Introduction to Linear Operator Theory

MONOGRAPHS AND TEXTBOOKS IN
PURE AND APPLIED MATHEMATICS

1. *K. Yano,* Integral Formulas in Riemannian Geometry (1970) *(out of print)*
2. *S. Kobayashi,* Hyperbolic Manifolds and Holomorphic Mappings (1970) *(out of print)*
3. *V. S. Vladimirov,* Equations of Mathematical Physics (A. Jeffrey, editor; A. Littlewood, translator) (1970)
4. *B. N. Pshenichnyi,* Necessary Conditions for an Extremum (L. Neustadt, translation editor; K. Makowski, translator) (1971)
5. *L. Narici, E. Beckenstein, and G. Bachman,* Functional Analysis and Valuation Theory (1971)
6. *D. S. Passman,* Infinite Group Rings (1971)
7. *L. Dornhoff,* Group Representation Theory (in two parts). Part A: Ordinary Representation Theory. Part B: Modular Representation Theory (1971, 1972)
8. *W. Boothby and G. L. Weiss (eds.),* Symmetric Spaces: Short Courses Presented at Washington University (1972)
9. *Y. Matsushima,* Differentiable Manifolds (E. T. Kobayashi, translator) (1972)
10. *L. E. Ward, Jr.,* Topology: An Outline for a First Course (1972) *(out of print)*
11. *A. Babakhanian,* Cohomological Methods in Group Theory (1972)
12. *R. Gilmer,* Multiplicative Ideal Theory (1972)
13. *J. Yeh,* Stochastic Processes and the Wiener Integral (1973) *(out of print)*
14. *J. Barros-Neto,* Introduction to the Theory of Distributions (1973) *(out of print)*
15. *R. Larsen,* Functional Analysis: An Introduction (1973)
16. *K. Yano and S. Ishihara,* Tangent and Cotangent Bundles: Differential Geometry (1973)
17. *C. Procesi,* Rings with Polynomial Identities (1973)
18. *R. Hermann,* Geometry, Physics, and Systems (1973)
19. *N. R. Wallach,* Harmonic Analysis on Homogeneous Spaces (1973)
20. *J. Dieudonné,* Introduction to the Theory of Formal Groups (1973)
21. *I. Vaisman,* Cohomology and Differential Forms (1973)
22. *B.-Y. Chen,* Geometry of Submanifolds (1973)
23. *M. Marcus,* Finite Dimensional Multilinear Algebra (in two parts) (1973, 1975)
24. *R. Larsen,* Banach Algebras: An Introduction (1973)
25. *R. O. Kujala and A. L. Vitter (eds),* Value Distribution Theory: Part A; Part B. Deficit and Bezout Estimates by Wilhelm Stoll (1973)
26. *K. B. Stolarsky,* Algebraic Numbers and Diophantine Approximation (1974)
27. *A. R. Magid,* The Separable Galois Theory of Commutative Rings (1974)
28. *B. R. McDonald,* Finite Rings with Identity (1974)
29. *J. Satake,* Linear Algebra (S. Koh, T. Akiba, and S. Ihara, translators) (1975)

30. *J. S. Golan,* Localization of Noncommutative Rings (1975)
31. *G. Klambauer,* Mathematical Analysis (1975)
32. *M. K. Agoston,* ALgebraic Topology: A First Course (1976)
33. *K. R. Goodearl,* Ring Theory: Nonsingular Rings and Modules (1976)
34. *L. E. Mansfield,* Linear Algebra with Geometric Applications: Selected Topics (1976)
35. *N. J. Pullman,* Matrix Theory and Its Applications (1976)
36. *B. R. McDonald,* Geometric Algebra Over Local Rings (1976)
37. *C. W. Groetsch,* Generalized Inverses of Linear Operators: Representation and Approximation (1977)
38. *J. E. Kuczkowski and J. L. Gersting,* Abstract Algebra: A First Look (1977)
39. *C. O. Christenson and W. L. Voxman,* Aspects of Topology (1977)
40. *M. Nagata,* Field Theory (1977)
41. *R. L. Long,* Algebraic Number Theory (1977)
42. *W. F. Pfeffer,* Integrals and Measures (1977)
43. *R. L. Wheeden and A. Zygmund,* Measure and Integral: An Introduction to Real Analysis (1977)
44. *J. H. Curtiss,* Introduction to Functions of a Complex Variable (1978)
45. *K. Hrbacek and T. Jech,* Introduction to Set Theory (1978)
46. *W. S. Massey,* Homology and Cohomology Theory (1978)
47. *M. Marcus,* Introduction to Modern Algebra (1978)
48. *E. C. Young,* Vector and Tensor Analysis (1978)
49. *S. B. Nadler, Jr.,* Hyperspaces of Sets (1978)
50. *S. K. Sehgal,* Topics in Group Rings (1978)
51. *A. C. M. van Rooij,* Non-Archimedean Functional Analysis (1978)
52. *L. Corwin and R. Szczarba,* Calculus in Vector Spaces (1979)
53. *C. Sadosky,* Interpolation of Operators and Singular Integrals: An Introduction to Harmonic Analysis (1979)
54. *J. Cronin,* Differential Equations: Introduction and Quantitative Theory (1980)
55. *C. W. Groetsch,* Elements of Applicable Functional Analysis (1980)
56. *I. Vaisman,* Foundations of Three-Dimensional Euclidean Geometry (1980)
57. *H. I. Freedman,* Deterministic Mathematical Models in Population Ecology (1980)
58. *S. B. Chae,* Lebesgue Integration (1980)
59. *C. S. Rees, S. M. Shah, and Č. V. Stanojević,* Theory and Applications of Fourier Analysis (1981)
60. *L. Nachbin,* Introduction to Functional Analysis: Banach Spaces and Differential Calculus (R. M. Aron, translator) (1981)
61. *G. Orzech and M. Orzech,* Plane Algebraic Curves: An Introduction Via Valuations (1981)
62. *R. Johnsonbaugh and W. E. Pfaffenberger,* Foundations of Mathematical Analysis (1981)

63. *W. L. Voxman and R. H. Goetschel, Jr.,* Advanced Calculus: An Introduction to Modern Analysis (1981)
64. *L. J. Corwin and R. H. Szczarba,* Multivariable Calculus (in press)
65. *V. I. Istrăţescu,* Introduction to Linear Operator Theory (1981)
66. *R. D. Järvinen,* Finite and Infinite Dimensional Linear Spaces: A Comparative Study in Algebraic and Analytic Settings (in press)
67. *J. K. Beem and P. E. Ehrlich,* Global Lorentzian Geometry (in press)

Other Volumes in Preparation

Introduction to Linear Operator Theory

VASILE I. ISTRĂŢESCU

MARCEL DEKKER, INC. New York and Basel

Library of Congress Cataloging in Publication Data

Istrăţescu, Vasile I.
Introduction to linear operator theory.

(Pure and applied mathematics ; 65)
Includes bibliographical references and indexes.
1. Linear operators. 2. Hilbert space. 3. Banach spaces. I. Title.
QA329.2.I8915 515.7'246 80-29511
ISBN 0-8247-6896-5

MARCEL DEKKER, INC.
270 Madison Avenue, New York, New York 10016

Current printing (last digit):
10 9 8 7 6 5 4 3 2 1

PRINTED IN THE UNITED STATES OF AMERICA

To my parents,
Paraschiva Istrăţescu and Ion Istrăţescu

PREFACE

During the last 15 or 20 years much progress has been made in the theory of non-self-adjoint operators on Hilbert and Banach spaces. The present volume is intended to provide an introduction to the subject.

The first four chapters are devoted to standard material on linear functional analysis. However, whenever possible a unified approach is used, and this is the case, for example, for the three basic theorems of linear functional analysis which are treated consequences of a unique theorem. Chapter 5 is concerned with the general spectral representation of operators on Hilbert spaces and depends on the Banach algebra theory developed in Chapter 4.

Chapter 6 is concerned with the basic notion of this book: the numerical range. First this is considered for Hilbert spaces and next for Banach spaces. Various classes of operators connected with the numerical range are also discussed.

The classes of nonnormal operators have a long history, and the problem of deciding when an operator is normal (also hermitian, unitary) forms the content of Chapters 7 and 8.

As is well known, the class of hermitian operators has many important applications in various branches of mathematics and physics; thus related classes of operators for which many properties of hermitian operators are preserved are of great interest. Chapter 9 gives an account of results involving such classes as well as some applications (for example, a simple proof of an interpolation theorem of Lions-Peetre).

In Chapter 10 the famous invariant subspace problem is discussed and some structure theorems are presented.

The Weyl spectrum of an operator is discussed in Chapter 11, as well as some applications. The elements of the von Neumann algebras are also given.

Chapter 12 is concerned with an important and useful notion: analytic and quasi-analytic vectors; also some applications are given. In Chapter 13 the Banach space version of the famous Schwarz theorem from complex function theory is presented. In Chapter 14 results on maximum theorems for operator-valued functions are given.

In the last chapter, I present some ergodic theorems for classes of operators containing the quasi-compact operators. The results presented are connected with the operator theoretic treatment of Markov processes as given by Kakutani-Yosida and refined by many authors. I have tried to indicate the origin of the various results, and the references (which in turn contain references to many earlier results) may be used to obtain further information. When I make no ascription, I am not claiming originality.

A part of this book has its origin in a course given by the author at the Centro Linceo Interdisciplinare di Scienze Matematiche e loro Applicazioni of the Accademia Nazionale dei Lincei (Rome, Italy).

For interesting and helpful conversations I am indebted to many friends. For discussions and constant encouragement I am indebted to Professor G. Köthe.

Vasile I. Istrăţescu

CONTENTS

PREFACE v

1 PRELIMINARIES: SET THEORY AND GENERAL TOPOLOGY 1

1.1 The Algebra of Sets 1
1.2 Partially Ordered Sets 4
1.3 Topology and Topological Spaces 6
1.4 Baire's Theorem 16

2 BANACH SPACES 19

2.1 Linear Spaces 19
2.2 Linear Independence 21
2.3 Sets in Linear Spaces 22
2.4 Classes of Spaces: Isomorphic Spaces, Quotient Spaces, and Complementary Spaces 25
2.5 Seminorms and Norms on Linear Spaces 27
2.6 Linear Topological Spaces 29
2.7 Banach Spaces 30
2.8 Linear Operators on Banach Spaces 32
2.9 Uniformly Convex and Rotund Banach Spaces: Some Generalizations 35
2.10 The Hahn-Banach Extension Theorem 39
2.11 Extension Theorems for Complex Banach Spaces 43
2.12 Three Basic Theorems of Linear Analysis 45
2.13 Convergence in Banach Spaces 56
2.14 The Adjoint of an Operator 63
2.15 The Spectrum of an Operator 65
2.16 The Local Spectrum of an Operator 70
2.17 Analytic Representation of the Dual of Some Banach Spaces 76
2.18 Measures of Noncompactness and Classes of Mappings on Banach Spaces 84

3 HILBERT SPACES 109

3.1 Inner Products on Linear Spaces 109
3.2 Orthonormal Bases and the Bessel Inequality
3.3 Separable Hilbert Spaces: Gram-Schmidt Orthogonalization Method 119
3.4 Orthogonal Subspaces of a Hilbert Space 125
3.5 The Dual of a Hilbert Space 126
3.6 Classes of Bounded Linear Operators on Hilbert Spaces 129

4 BANACH ALGEBRAS 142

4.1 Definitions and Some Examples 142
4.2 The Spectrum of an Element in a Banach Algebra with Unit 147
4.3 Representation Theorems for Commutative Banach Algebras 151
4.4 Structure Theorems for Commutative Banach Algebras 154
4.5 Representation Theorems for Noncommutative Banach Algebras 167

5 SPECTRAL REPRESENTATION OF OPERATORS ON HILBERT SPACES 174

5.1 Semispectral and Spectral Families of Radon Measures 174
5.2 Measurability and Integrability with Respect to Spectral Families 179
5.3 A Representation Theorem for L^{∞} 181
5.4 Spectral Decomposition of Some Classes of Operators 188
5.5 Some Remarks on the Spectral Mapping Theorem for Hermitian and Normal Operators 191

6 THE NUMERICAL RANGE 200

6.1 The Numerical Range for Bounded Linear Operators on Hilbert Spaces 200
6.2 The Numerical Range and the Spectrum 206
6.3 The Numerical Range and Its Closure 214
6.4 The Essential Numerical Range for Bounded Linear Operators on Hilbert Spaces 217
6.5 The Maximal Numerical Range of a Bounded Operator on a Hilbert Space 219
6.6 The Extreme Points of the Numerical Range for Hyponormal Operators and (WN) Operators 221
6.7 The Numerical Range and Some Classes of Operators 223
6.8 The Numerical Range and Tensor Products 224

6.9 The Numerical Range for Bounded Linear Operators on Banach Spaces 226
6.10 The Exponential Function on the Set of All Bounded Linear Operators on a Banach Space 233
6.11 The Numerical Radius, the Spectral Radius, and the Norm of a Bounded Linear Operator on a Banach Space 236
6.12 Hermitian and Normal Operators on Banach Spaces 238
6.13 Normal Operators on Banach Spaces 244
6.14 Classes of Elements in Banach Algebras with Unit: The Vidav-Palmer Theorem 246
6.15 Some Properties of Hermitian and Normal Elements of a Banach Algebra 255
6.16 The Numerical Radius and the Iterates of an Element 260
6.17 The Numerical Range of Elements of Locally m-Convex Algebras 262

7 NONNORMAL CLASSES OF OPERATORS 265

7.1 Classes of Nonnormal Operators 266
7.2 Spectral Sets and Dilations of Operators 275
7.3 Operators with G_1 Property and Some Generalizations 286
7.4 Operators with Property Re $\sigma(T) = \sigma(\text{Re } T)$ 297
7.5 The Class $\bar{R}_1$ 302
7.6 Other Classes of Bounded Operators 307

8 CONDITIONS IMPLYING NORMALITY 310

8.1 Conditions Implying Hermitianity 310
8.2 Conditions Implying Unitarity 317
8.3 Conditions Implying Normality 321

9 SYMMETRIZABLE OPERATORS: GENERALIZATIONS AND APPLICATIONS 344

9.1 Symmetrizable Operators on Hilbert Spaces 344
9.2 Symmetrizable Elements in Banach Algebras 349
9.3 Inner Products on Banach Spaces: Symmetrizable Operators and Some Generalizations 352
9.4 Some Applications of Symmetrizable Operators and Quasi-Normalizable Operators 369
9.5 Further Results on Symmetrizable Operators on Hilbert Spaces 374

10 INVARIANT SUBSPACES AND SOME STRUCTURE THEOREMS 380

10.1 Invariant Subspaces: Some Existence Theorems 380
10.2 Reducing Invariant Subspaces 398
10.3 Some Structure Theorems 408

11 THE WEYL SPECTRUM OF AN OPERATOR 412
11.1 Preliminaries and Some General Results 412
11.2 Weyl's Theorem 420
11.3 Weyl's Theorem for Some Classes of Operators 426
11.4 The Weyl Spectrum of an Element in a von Neumann Algebra 433
11.5 The von Neumann Theorem 439

12 ANALYTIC AND QUASI-ANALYTIC VECTORS 444

12.0 Introduction 444
12.1 Self-Adjoint Operators 446
12.2 Classes of Vectors for an Operator 452
12.3 Analytic and Quasi-Analytic Vectors and Essentially Self-Adjoint Operators 455
12.4 Quasi-Analytic Vectors and Semigroups of Operators 463
12.5 Analytic and Quasi-Analytic Elements in Commutative Banach Algebras 466

13 SCHWARZ NORMS 468

13.1 Schwarz Norms 469
13.2 A New Class of Schwarz Norms 476
13.3 Schwarz Norms on Banach Spaces 478

14 MAXIMUM THEOREMS FOR OPERATOR-VALUED HOLOMORPHIC FUNCTIONS 483

14.1 Holomorphic Functions 483
14.2 Subharmonic Functions 485
14.3 Maximum Theorems for the Norm 498
14.4 Maximum Theorems for the Spectral Radius and for the Essential Spectral Radius 508
14.5 Maximum Theorems for Other Operator-Valued Holomorphic Functions 528

15 UNIFORM ERGODIC THEOREMS FOR SOME CLASSES OF OPERATORS 537

15.1 Classes of Operators 537
15.2 Applications to Markov Processes 545

APPENDIX. C_p CLASSES 549

REFERENCES 553

SYMBOL INDEX 573

SUBJECT INDEX 575

Introduction to Linear Operator Theory

Chapter 1

PRELIMINARIES: SET THEORY AND GENERAL TOPOLOGY

1.1 THE ALGEBRA OF SETS

In what follows we assume that the reader is familiar with the notion of a set. We give several examples of sets.

1.1.1 EXAMPLE The letters of the English alphabet form a set.

1.1.2 EXAMPLE The rational numbers in the interval $[0, 1] = \{x \mid 0 \leq x \leq 1\}$ form a set.

1.1.3 EXAMPLE The numbers of the form 2n, where n is an integer, also form a set.

We consider a set A as being given when we can identify its elements; for example, the set E has the elements a, b, c, ..., or, in brief,

$$E = \{a, b, c, \ldots\}$$

Another way is to identify the elements of a set by a property P, or in brief,

$$E = \{a \mid a \text{ has the property } P\}$$

Generally speaking, we adhere to the standard notational conventions: we denote by lowercase letters a, b, c, ... the elements of a set E and we write this as $a \in E$, $b \in E$, $c \in E$, ... Sets are denoted by uppercase letters A, B, C, ..., X, Y, Z and families of sets by $\mathcal{A}$, $\mathcal{B}$, $\mathcal{C}$, ... (script uppercase letters). We write $\emptyset$ for the empty set, and for any element a, {a} denotes the set which has as an element only the element a.

1.1.4 DEFINITION Let A and B be two sets. We say that A is a *subset* of B if for any a ∈ A we have a ∈ B. We write this as A ⊆ B. If there exists an element b ∈ B, b not in A, then we say that A is a *proper* subset of B.

If an element a is not in a set A, we write this as a ∉ A.

1.1.5 DEFINITION If A and B are two sets, we say that A = B if A ⊆ B and B ⊆ A; in the contrary case, we say that A and B are distinct sets.

1.1.6 DEFINITION If A and B are two sets, then A ∪ B denotes the set of all elements which are in A or in B; A ∩ B denotes the set of all elements which are in A and in B.

The set A ∪ B is called the *union* of the sets A and B; A ∩ B is called the *intersection* of the sets A and B.

1.1.7 REMARK Similar definitions can be given for the case when we have a family of sets, $A = \{A_\alpha\}_{\alpha \in I}$ for $\bigcup_\alpha A_\alpha$ and $\bigcap_\alpha A_\alpha$.

1.1.8 EXAMPLE If A = {1, 2, 3} and B = {2, 5}, then

A ∪ B = {1, 2, 3, 5} A ∩ B = {2}

If A = {1, 2} and B = {3, 5}, then

A ∪ B = {1, 2, 3, 5} A ∩ B = ∅

1.1.9 DEFINITION For any set A we note $P(A)$, the family of all subsets of A.

1.1.10 EXAMPLE For A = {2} we have

$P(A) = \{\emptyset, \{2\}\}$

1.1.11 PROPOSITION If A, B, and C are sets, then the following relations hold:

1. A ∪ B = B ∪ A
2. A ∪ A = A
3. A ∪ ∅ = A

4. $A \cup (B \cup C) = (A \cup B) \cup C$

5. $A \subseteq A \cup B$

6. $A \subseteq B$ if and only if $A \cup B = B$

1.1.12 REMARK The reader can prove similar properties for the intersection; for example

1'. $A \cap B = B \cap A$

2'. $A \cap A = A$

1.1.13 PROPOSITION If A, B, and C are arbitrary sets, then

$$A \cap (B \cup C) = (A \cap B) \cup (A \cap C)$$

$$A \cup (B \cap C) = (A \cup B) \cap (A \cup C)$$

Proof. We prove only the first assertion; the second relation can be proved in a similar way.

Let $x \in A \cap (B \cup C)$. Then $x \in A$ and x is also in B or in C.

Suppose, for example, that $x \in B$. The case $x \in C$ is similar. In this case $x \in A \cap B$ and thus $x \in (A \cap B) \cup (A \cap C)$. In this way, $A \cap (B \cup C) \subseteq (A \cap B) \cup (A \cap C)$. Let us now suppose that x is in $(A \cap B) \cup (A \cap C)$ and for simplicity x is in $A \cap B$. In this case it is clear that x is in $(A \cup B) \cap (A \cup C)$. The proposition is proved.

1.1.14 DEFINITION Two sets A and B are called *disjoint* if $A \cap B = \emptyset$; if $\mathcal{A}$ is a family of sets such that for any sets A, B in $\mathcal{A}$ we have $A \cap B = \emptyset$, we say that $\mathcal{A}$ has *pairwise disjoint* sets.

Suppose that we have a set T and we consider $\mathcal{P}(T)$. For any $A \in \mathcal{P}(T)$ we can define a new set, called the *complement* of A, by the relation,

$$C_A = \{x \mid x \in T, x \notin A\}$$

It is easy to see that the following properties hold:

1. $C_{A \cup B} = C_A \cap C_B$

2. $C_{A \cap B} = C_A \cup C_B$

Also it is clear that these relations hold for the case of families of sets. Since the proof is easy we omit this.

The set $A \Delta B$ (the symmetric difference of the sets A and B) is defined by

$$A \Delta B = \{x \mid x \in A \cup B, x \notin A \cap B\}$$

It is obvious that $A \Delta B = B \Delta A$.

If T is a given set, then in $\mathcal{P}(T)$ we have several types of families of sets. Among these we mention two which are very useful in measure theory: the ring of sets and the σ-ring (σ-algebra).

1.1.15 DEFINITION A family of sets $\mathcal{R} \subseteq \mathcal{P}(T)$ is called a *ring* of sets if for any A and B in $\mathcal{R}$ the following sets are also in $\mathcal{R}$:

$A \cup B \qquad A \cap C_B$

1.1.16 DEFINITION A ring of sets $\mathcal{B}$ is called a *σ-algebra* if for any $A_n \in \mathcal{B}$, $\bigcup_n A_n$ is again an element of $\mathcal{B}$.

1.2 PARTIALLY ORDERED SETS

Let A be a nonempty set.

1.2.1 DEFINITION A *relation* on A is a collection of ordered pairs (x, y) of elements of A.

1.2.2 DEFINITION A *partially ordered set* is a nonempty set with a relation denoted "≤" such that the following properties hold:

1. If $a \le b$ and $b \le c$, then $a \le c$ (transitivity).
2. $a \le a$ for all $a \in A$ (reflexivity).
3. If $a \le b$ and $b \le a$, then $a = b$ (antisymmetry).

1.2.3 DEFINITION A *totally ordered set* is any partially ordered set A (with the relation ≤) with the property that for any pair a, b of elements of A, we have $a \le b$ or $b \le a$.

1.2.4 EXAMPLE The set of all real numbers R is a totally ordered set. The relation "≤" is defined as follows: We say that two real numbers are in the relation $a \le b$ if the difference $b - a$ is a positive number.

The set of all complex numbers is a partially ordered set.

Here we have several possible ways to define the relation $\leq$; for example, we can define the relation "$\leq$" as follows: Two complex numbers z_1 and z_2 are in the relation $z_1 \leq z_2$ if $\text{Re } z_1 \leq \text{Re } z_2$, where Re z is the real part of z.

If T is any set, then we can define in $P(T)$ a relation in the following way: If A and B are elements of $P(T)$, then we say that $A \leq B$ if A is a subset of B. In this case $P(T)$ is a partially ordered set. In this case we say also that we have an ordering by *inclusion*.

1.2.5 DEFINITION If a is a partially ordered set and $A_1 \subset A$, then an element $a \in A$ is said to be an *upper bound* for A_1 if $a_1 \leq a$ for all $a_1 \in A_1$.

The element $\tilde{a}$ is called a *least upper bound* of A_1

1. If $\tilde{a}$ is an upper bound of A_1.
2. If a_1 is another upper bound of A_1, then $\tilde{a} \leq a_1$.

The least upper bound is denoted, generally, by lub A_1.

The element $b \in A$ is called a *lower bound* of A_1 if $b \leq a_1$ for all $a_1 \in A_1$ and an element $\tilde{b} \in A$ is called the *greatest lower bound*

1. If $\tilde{b}$ is a lower bound.
2. If b_1 is another lower bound, then $b_1 \leq \tilde{b}$.

The greatest lower bound is denoted, generally, by glb A_1.

1.2.6 DEFINITION An element x of a partially ordered set A is called *maximal* if $x \leq y$ implies $y \leq x$.

Similarly we can define the notion of *minimal* element.

1.2.7 DEFINITION A *chain* in a partially ordered set is any subset C of A such that the relation order "$\leq$" of A, restricted to C, gives that C with this relation is a totally ordered set.

One of the most important axioms in set theory is the axiom of E. Zermelo and is called the *axiom of choice*. There exist several equivalent formulations of this axiom and here we quote without proof only two. First we define the notion of cartesian product. For I, any set, we suppose that for each $i \in I$ there exists a set

A_i. The cartesian product $\Pi_{i \in I} A_i$ is the set of all functions f defined on I such that,

$$f(i) = f_i \in A_i$$

for each $i \in I$; f_i is called the i coordinate of f.

Now we are ready to give the two formulations of the choice axiom.

1.2.8 CHOICE AXIOM The cartesian product of any nonvoid family of nonvoid sets is a nonvoid set.

1.2.9 ZORN'S LEMMA If A is a partially ordered set such that for every chain there exists an upper bound, then A has a maximal element.

In several sections of this book we apply these assertions; we mention here the application in the proof of the Hahn-Banach theorem.

1.3 TOPOLOGY AND TOPOLOGICAL SPACES

In analysis we use intensively the notion of convergence. For example, we say that a sequence of complex numbers or real numbers (or a sequence of real- or complex-valued functions) converge.

Also the notion of convergence is used to characterize certain important classes of functions. For example, a function F: $[0, 1] \to R$ is continuous if and only if for each $s \in [0, 1]$ and any $s_n \to s$, $f(s_n) \to f(s)$.

These situations and others lead to an axiomatic treatment of the notion of convergence, and one of the basic settings in which this is best realized is in metric spaces.

1.3.1 DEFINITION A metric space (X, ρ) is a pair, where X is a nonempty set and ρ is a real-valued function on $X \times X$ with the following properties:

1. $\rho(x, y) = 0$ if and only if $x = y$
2. $\rho(x, y) = \rho(y, x)$ for all $x, y \in X$

3. $\rho(x, y) \leq \rho(x, z) + \rho(z, y)$ for all $x,y,z \in X$

The inequality in property 3 is called the *triangle inequality.* It is trivial to see that on any set there exists a metric ρ,

$$\rho(x, y) = \begin{cases} 1 & \text{if } x \neq y \\ 0 & \text{if } x = y \end{cases}$$

In what follows we suppose that ρ is not such a metric.

For any x and y the number $\rho(x, y)$ is called the *distance* between x and y.

1.3.2 EXAMPLE

1. Let R^n be the set of all elements of the form

$x = (x_1, x_2, x_3, \ldots, x_n)$

where x_i are real numbers. Let $y = (y_1, y_2, y_3, \ldots, y_n) \in R^n$; then we define

$$\rho_p(x, y) = (\Sigma\, |x_i - y_1|^p)^{1/p} \qquad 1 \leq p < \infty$$

and for $p = \infty$,

$$\rho_\infty(x, y) = \sup_i \{|x_i - y_1|\}$$

It is well known that ρ_p are metrics and thus (R^n, ρ_p) is a metric space, called the *n-dimensional euclidean space.*

2. Let $C_{[0,1]}$ be the set of all continuous real-valued functions on [0, 1]. If we define, for any $f,g \in C_{[0,1]}$

$$\rho(f, g) = \sup_{x\in[0,1]} \{|f(x) - g(x)|\}$$

then it is easy to verify that $(C_{[0,1]}, \rho)$ is a metric space.

3. Let ℓ^p be the set of all infinite sequences $x = (x_1, x_2, x_3, \ldots)$ such that $\Sigma\, |x_i|^p < \infty$ and for any $x,y \in \ell^p$ we define

$$\rho_p(x, y) = (\Sigma\, |x_i - y_i|^p)^{1/p}$$

Using the inequalities of Hölder and Minkowski, we can see that (ℓ^p, ρ_p) is a metric space.

Also for $p = \infty$, we define

$$\rho_\infty(x, y) = \sup_i |x_i - y_i|$$

and $(\ell^\infty, \rho_\infty)$ is a metric space.

We define now a very important notion in the theory of metric space: the notion of a neighborhood of a point.

Let (X, ρ) be a metric space and $x_0 \in X$ be an arbitrary point.

1.3.3 DEFINITION A set $V \subset X$ is called a *neighborhood* of x_0, and there exists $\varepsilon > 0$ such that the set

$$S_\varepsilon(x_0) = \{x \mid \rho(x, x_0) < \varepsilon\}$$

is contained in V.

With this notion we define two dual notions: open sets and closed sets.

1.3.4 DEFINITION A set G in a metric space (X, ρ) is called an *open set* if for each point $x \in G$ there exists a neighborhood V_x of x such that $V_x \subset G$.

1.3.5 DEFINITION A set F in a metric space (X, ρ) is called a *closed set* if the set

$$C_F = \{x \mid x \in X, x \notin F\}$$

is open.

From the definition of open sets and closed sets, it is easy to see that the following assertions are true.

1.3.6 PROPOSITION

1. If $\{G_i\}_{i \in I}$ is a family of open sets, then $\bigcup G_i$ is an open set.
2. If $\{F_i\}_{i \in I}$ is a family of closed sets, then $\bigcap F_i$ is a closed set.
3. The intersection of a finite family of open sets is also an open set.

Simple examples show that the intersection of an infinite family of open sets is not open. Indeed, let $X = [0, 1]$ with the metric

$$\rho(x, y) = |x - y|$$

and the sets

$$G_n = \{x \mid x < \frac{1}{n}\}$$

and it is clear that the intersection is $\{0\}$, which is a closed set.

The notion of convergence can be defined in the setting of metric spaces, following the model of real and complex numbers.

1.3.7 DEFINITION A sequence $\{x_n\}$ in a metric space (X, ρ) is called *convergent* to an element $x \in X$ if for any $\varepsilon > 0$ there exists an integer N_ε such that for any $n \geq N_\varepsilon$ the following inequality holds:

$$\rho(x, x_n) \leq \varepsilon$$

We write this as $\lim x_n = x$.

It is clear from the definition that the convergence defined in Definition 1.3.7 coincides, in the case of metric spaces of real and complex numbers, with the usual notion, and also in the space of continuous functions on [0, 1].

1.3.8 DEFINITION A metric space (X, ρ) is called *complete* if for any Cauchy sequence $\{x_n\}$ [i.e., a sequence $\{x_n\}$ with the property that for any $\varepsilon > 0$ there exists an integer N_ε such that for all $n, m \geq N_\varepsilon$ the following inequality holds: $\rho(x_n, x_m) \leq \varepsilon$], there exists $x \in X$ such that $\lim x_n = x$.

1.3.9 REMARK As is well known, all the metric spaces considered in Example 1.3.2 are complete metric spaces. An example of a non-complete metric space is the set of all rational numbers.

If (X, ρ) is a metric space and X_1 is a subset of X, then we can define a metric on X_1 using the metric ρ,

$$\rho_1(x, y) = \rho(x, y)$$

The metric ρ_1 is called the *induced* metric. Also, if (X_1, ρ_1) and (X_2, ρ_2) are two metric spaces, then the cartesian product is also a metric space. The metric can be defined in various ways,

$$\rho_p(\{x_1, y_1\}, \{x_2, y_2\}) = [\rho_1(x_1, x_2)^p + \rho_2(y_1, y_2)^p]^{1/p}$$

and this metric space is complete if and only if (X_i, ρ_i) are complete metric spaces.

The above considerations lead us to the fundamental notion of general topology: the notion of topological space.

1.3.10 DEFINITION A collection $\mathcal{D}$ of subsets of a set X is called a *topology* on X if the following conditions are satisfied:

1. For any family of sets in $\mathcal{D}$, their union is again in $\mathcal{D}$.
2. If $G_1, G_2, \ldots, G_n \in \mathcal{D}$, then $\bigcap_{i=1}^{n} g_i \in \mathcal{D}$.
3. X is in $\mathcal{D}$.

We say that the pair $(X, \mathcal{D})$ is a *topological space* or that X is a topological space.

The members of $\mathcal{D}$ are called *open sets*.

We give now some examples of topologies.

1.3.11 EXAMPLE

1. For any set X let $\mathcal{D} = \mathcal{P}(X)$. Then $\mathcal{D}$ is obviously a topology on X and is called the *discrete topology*.

2. For any set X let $\mathcal{D} = \{X, \emptyset\}$. Clearly this is a topology, which is called the *indiscrete topology*.

3. If (X, ρ) is a metric space, then the family of all open sets of X is a topology.

In connection with the topology considered in part 3 of Example 1.3.11, if $(X, \mathcal{D})$ is a topological space, we say that the topology is metrizable, or simply that the topological space $(X, \mathcal{D})$ is metrizable, if there exists a metric on X such that the open sets as were defined in Definition 1.3.4 with respect to this metric are exactly the members of $\mathcal{D}$.

1.3.12 DEFINITION If $(X, \mathcal{D})$ is a topological space and X_1 is a subset of X, we say that a subset of X_1 is X_1 open if it is the intersection of an open set in X with X_1.

It is easy to see that all X_1 open sets form a topology on X which is called the *induced topology* of X on X_1.

In the family of all topologies on a set X we can introduce a partial order by the following relation: we say that two topologies $\mathcal{D}_1$ and $\mathcal{D}_2$ are in the relation $\mathcal{D}_1 \leq \mathcal{D}_2$ if $\mathcal{D}_1 \subset \mathcal{D}_2$. In other words, we say that $\mathcal{D}_1$ is weaker than $\mathcal{D}_2$ or that $\mathcal{D}_2$ is stronger than $\mathcal{D}_1$.

From this definition it is clear that the indiscrete topology is the weakest topology and the discrete topology is the strongest topology on X.

We define now an important notion: the continuous function.

1.3.13 DEFINITION If X and Y are two topological spaces and $f\colon X \to Y$ is a function, then we say that f is *continuous* at $x \in X$ if for any neighborhood W of f(x) in Y, there exists a neighborhood V of x such that for all $\tilde{x} \in V$, $f(\tilde{x}) \in W$. The function f is said to be *continuous on X* if it is continuous at each $x \in X$.

In terms of continuous functions the order relation between topologies can be characterized as follows:

1.3.14 PROPOSITION Let $\mathcal{D}_1$ and $\mathcal{D}_2$ be two topologies on a set X. Then $\mathcal{D}_1$ is weaker than $\mathcal{D}_2$ if and only if $f\colon X \to X$, defined by $f(x) = x$, is a continuous function from the topological space $(X, \mathcal{D}_2)$ to the topological space $(X, \mathcal{D}_1)$.

Proof. Since this is easy we omit it.

Suppose that we have a topological space X and $x \in X$ arbitrary.

1.3.15 DEFINITION We say that the point x is in the *interior* of a set S if there exists an open set G such that $x \in G \subseteq S$. S also is called a neighborhood of x.

It is obvious that a set G in a topological space is open if and only if it is a neighborhood of each of its points.

1.3.16 DEFINITION A *local base* at x, $x \in X$, where X is a topological space, is a collection C of neighborhoods of x such that given any neighborhood V of x there exists $G_1 \in C$ such that $G_1 \subset V$.

1.3.17 DEFINITION A topological space X is said to satisfy the *first axiom of countability* if it has a countable local basis at every $x \in X$.

One way to construct topologies on a set is to give a collection of sets with some special properties. Such a collection is considered in the following definition.

1.3.18 DEFINITION A *subbase* for a topology $\mathcal{D}$ on a set X is a collection C of subsets such that

1. If $G \in C$, G is open.
2. If x is any point in X and V is any neighborhood of x, there exist $G_1, G_2, \ldots, G_n \in C$ such that $x \in \bigcap_{i=1}^{n} G_i \subset V$.

1.3.19 EXAMPLE In any metric space (X, ρ) a base for the topology is the family C of subsets of the form

$$S_\varepsilon(x) = \{y \mid \rho(x, y) < \varepsilon\}$$

The following theorem shows the connection between topology and subbases.

1.3.20 THEOREM If X is a nonempty set and C a collection of subsets whose union is X, then there exists a unique topology on X such that C is a subbase for $\mathcal{D}$. This topology is the weakest topology including C.

Proof. If C is given, then we construct $\mathcal{D}$ as follows. First we set in $\mathcal{D}$ the empty set and X, and also the unions of all finite intersections of elements in C. Clearly $\mathcal{D}$ is a topology and C is a subbase for $\mathcal{D}$. Thus the existence is proved. The uniqueness is proved as follows: Suppose that C is a subbase for two topologies $\mathcal{D}$ and $\tilde{\mathcal{D}}$. Since, by definition of subbase, C contains sets such that if x is any point of X and G a neighborhood of x in $(X, \tilde{\mathcal{D}})$ we have

$$x \in G_1 \cap G_2 \cap G_3 \cap \cdots \cap G_n \subset G$$

Clearly, $G_1 \cap G_2 \cap G_3 \cap \cdots \cap G_n$ is open and so it is a neighborhood of x in the topology $\mathcal{D}$. Thus any set which is open in the $\tilde{\mathcal{D}}$ topology is open in the $\mathcal{D}$ topology. By symmetry, we obtain that every set which is open in the $\mathcal{D}$ topology is open in the $\tilde{\mathcal{D}}$ topology and thus $\mathcal{D} = \tilde{\mathcal{D}}$.

The following definition gives a way to construct new topologies.

1.3.21 DEFINITION If $\{\mathcal{D}_i\}_{i \in I}$ is a family of topologies on a set X, then the topology generated by $\{\cup \mathcal{D}_i, i \in I\}$ in the sense of Theorem 1.3.20 (i.e., $\{\mathcal{D}_i\} = C$) is called $\vee \mathcal{D}_i = \sup \mathcal{D}_i$.

A related and important notion is that of a base for a topology.

1.3.22 DEFINITION If $(X, \mathcal{D})$ is a topological space, a family of nonempty subsets is called a *basis* for $\mathcal{D}$ if any G in $\mathcal{D}$ is of the form $G = \cup G_i$ with G_i in the family.

1.3.23 EXAMPLE If (X, ρ) is a metric space, then a basis for the topology of X is

$$B = \{S_\varepsilon(x) = \{y \mid \rho(x, y) < \varepsilon\}, \varepsilon > 0$$

As a corollary of this example, we obtain that the topology on the spaces R^n, C_I, ℓ^p has a basis, and also it is easy to see that we consider only $\varepsilon > 0$ as rational. Thus the base is countable. Also this remark applies for all separable metric spaces (i.e., metric spaces in which there exists a countable set $\{x_n\}$ such that for any $x \in X$ there exists a subsequence $\{x_{n_i}\}$ converging to x).

1.3.24 DEFINITION A topological space is said to satisfy the second axiom of countability if the topology on X has a countable basis.

1.3.25 COROLLARY Every separable metric space satisfies the second axiom of countability.

An important class of topological space is the class of separate topological spaces. Since there exist several notions of separation we quote here only two, which will be used further.

1.3.26 DEFINITION A topological space is called a *Hausdorff space* if for any $x, y \in X$, $x \neq y$ there exist open sets V_x and V_y such that $V_x \cap V_y = \emptyset$, $x \in V_x$, $y \in V_y$.

1.3.27 DEFINITION The topological space is called *completely regular* if for each closed set F and any $x \in X$, $x \notin F$ there exists a continuous function $f\colon X \to [0, 1]$ such that the restriction to F is zero and $f(x) = 1$.

1.3.28 EXAMPLE Every metric space is a Hausdorff space and also a completely regular space.

We consider now the notion of convergence in topological spaces.

1.3.29 DEFINITION A directed set is any partially ordered set such that for any a and b in the set there exists another element in the set, say c, such that

$$a \le c \qquad b \le c$$

We say that the set is *directed* by the relation "$\le$."

1.3.30 DEFINITION A *net* in a topological space X is any function on a directed set with values in X.

The nets are denoted by $\{x_\delta\}$ where $\delta \in S$ and S is a directed set. If S is N (the set of integers), then we denote the corresponding net as $\{x_n\}$ and we call $\{x_n\}$ a *sequence*.

1.3.31 DEFINITION If $\{x_\delta\}$ is a net with a directed set S, then we say that $\{x_\delta\}$ is *eventually* in X_1 if there exists a set S_0 in S such that $x \in X_1$ for all $\delta \ge \delta_0$, and we say that $\{x_\delta\}$ is *frequently* in X_1 if for any $\delta_0 \in S$ there exists $\delta \ge \delta_0$ such that $x_\delta \in X_1$.

1.3.32 DEFINITION Let $\{x_\delta\}$ be a net in a topological space $(X, \mathcal{D})$. Then we say that $\{x_\delta\}$ *converges* to x if for any neighborhood V of x, x_δ is eventually in V. If $\{x_\delta\}$ converges to a unique x, we write $\lim x_\delta = x$.

We can use the nets to obtain a characterization of continuous functions. We state the results without proof.

1.3.33 THEOREM If X and Y are two topological spaces, then $f: X \to Y$ is continuous if and only if for each $x \in X$ and any net $\{x_\delta\}$, $\lim x_\delta = x$, $\lim f(x_\delta) = f(x)$.

Another useful notion is the notion of filter; we give only the definition.

1.3.34 DEFINITION A filter F on a set X is a collection of subsets such that

1. $\emptyset \notin f$.
2. $A,B \in F \to A \cap B \in F$.
3. If C is an arbitrary set which contains a member of f, then C is in F.

Suppose that X and Y are two topological spaces and $F: X \to Y$ is any function.

1.3.35 DEFINITION The topology generated by the sets $\{f^{-1}(G)\}$, where G is open in Y, is called the *weak topology generated by the function* f and is denoted by w_f. If we have $f_i: X \to Y_i$, $i \in I$, then the topology generated by $\{f_i^{-1}(G_k) \mid G_i \text{ open in } Y_i\}$ is called the *weak topology generated by* $\{f_i\}_{i\in I}$ and is denoted by $V\{w_{f_i}\}_{i\in I}$.

An important application of the topologies defined as in 1.3.35 is the notion of product topology.

1.3.36 DEFINITION If $\Pi_i X_i$ is product of topological spaces, then the product topology on $\Pi_i X_i$ is the topology generated by the family of functions $f_i(\{x_i\}) = x_i$.

In a topological space there exist some very important classes of sets. As we know, examples of such classes are open sets and also closed sets. We consider now another class: the class of compact sets.

1.3.37 DEFINITION A set C in a topological space is called *compact* if and only if from any family of open sets $\{G_i\}_{i\in I}$, $\bigcup G_i \supseteq C$, there exists a finite family $\{G_i\}_{i=1}^{n}$ such that $\bigcup_{i=1}^{n} G_i \supseteq C$.

A characterization of compact sets can be given using the notion of *finite intersection property*. We say that a family $\mathcal{A}$ of subsets of a set A and $A_1 \subseteq A$ have the *finite intersection property* relative to A_1 if A_1 meets any finite subcollection of $\mathcal{A}$, then A_1 meets all members of $\mathcal{A}$.

The characterization of compact sets is as follows.

1.3.38 PROPOSITION C is compact if and only if for any family of closed sets the finite intersection property is satisfied.

1.4 BAIRE'S THEOREM

One of the most important theorems concerning complete metric spaces, which has numerous applications, is the theorem of Baire.

We give it here and the reader is warned that we apply this theorem many times and especially in the proof of a lemma, from which follows in a simple way three basic theorems of linear analysis.

1.4.1 THEOREM (Baire) [See Rudin (1972).] If (X, ρ) is a complete metric space and $\{A_n\}_{n=1}^{\infty}$ is a countable family of open dense sets in X, then $\bigcap A_n$ is dense in X.

Proof. Suppose that for each i, A_i is open and dense in X. Let $r > 0$ and define

$$S_r(x) = \{y, \rho(x, y) < r\}$$

and $\overline{S_r(x)}$ denotes the closure of this set. Since A_1 is dense in X and also open, then for any nonvoid open set V we find x_1 and r_1 such that

$$\overline{S_{r_1}(x_1)} \subseteq V \cap A_1 \qquad 0 < r_1 < 1$$

We define now a sequence of sets $\{\overline{S_{r_n}(x_n)}\}$ in the following way: If $x_1, x_2, \ldots, x_{n-1}$ are always chosen, as well as $r_1, r_2, \ldots, r_{n-1}$, since A_n is dense and open, the set

$$A_n \cap S_{r_{n-1}}(x_{n-1})$$

is also open and nonvoid. Then there exists x_n and $r_n > 0$ such that

$$\overline{S_{r_n}(x_n)} \subseteq A_n \cap S_{r_{n-1}}(x_{n-1}) \qquad 0 < r_n < \frac{1}{n}$$

In this way we obtain a sequence of elements of X, $\{x_n\}$, and a sequence of positive numbers $\{r_n\}$, and we remark that this sequence is a Cauchy sequence. Indeed, let $\varepsilon > 0$ and N_ε such that for $n > N_\varepsilon$, $2/n < \varepsilon$. Let $n,m > N_\varepsilon$, and since we may suppose without loss of generality that $m > n$, we have

$$\rho(x_n, x_m) < 2r_n < \frac{2}{n}$$

since x_n and x_m are in $S_{r_n}(x_n)$. Let $\lim x_n = x$. Since $x \in S_{r_n}(x_n)$ for $m > n$, we obtain that x is in $\overline{S_{r_n}(x_n)}$ for all n and thus x is in $\cap A_n$ and in V. Thus for any V the intersection of V with $\cap A_n$ is a nonvoid set and this proves that $\cap A_n$ is dense.

The Baire theorem can be used to obtain a structure theorem for complete metric spaces. For this we need the following notions.

1.4.2 DEFINITION A set M in a metric space is called *nowhere dense* if its closure does not contain open sets which are nonvoid. A set in a metric space is of the *first category* if it is a union of a countable family of nowhere dense subsets of X.

The theorem of Baire can be formulated as follows.

1.4.3 THEOREM A complete metric space is not a set of the first category.

Chapter 2

BANACH SPACES

2.1 LINEAR SPACES

Let K be the field of real or complex numbers. The real numbers will be denoted by R and the complex numbers by C.

2.1.1 DEFINITION A linear space over the field K is a set L whose elements are called *points* and on the sets $L \times L$ and $K \times L$ there exist two functions having values in L and denoted "+" and "•" with the following properties for all $x, y, z \in L$ and $\alpha, \beta \in K$:

1. $x + (y + z) = (x + y) + z$.
2. $x + y = y + x$.
3. There exists in L one element called the zero element and denoted by 0 such that for all $x \in L$, $x + 0 = x$.
4. For any $x \in L$ there exists a point $(-x) = x \in L$ such that $x + (-x) = 0$.
5. $\alpha \cdot (x + y) = \alpha \cdot x + \alpha \cdot y$.
6. $(\alpha + \beta) \cdot x = \alpha \cdot x + \beta \cdot x$.
7. $(\alpha\beta) \cdot x = \alpha \cdot (\beta \cdot x)$.
8. $1 \cdot x = x$.

For simplicity we write $\alpha \cdot x = \alpha x$.

It is easy to see that the following assertion holds:

2.1.2 PROPOSITION If L is a linear space, then

1. 0 is the unique element of L satisfying 3.
2. The element $-x$ is also unique.

The proof being simple we omit it.

2.1.3 DEFINITION A *linear subspace* of a linear space L is a subset $L_1 \subset L$ such that it is a linear space with "+" and " " of L. We give now several examples of linear spaces. Other examples will be given later in the book.

2.1.4 EXAMPLE

1. Let R^n be the set of all sequences $x = (x_1, x_2, \ldots, x_n)$ with $x_i \in R$ and the functions "+" and "•" defined as follows:

$$x + y = (x_1 + y_1, x_2 + y_2, \ldots, x_n + y_n)$$

$$\alpha \cdot x = (\alpha x_1, \alpha x_2, \ldots, \alpha x_n)$$

where $y = (y_1, y_2, \ldots, y_n)$. R^n is a linear space over R.

2. Let C^n be the set of all sequences of complex numbers $z = (z_1, z_2, \ldots, z_n)$ with the functions "+" and "•" defined as in the above example. Then C^n is a linear space over C.

3. Let m be the set of all sequences $x = (x_1, x_2, \ldots)$ such that the sequence $\{x_n\}$ is bounded. This is also a linear space with "+" and "•" defined as follows:

$$x + y = (x_1 + y_1, x_2 + y_2, \ldots)$$

$$\alpha \cdot x = (\alpha x_1, \alpha x_2, \ldots)$$

where $y = (y_1, y_2, \ldots)$.

4. The set of all convergent sequences, denoted by c, is a linear subspace of m.

5. The set of all convergent sequences with the limit zero, denoted by c_0, is a linear space; moreover, it is a linear subspace of m and c.

6. The set ℓ^p of all sequences $x \in m$ such that $\Sigma |x_i|^p < \infty$, $1 \le p < \infty$, is a linear subspace. This follows applying the inequality of Minkowski.

7. The space C_I of all continuous functions on the interval [0, 1] with "+" and "•" defined as

$$(f + g)(t) = f(t) + g(t)$$
$$(\alpha f)(t) = \alpha \cdot f(t)$$

is a linear space.

8. Let X be a locally compact space (i.e., a topological space such that for any point $x \in X$ there exists a neighborhood whose closure is a compact set), and let μ be a Radon measure on X. Then the set of all equivalence classes of functions such that $\int |f(t)|^p \, d\mu < \infty$, is a linear space, denoted by $L^p(X, \mu)$.

9. Let C^α be the set of all real functions on [0, 1] such that

$$m(f) = \sup \frac{|f(t) - f(s)|}{|t - s|^\alpha} < \infty$$

Then with "+" and "$\cdot$" defined as in 7, C^α is a linear space (the space of functions satisfying the Lipschitz condition or order α).

2.1.5 REMARK The linear space over K = R is called a *real linear space* and when K = C, L is called a *complex linear space*.

2.2 LINEAR INDEPENDENCE

Let L be a linear space over K and A be a subset of L. A linear combination of elements of A is a vector of the form $\Sigma_{i=1}^n \lambda_i a_i$, where $\lambda_i \in K$ and $a_i \in A$.

2.2.1 DEFINITION A finite set A is *linearly independent* if for any linear combination of elements of A we have $\Sigma \lambda_i a_i = 0$, then necessarily $\lambda_i = \cdots \lambda_n = 0$. Of course, here $A = \{a_1, \ldots, a_n\}$.

A set A is called linearly independent if any finite subset is linearly independent.

2.2.2 DEFINITION A set A is called *linearly dependent* if it is not linearly independent.

2.2.3 EXAMPLE

1. In the space R^n, let $A = \{e_1, e_2, \ldots, e_n\}$, where $e_i = (0, 0, \ldots, 0, 1_i, 0, \ldots)$. Then it is easy to see that A is a linearly independent set.

2. In the space $C_{[0,1]}$ the set $A = \{t^i\}_{i=0}^{\infty}$ is linearly independent.

3. In the space C^n the set $\{z_1, z_2\}$, where $z_1 = (1, 0, \ldots, 0, 1)$ and $z_2 = (-1, 0, \ldots, 0, -1)$, is a linearly dependent set.

2.2.4 DEFINITION If A is a subset of a linear space L, then the set of all linear combinations of elements of A is called the *span* of A or the *space generated by* A.

2.2.5 REMARK Sometimes we consider only combinations with coefficients in R; we call this the *linear real span* of A or the *real linear space generated by* A.

From the definition of linear space and the order relation in the family of all linear subspaces of L (this order is by inclusion), we have the following proposition.

2.2.6 PROPOSITION The linear span of a set is the smallest linear subspace which contains A.

If A and B are two subsets in a linear space L, then $A \oplus B$ denotes the linear span of $A \cup B$. It is clear that $A \oplus B = B \oplus A$ and that the definition can be extended to a family of subsets $\{A_i\}$ of L: $\{A_i\}$, $\oplus A_i$ is the linear span of $\cup A_i$.

2.3 SETS IN LINEAR SPACES

Suppose that L is a linear space and $\mathcal{P}(L)$ is the family of all subsets of L. We define now several classes of elements of $\mathcal{P}(L)$ and some operations on them.

2.3.1 DEFINITION If A and B are two subsets in a linear space L, we define

1. $A + B = \{a + b \mid a \in A, b \in B\}$.
2. $sA = \{sa \mid a \in A\}$, where s is a scalar in K.
3. $-A = \{-a \mid a \in A\}$.
4. $a + A = \{a\} + A$ (the translation of A with the vector a).

5. A is *symmetric* if $A = -A$, and A is *balanced* if $tA \in A$ for all t, $|t| \leq 1$.
6. A is *convex* if $tx + (1 - t)y \in A$ for all $t \in [0, 1]$ and $x,y \in A$.
7. A is an *absorbing set* if for any $x \in L$ there exists a scalar $\varepsilon_x > 0$ such that $tx \in A$ for all t, $|t| \leq \varepsilon_x$.
8. A line throughout x and y is the set $\{sx + (1 - s)y\}_{s \in R}$ and a segment $[x,y] = \{sx + (1 - s)y \mid s \in [0, 1]\}$.

The following examples illustrate the above definition.

2.3.2 EXAMPLE

1. Let $L = R^n$ and

$$A = \{s \mid \Sigma |x_i| < 1, x = (x_1, \ldots, x_n)\}$$

Then it is easy to see that A is a symmetric, balanced, absorbing, and convex set.

2. In the space R^2 the set

$$A = \{x = (x_1, x_2) \mid x_i \geq 0\}$$

is a convex set, but not symmetric and not absorbing.

We define now an important notion related to convex sets.

By a *weighted mean* of a set A in a linear space L we mean any linear combination $\Sigma_{i=1}^n \lambda_i a_i$, $\lambda_i \geq 0$, $\Sigma_{i=1}^n \lambda_i = 1$.

2.3.3 PROPOSITION If C is a convex set, then it contains all of its weighted means.

Proof. Let $x = \Sigma_{i=1}^n \lambda_i a_i$, $\lambda_i \geq 0$, $\Sigma_{i=1}^n \lambda_i = 1$. It is clear that the assertion of the proposition is true for $n = 2$ by the definition of convexity. Suppose now the assertion is true for all $k \leq n$ and we prove it for $n + 1$. Indeed, we have

$$\tilde{x} = \sum_{i=1}^{n+1} s_i a_i = \sum_{i=1}^{n} s_i a_i + s_{n+1} a_{n+1} = s_{n+1} a_{n+1}$$

$$+ (1 - s_{n+1}) \sum \frac{s_i}{1 - s_{n+1}} a_i$$

and if we set

$$s_i' = \frac{s_i}{1 - s_{n+1}}$$

then $\Sigma s_i' = 1$ and the element $z_1 = \Sigma_{i=1}^{n}(s_i/1 - s_{n+1})a_i \in C$. Since we have $\tilde{x} = s_{n+1}a_{n+1} + (1 - s_{n+1})z$, then $\tilde{x} \in C$; hence the proposition is proved.

2.3.4 DEFINITION If A is a set in a linear space L, then the *convex hull* of A, denoted by conv A, is the set of all its weighted means.

In $\mathcal{P}(L)$, as we know, we consider an order, using "inclusion," and with this order we have the following proposition.

2.3.5 PROPOSITION The set conv A is the smallest convex set containing A.

Proof. It is obvious that conv A contains A. We show now that conv A is a convex set. Indeed, if $x = \Sigma s_i a_i$ and $y = \Sigma r_i b_i$ with s_i, r_i positive numbers such that $1 = \Sigma_{i=1}^{n} s_i = \Sigma_{i=1}^{m} r_i$, then

$$tx + (1 - t)y = \sum_{1}^{n+m} s_i'\tilde{x}_i \qquad s_i' \geq 0, \sum s_i' = 1$$

and so conv A is a convex set.

2.3.6 PROPOSITION If A and B are convex sets, then A + B is also a convex set.

2.3.7 EXAMPLE

1. If L = C and $A = \{z \mid |z| = 1\}$, then conv A is the set $\text{conv } A = \{z \mid |z| \leq 1\}$
2. If $A = \{0, 1\}$ and L = C, then conv A = [0, 1].

A related notion to a linear subspace of a linear space is the notion of an affine set.

2.3.8 DEFINITION An *affine set* is a translation of a subspace of L; i.e., there exists a linear subspace $L_1 \subset L$ such that $A = a + L_1$ for some $a \in L$.

2.3.9 EXAMPLE In R^2 an affine set is a translation of a line.

2.3.10 DEFINITION A cone in a linear space is a subset K with the property that for all $s > 0$, $sK \subset K$.

2.3.11 EXAMPLE

1. In R^2 the set $K = \{x = (x_1, x_2) \mid x_i > 0\}$ is a cone.
2. In $C_{[0, 1]}$ the set $K = \{f \mid f \geq 0\}$ is a cone.
3. Any linear subspace is a cone.

2.4 CLASSES OF SPACES: ISOMORPHIC SPACES, QUOTIENT SPACES, AND COMPLEMENTARY SPACES

Let L be a linear space over K and let L_1 and L_2 be two subspaces of L.

2.4.1 DEFINITION The subspaces L and L are called *complementary* subspaces if

1. $L_1 + L_2 = L$
2. $L_1 \cap L_2 = \{0\}$

2.4.2 EXAMPLE

1. Let $L = R^2$ and $L_1 = \{x \mid x = (x_1, 0)\}$, $L_2 = \{x \mid x = (0, x_2)\}$ and it is clear that $L = L_1 + L_2$ and $L_1 \cap L_2 = 0$.

2. Let $L = L^p[-1, 1]$ and $L_1 = \{f \mid f = 0$ a.e. on $[-1, 0]\}$, $L_2 = \{f \mid f = 0$ a.e. on $[0, 1]\}$. It is easy to see that L_1 and L_2 are complementary subspaces.

2.4.3 PROPOSITION If L is a linear space and L_1 is a linear subspace of L, then L_1 has a complementary subspace in the sense that there exists a linear subspace L_2 of L such that L_1 and L_2 are complementary subspaces.

Proof. Using Zorn's lemma, it is easy to see that there exists a linear independent set S_1 in L_1 which spans L_1; we call such a set the *Hammel basis* for L_1.

Also we can extend this to a Hammel basis for L. We denote by L_2 the span of $S - S_1$, where S is the Hammel basis for L. Clearly,

L_1 and L_2 satisfies the proposition.

For other results concerning a related notion see Section 2.18.

2.4.4 DEFINITION Let L be a linear space and L_1 be a linear subspace of L. The *quotient space* L/L_1 is the set whose elements are translates of L_1, i.e., elements of the form $a + L_1$, where $a \in L$.

The functions "+" and "·" are defined as follows:

1. $(a_1 + L_1) + (a_2 + L_1) = a_1 + a_2 + L_1$
2. $s \cdot (a + L_1) = sa + L_1$

In this way L/L_1 is a linear space. From the definition it is clear that the zero of L/L_1 is the subspace L_1.

A natural homomorphism of L onto L/L_1 can be defined by

$$\phi(a) = a + L_1$$

and it is obvious that $\phi^{-1}(0) = L_1$ and that this is one-to-one.

2.4.5 DEFINITION Two linear spaces L and $\tilde{L}$ on the same K are *isomorphic* if there exists a function $f: L \to \tilde{L}$ such that

1. $f(x + y) = f(x) + f(y)$
2. $f(sx) = sf(x)$
3. $f(L) = \tilde{L}$
4. $f(x) = 0$ if and only if $x = 0$

for all $x,y \in L$ and $s \in K$.

From the above remark we have the following.

2.4.6 PROPOSITION The spaces L_2 and L/L_1 are isomorphic.

2.4.7 DEFINITION A *maximal subspace* M of a linear space is any subspace of codimension 1.

2.4.8 DEFINITION A *hyperplane* is a translation of a maximal subspace.

2.4.9 PROPOSITION If $f: L \to K$ has the properties

1. $f(x + y) = f(x) + f(y)$
2. $f(sx) = sf(x)$

and is not identically zero, then $H_s = \{x \mid f(x) = s\}$ is a hyperplane.

Proof. Since the proof is easy, we omit it.

2.5 SEMINORMS AND NORMS ON LINEAR SPACES

Let L be a linear space over K.

2.5.1 DEFINITION A *seminorm* on a linear space L is any function $p\colon\ L \to R$ such that

1. $p(x) \geq 0$
2. $p(x + y) \leq p(x) + p(y)$ (triangle inequality)
3. $p(sx) = |s|p(x)$

for all $x, y \in L$ and $s \in K$. A seminorm is a *norm* if and only if $p(x) = 0$ if and only if $x = 0$.

2.5.2 EXAMPLE

1. Let $L = C_{[0,1]}$ and $t_0 \in [0, 1]$ be a fixed point. In this case it is easy to see that

$$p(f) = f(t_0)$$

is a seminorm on L.

2. Let $L = R^n$. Then the following functions are norms:

(a) $x \mapsto (\Sigma |x_i|^p)^{1/p}$ $x = (x_1, x_2, \ldots, x_n)$, $1 \leq p < \infty$

(b) $x \mapsto \max_i |x_i|$

(c) Let $L = L^p$ and consider the functions

$$f \mapsto \left\{\int | f(t) \, d\mu \, (t)\right\}^{1/p} \qquad 1 \leq p < \infty$$

All the above assertions follows from the Hölder and Minkowski inequalities.

The following result gives a connection between the seminorms and a class of convex sets.

2.5.3 DEFINITION A *balloon* is any convex, balanced, and absorbing set in a linear space.

2.5.4 PROPOSITION If p is a seminorm in a linear space, then the sets

$$B_p = \{x \mid p(x) < 1\} \qquad \tilde{B}_p = \{x \mid p(x) \le 1\}$$

are balloons.

The following result is in the opposite direction.

2.5.5 PROPOSITION For any balloon B there exists a seminorm p such that $B_p \subset B \subset \tilde{B}_p$.

Proof. Since B is a balloon, we define

$$p(x) = \inf\{t > 0 \mid x \in tB\}$$

and we show that p is a seminorm satisfying the proposition.

Indeed since B is a balloon for any x and $s \in K$, we have that then $sx \in |s|tB$ and this gives $p(sx) = |s|p(x)$. The property $p(x) \ge 0$ is trivial. We prove now the triangle inequality. Let x and y be in L and for $\varepsilon > 0$ we find that

$$x \in (p(x) + \varepsilon)B \qquad y \in (p(y) + \varepsilon)B$$

and thus

$$(x + y) \in (p(x) + p(y) + 2\varepsilon)B$$

and since ε is arbitrary the triangle inequality follows. Since B is absorbing the function p is well defined and from the definition of p the stated properties are clear.

Using Proposition 2.5.5, we can prove the following.

2.5.6 PROPOSITION If B is a balloon, then there exists a balloon B_1 such that $B_1 + B_1 \subseteq B$.

Proof. It is sufficient to take

$$B = \{x \mid p(x) < \tfrac{1}{2}\}$$

where p is the seminorm associated to B as in Proposition 2.5.5.

2.5.7 REMARK Sometimes p is called the *gauge* of B or the *Minkowski functional* of B.

2.6 LINEAR TOPOLOGICAL SPACES

Let L be a linear space and $\mathcal{D}$ be a topology on L.

2.6.1 DEFINITION $\mathcal{D}$ is called a *linear topology* for L if the functions "+" and "•" are continuous.

The topological space (L, $\mathcal{D}$), where $\mathcal{D}$ is a linear topology for L, is called a *linear topological space*.

2.6.2 EXAMPLE If L is a linear space and is the (weak) topology generated by a linear functional, then (L, $\mathcal{D}$) is obviously a linear topological space.

2.6.3 EXAMPLE If L is a linear space of all C^∞ functions on [0, 1] and the topology is generated by the family of seminorms,

$$p_n(f) = \sup_t \sum_{i=1}^{n} \frac{|f^i(t)|}{i!} \qquad n = 0, 1, 2, \ldots$$

then (L, $\mathcal{D}$) is a linear topological space.

Of central importance in linear analysis is a subclass of linear topological spaces, namely, the class of locally convex spaces. This class is considered in the following.

2.6.4 DEFINITION A *locally convex space* is a linear topological space (L, $\mathcal{D}$) with the property that for every neighborhood V of zero, there is a convex neighborhood V_c of zero such that $V_c \subset V$.

It is easy to see that in Examples 2.6.3 and 2.6.4 we have in fact locally convex spaces. As another important example we quote the following: Let $\{L_i \mathcal{D}_i\}_{i \in I}$ be a family of locally convex spaces and consider the product space $\Pi_i L_i$ with the product topology. Then it follows from the definition of product topology that $\Pi_i L_i$ with the product topology is a locally convex space.

Also the following example is of interest: Let $p \in (0, \infty)$ and denote by ℓ^p the space of all sequences $x = (x_i)_{i=1}^{\infty}$ with the topology generated by the seminorms

$$p_n(x) = \left(\sum_{i=1}^{n} |x_i|^p\right)^{1/p}$$

If we consider the space of all sequences x with the seminorms $\{p_n\}$ defined formally as above but $p \in (0, 1)$, then the corresponding space is a linear topological space but not locally convex.

We define now an important notion in linear topological spaces: that of bounded set.

2.6.5 DEFINITION A set M in a linear topological space is called *bounded* if for any neighborhood V of zero there exists $d > 0$ such that

$$\varepsilon M \subset V$$

for all $0 < \varepsilon < d$.

Equivalently, we have the Banach definition: M is bounded if and only if for every sequence $\varepsilon_n \to 0$, $\varepsilon_n \in K$, and $\{m_n\} \subset M$, then $\varepsilon_n m_n \to 0$.

2.7 BANACH SPACES

An important class of locally convex spaces is the class of Banach spaces. This class of spaces was considered by Banach (1922), Hahn (1922, and Wiener (1922). Their definition is as follows.

2.7.1 DEFINITION A linear space L is called a *normed space* if there exists a norm on L.

We denote the norm on a normed space simply by

$$x \mapsto \|x\|$$

It is easy to see that every norm generates a metric on L by

$$\rho(x, y) = \|x - y\|$$

2.7.2 DEFINITION A *Banach space* L is any normed space such that (L, ρ) is a complete metric space.

2.7.3 EXAMPLES

1. The space L of all continuous real-valued functions on the interval [0, 1] with the norm

$$f \mapsto \|f\| = \sup\{|f(t)| \mid t \in [0, 1]\}$$

is a Banach space.

2. For $p \in (1, \infty)$ the space ℓ^p of all sequences $x = (x_i)$,

$$(\Sigma\, |x_i|^p)^{1/p} = \|x\|$$

is a Banach space.

3. For $p = \infty$, the space ℓ^∞ of all sequences $x = (x_n)$ with the norm

$$\|x\| = \sup_i |x_i|$$

is a Banach space.

4. The space of all continuous functions on [0, 1] with the norm

$$f \mapsto \int_0^1 |f(t)|\, dt$$

is a normed space (this is an example of a normed space which is not a Banach space).

As is well known, if (X, ρ) is a metric space, then there exists a new metric space $(\tilde{X}, \tilde{\rho})$ such that

1. X is isometric with a dense set in $\tilde{X}$.
2. $(\tilde{X}, \tilde{\rho})$ is complete.

In the case of normed spaces the same method gives the following.

2.7.4 PROPOSITION Every normed space can be embedded isometrically in a Banach space, and the normed space is isometric with a dense subset of the Banach space.

2.8 LINEAR OPERATORS ON BANACH SPACES

Let L_1 and L_2 be two Banach spaces over the same scalar field K.

2.8.1 DEFINITION A linear operator $T: L_1 \to L_2$ is any function on L_1 with values in L_2 with the properties:

1. $T(x + y) = Tx + Ty$
2. $T(ax) = aTx$

for all $x,y \in L_1$ and $a \in K$.

2.8.2 DEFINITION The linear operator $T: L \to L$ is called continuous if T is continuous as a function.

2.8.3 EXAMPLE

1. In the space $C_{[0,1]}$ the operator

$$(Tf)(x) = \int_0^x f(t)\, dt$$

is continuous. This is the so-called Volterra operator.

2. Again on the space $C_{[0,1]}$ the operator

$$(Tf)(x) = f(t_0)$$

for some fixed $t_0 \in [0, 1]$ is a continuous operator.

We give now a necessary and sufficient condition for an operator to be continuous.

2.8.4 THEOREM The operator $T: L_1 \to L_2$, L_1 and L_2 are Banach spaces, is continuous if and only if there exists $0 < M < \infty$ such that

$$\|Tx\| \le M\|x\|$$

2.8.5 REMARK The norm of x is computed in the space L_1 and the norm of Tx is computed in the space L_2.

To prove Theorem 2.8.4 we first give the following lemma.

2.8.6 LEMMA A linear operator $T: L_1 \to L_2$ is continuous if and only if it is continuous at some point of L_1.

Proof. It is obvious that the condition is necessary. Suppose now that T is continuous at some point $x_0 \in L_1$. Since for any sequence $\tilde{x}_n \to \tilde{x}_0$ we have that

$$(\tilde{x}_n - \tilde{x}_0) + x_0 \to x_0$$

then

$$T(\tilde{x}_n - \tilde{x}_0 + x_0) \to Tx_0$$

and thus $T\tilde{x}_n \to T\tilde{x}_0$, which proves the lemma.

We now prove Theorem 2.8.4. It is obvious that the condition is sufficient. We prove that it is necessary. Indeed, in the contrary case, we find a sequence $\{x_n\}$, $\|x_n\| = 1$, such that

$$\|Tx_n\| \geq n$$

and thus the sequence $\{(1/n)x_n\}$ converges to zero, but

$$\lim \left\|T\left(\frac{1}{n} x_n\right)\right\| \geq n \frac{1}{n} = 1$$

which represents a contradiction, and the theorem is proved.

2.8.7 DEFINITION The norm of a continuous linear operator $T: L_1 \to L_2$ is the number

$$\|T\| = \sup\{\|Tx\| \mid \|x\| = 1\}$$

2.8.8 PROPOSITION On the linear space of all continuous linear operators defined on L_1 with values in L_2, the function

$$T \mapsto \|T\|$$

is a norm.

The set of all continuous operators $T: L_1 \to L_2$ is denoted by $L(L_1, L_2)$.

2.8.9 PROPOSITION If L_2 is a Banach space, then $L(L_1, L_2)$ is a Banach space.

Proof. Since all the assertions are clear except completeness, we give now a proof of this property. Let $\{T_n\}$ be a Cauchy sequence in $L(L_1, L_2)$ and $\varepsilon > 0$. Since

$$\sup_{\|x\|=1} \|(T_n - T_m)x\| \leq \varepsilon \qquad (*)$$

for all $n, m \geq N_\varepsilon$, we obtain that for each x, $\|x\| = 1$, the sequence $\{T_n x\}$ is a Cauchy sequence. Let

$$Tx = \lim T_n x$$

which can be defined in an obvious way for all x. Clearly, this is a linear operator. We show now that $T \in L(L_1, L_2)$ and $\lim T_n = T$.

From the relation (*) we obtain for $m \to \infty$ that

$$\sup_{\|x\|=1} \|(T_n - T)x\| < \varepsilon$$

and thus $T_n \to T$ and since for $n \geq N_\varepsilon$, $\|T_n - T\| \leq \varepsilon$, it follows that $T \in L(L_1, L_2)$.

2.8.10 DEFINITION If L_1 is a normed space and $L_2 = R$ or C is the field of scalars of L_1, the space $L(L, L)$ is called the *dual* of L_1 and is denoted by L_1^*.

Similarly we can define $(L_1^*)^* = L_1^{**}$ and by induction, $L_1^{*n} = (L^{*(n-1)})^*$.

In the case $L_1 = L_2 = L$, $L(L, L) = L(L)$ has some special properties which are given in the following.

2.8.11 PROPOSITION $L(L)$ is a Banach algebra with unit.

For definition and properties of Banach algebras, see Chapter 4.

The identity in $L(L)$ is the operator

$$Ix = x$$

for all $x \in L$.

2.8.12 DEFINITION An operator $T \in L(L)$ is called *invertible* if there exists an element, noted by $T^{-1} \in L(L)$, such that

$$T^{-1}T = TT^{-1} = I$$

2.8.13 THEOREM An operator $T \in L(L)$ is invertible if and only if there exists $m > 0$ such that for all $x \in L$,

$$\|Tx\| \geq m\|x\|$$

The condition is obviously necessary. Indeed, if T^{-1} exists, we have

$$\|x\| = \|TT^{-1}x\| \leq \|T^{-1}\| \cdot \|Tx\|$$

and thus

$$\|Tx\| \geq \frac{1}{\|T^{-1}\|}\|x\|$$

We prove now that it is also sufficient. Indeed, from the condition it follows that T is one-to-one. On the space $TL = \{y \mid y = Tx \text{ for some } x \in L\}$ we define $T^{-1}y = x$, and clearly we have

$$\|T^{-1}y\| \leq \frac{1}{m\|y\|}$$

which is continuous as an operator on TL with values in L.

2.8.14 REMARK The above proof shows that a more general result holds.

2.8.15 THEOREM If $T: L \to L$ is a continuous operator and for some $m > 0$, $\|Tx\| \geq \|x\| \cdot m$, then there exists a continuous operator $S: TL_1 \to L_1$ such that

$$STx = x$$

for all $x \in L_1$.

2.9 UNIFORMLY CONVEX AND ROTUND BANACH SPACES: SOME GENERALIZATIONS

In applications it is of some interest if the Banach space under consideration has some additional properties. Such properties are shared by the classes of classical Banach spaces L^p and ℓ^p for $p \in (1, \infty)$.

We begin with the so-called uniformly convex spaces considered by Clarkson (1936).

2.9.1 DEFINITION A Banach space L is called *uniformly convex* if for each $\varepsilon > 0$ there exists $\delta(\varepsilon) > 0$ such that for all $x,y \in L$, $\|x\| \le 1$, $\|y\| \le 1$, and $\|x + y\| \ge 2 - \delta(\varepsilon)$, then $\|x - y\| \le \varepsilon$.

The following is an important property of uniformly convex spaces.

2.9.2 PROPOSITION If L is a uniformly convex Banach space and $\{x_n\}$ is a sequence with the following properties:

1. $\|x_n\| = 1$
2. $\lim_{n,m\to\infty} \|x_n + x_m\| = 2$

then $\{x_n\}$ is a Cauchy sequence.

Proof. First we recall that condition 2 means that for $\varepsilon > 0$ there exists N_ε such that for $n,m \ge N_\varepsilon$, $\big|\|x_n + x_m\| - 2\big| \le \varepsilon$.

If for all n, $\|x_n\| \le 1$, for N_ε we have that $\|x_n - x_m\| < \varepsilon$ since L is uniformly convex.

Now if $\|x_n\| \to 1$, then we define $y_n = x_n/\|x_n\|$, and we remark that the following relations hold:

1. $\|y_n\| = 1$
2. $\|y_n + y_m\| \to 2$

Indeed since 1 is obvious, we prove only 2. We have

$$2 \ge \|y_n + y_m\| = \left\| \frac{x_n + x_m}{\|x_n\|} + \frac{\|x_n\| - \|x_m\|}{\|x_n\| \cdot \|x_m\|} \right\|$$

$$\ge (\|x_n + x_m\| - \big|\|x_n\| - \|x_m\|\big|)\|x_n\| \to 2$$

and 2 follows. We conclude that $\{y_n\}$ is a Cauchy sequence and this obviously implies that $\{x_n\}$ is a Cauchy sequence.

As a generalization of uniformly convex spaces, we give Lovaglia's notion (1955) of "locally uniformly convex spaces."

2.9.3 DEFINITION A Banach space L is called *locally uniformly convex* if for each $\varepsilon > 0$ and for each x with $\|x\| = 1$ there exists $\delta_x(\varepsilon) > 0$ such that for all y, $\|y\| \le 1$ and $\|x + y\| > 2 - \delta_x(\varepsilon)$, the inequality

$$\|x - y\| < \varepsilon$$

holds.

It is obvious that every uniformly convex space has the property in Definition 2.9.3. An interesting property of locally uniformly convex spaces (l.u.c.) refers to the convergence of a special sequence of elements.

2.9.4 DEFINITION A sequence of elements of a Banach space *converges weakly* to x if for all x^* in the dual space,

$$\lim x^*(x_n) = x^*(x)$$

We write this more concisely as $x_n \rightharpoonup x$.

2.9.5 THEOREM If $\{x_n\}$ is a sequence of elements in a l.u.c. space and

1. $x_n \rightharpoonup x$
2. $\|x_n\| \to \|x\|$

then $\lim x_n = x$.

Proof. The proof of a more general result is given in 2.17.9.

2.9.6 THEOREM If C is a closed convex set in a uniformly convex space, then there exists a unique element in C of smallest norm.

Proof. If $0 \in C$ the assertion is clear. Suppose that this is not the case. Let

$$d = \inf\{\|x\| \mid x \in C\}$$

and obviously $d > 0$. Let $x_n \in C$ such that $\|x_n\| \to d$. Since C is convex, we have that for all n and m

$$\|x_n + x_m\| \ge 2d$$

and it is also clear that

$$\lim \|x_n + x_m\| = 2d$$

If we define $y_n = x_n/d$, then $\{y_n\}$ satisfies the conditions of Proposition 2.9.2. Thus $\{y_n\}$ is a Cauchy sequence; let $y = \lim y_n$. Clearly $\tilde{y} = dy$ is an element of C satisfying $\|\tilde{y}\| = d$.

The existence being proved we prove now the uniqueness. Suppose that there exist two points y and $\tilde{y}$ in C such that

$$\|\tilde{y}\| = \|y\| = d$$

and since C is a convex set, we have that

$$2 \leq \left\|\frac{y + y}{2d}\right\| \leq 2$$

and $\|(\tilde{y} - y)/(2d)\| \geq \varepsilon$, implies the relation that $\|y + \tilde{y}\| = 2d$ which is not true. Thus $y = \tilde{y}$.

Another interesting class of Banach spaces is the so-called rotund spaces.

2.9.7 DEFINITION A point x of a convex set C is called an extreme point if x does not belong to any segment joining two points of C.

2.9.8 DEFINITION A normed space is called *rotund* (or *strictly convex*) if any point of $S_{(0,1)} = \{x \mid \|x\| \leq 1\}$ of norm 1 is an extreme point.

2.9.9 PROPOSITION If C is a convex closed set in a rotund space, then there exists only one point of minimal norm.

Also it is easy to see that

2.9.10 PROPOSITION Every uniformly convex space is rotund.

As a consequence of Proposition 2.9.9, we have

2.9.11 PROPOSITION If $x^* \in L^*$ and L is rotund, then there exists at most one point x_0 such that

1. $\|x_0\| = 1$

2. $f(x_0) = 1 = \|f\|$

2.10 THE HAHN-BANACH EXTENSION THEOREM

Suppose that L is a Banach space over the reals and L_1 is a subspace which is (without loss of generality) closed.

Suppose that $x_1^* \in L_1^*$. Then the following problem appears naturally: there exists an element $F \in L^*$ such that its restriction to L_1 is exactly x_1^*.

If F exists, then it is called the *extension* of x_1^*. Another aspect of this problem is concerned with the relation between $\|F\|$ and $\|x_1^*\|$. If $\|x_1^*\| = \|F\|$, then we say that F is the norm-preserving extension of x*. A famous result of Hahn's is that such norm-preserving extensions exist. The generalization by Banach refers to arbitrary seminorms. More precisely, the result is as follows:

2.10.1 THEOREM If $x^* \in L_1^*$ and for some seminorm $p: L \to L$,

$$x^*(x) \leq p(x) \qquad x \in L_1$$

then there exists an extension F of x* such that $F(x) \leq p(x)$.

Proof. We suppose, without loss of generality, that x* is not identically zero. Let x_0 be an element of L not in L_1 and consider the subspace $\tilde{L}_1 = \{x_0, L_1\}$. We remark that every element of this space can be written in a unique form as

$$\tilde{x} = ax_0 + y \qquad y \in L_1$$

Indeed, in the contrary case,

$$0 = (a - b)x + z \qquad z \in L_1$$

which gives that $x_0 \in L_1$.

We define now the functional x_1^* on the subspace $\tilde{L}_1$ as follows:

$$x_1^*(\tilde{x}) = x^*(y) + ac$$

where c is a number which is to be determined in such a way that

$$x_1^*(\tilde{x}) \leq p(\tilde{x})$$

We show now that such a number exists. Since x* and p on L_1 have the property $x^*(x) \leq p(x)$ and since $\tilde{x} = ax_0 + x$, we have

$$x^*\left(\frac{x}{a}\right) + c \le p\left(\frac{x}{a} + x_0\right) \qquad a > 0$$

and

$$x^*\left(\frac{x}{a}\right) + c \ge -p\left(\frac{-x}{a} - x_0\right) \qquad a < 0$$

for all $x \in L_1$.

From these it follows that for c it will be sufficient to have

$$x^*(x) - p(x - x_0) \le c \le -x^*(y) + p(y + x_0)$$

for all $x,y \in L_1$. Now since

$$x^*(x) + x^*(y) = x^*(x + y) \le p(x + y) \le p(x - x_0 + x_0 + y)$$
$$\le p(x - x_0) + p(x_0 - y)$$

and if we define

$$m = \sup\{x^*(x) - p(x - x_0)\}$$
$$M = \inf\{-x^*(x) + p(x_0 + y)\}$$

then $m \le M$, and for c we can choose any number between m and M and the inequality

$$\tilde{x}_1^*(\tilde{x}) \le p(\tilde{x})$$

is satisfied.

Next, by applying Zorn's lemma, we show that F exists. Indeed let ε be the set of all extensions of x^* satisfying the inequality with respect to the seminorm. We can define a partial order in this set as follows: we say that two extensions f and g of x^* are in the relation $f < g$ if the following conditions are satisfied:

1. The subspaces L_f and L_g on which they are defined, respectively, satisfy the relation $L_f \subset L_g$.
2. For all $x \in L_f$, $f(x) = g(x)$.

Using this order relation it is an easy exercise to show that we can apply Zorn's lemma, and thus we obtain a maximal element, i.e., a maximal extension of x^*. If the space on which this maximal extension exists is not L, then we apply step 1 and obtain a new extension which clearly contradicts the fact that our extension is maximal.

The Hahn-Banach extension theorem is thus proved.

We give now several corollaries of this fundamental theorem.

2.10.2 COROLLARY If L is a real linear subspace and x_0 an arbitrary element of L, then there exists $x^* \in L^*$ such that

1. $x^*(x_0) = 1$
2. $\|x^*\| = \|x_0\|$

Proof. We consider the space $L_1 = \{ax_0\}$ and define on L_1

$$x^*(ax_0) = a$$

and the assertion follows from the Hahn-Banach theorem.

2.10.3 COROLLARY If L_1 is a closed subspace of a Banach space L, then there exists, for any $x_0 \notin L_1$, an element $F \in L^*$ such that

1. $(x) = 0$ for all $x \in L_1$
2. $F(x_0) = d = \inf\{\|x_0 - y\| \mid y \geqq L_1\}$
3. $\|F\| = 1$

Proof. We recall that the number d is the distance between the point x_0 and the set L_1. For the proof we consider the space generated by the element x_0 and the elements of L_1, $\tilde{L}_1$. We define the linear functional on $\tilde{L}_1$ by the formula

$$f(\tilde{x}) = ad$$

where $\tilde{x} = ax_0 + x$, $x \in L_1$. The definition is correct since $\tilde{x}$ can be written in a unique form as above. We show now that this functional is in $\tilde{L}_1^*$. Indeed, we have

$$\|\tilde{x}\| = \|ax_0 + x\| = \left\|-a\left(\frac{-1}{a}x - x_0\right)\right\| = |a|\left\|\frac{-1}{a}x - x_0\right\| \geq |a|d$$

and thus

$$|f(\tilde{x})| \leq |a|d \leq \|\tilde{x}\|$$

Thus f is in $\tilde{L}_1^*$ and by the Hahn-Banach theorem we obtain a norm-preserving extension of this to L; clearly this extension has the norm ≤ 1. We show that the norm of f is 1. Indeed, for any $\varepsilon > 0$, we find x_ε such that

$$\|x_\varepsilon - x_0\| \leq d + \varepsilon$$

and if we set

$$y_\varepsilon = \frac{x_\varepsilon - x_0}{\|x_\varepsilon - x_0\|}$$

we obtain

$$\|f\| \geq \frac{d}{\|x_\varepsilon - x_0\|} = \frac{d}{d + \varepsilon}$$

and this gives that the norm is 1.

2.10.4 REMARK The Hahn-Banach theorem asserts only the existence of an extension and nothing about uniqueness; general speaking, uniqueness is a rare event. For this topic see Phelps (1960).

The following example of Bronson and Steiner (1973) illustrates the power of the Hahn-Banach theorem.

2.10.5 EXAMPLE It is a simple matter to extend a linear functional on the entire space when it is known on a one-dimensional supspace.

Suppose now that we have an infinite dimensional space and let $x_0 \in L$. We consider the Hammel basis of L which contains x_0.

We define, for any $x \in L$,

$$f(x) = f(a_0 x_0 + \sum_i a_i x_i) = a_0$$

and the problem is the following: Is f in L*? We show that in general the answer is no. Indeed, let $L = \ell^1$ and $x_0 = (1, 0, 0, \ldots)$. Let $x_k = (10^k, 1/k, 1/k^2, \ldots)$, $k = 2, 3, \ldots$. Then clearly $\{x_k\}$ is a linear independent set. We consider the Hammel basis which contains this set. Consider $\{y_k\}$, where $y_k = (0, 1/k, 1/k^2, \ldots)$. It is clear that these are in ℓ^1 and

$$|f(y_k)| = |f(-10^k x_0 + x_k)| = 10^k > 10^k \|y_k\|$$

and thus f is unbounded.

2.11 EXTENSION THEOREMS FOR COMPLEX BANACH SPACES

In what follows we give some results about the extension problem in the case of complex spaces.

2.11.1 THEOREM If $f: L_1 \to C$ is a functional on a subspace of a complex space L having the property

$$|f(x)| \leq p(x) \qquad \text{for all } x \in L_1$$

where p is a seminorm, then there exists an extension $F: L \to C$ of f satisfying the property

$$|F(x)| \leq p(x) \qquad \text{for all } x \in L$$

For the proof we need the following lemma.

2.11.2 LEMMA Every linear functional F on a complex space L has the form

$$F(x) = U(x) - iU(ix)$$

where U is a real linear functional, i.e., a functional on L as a real space.

Proof. If $z = x + iy$ is a complex number, we set

$$x = \text{Re } z \qquad y = \text{Im } z$$

and thus

$$z = \text{Re } z - i \text{ Re } (iz)$$

From this we obtain that

$$F(x) = \text{Re } F(x) - i \text{ Re } F(ix)$$

and clearly Re F is a real linear functional on L. Conversely, if F has the form $U(x) - iU(ix)$, where U is a real linear functional, then F is a functional on L as a complex linear space. For this it is sufficient to show that

$$F(ix) = iF(x)$$

This can be proved as follows:

$$F(ix) = U(ix) - iU(-x) = U(x) + iU(x) = iF(x)$$

Since we need the extension only for the case $p(x) = \|x\|$, we prove that $\|F\| = \|U\|$.

Indeed it is obvious that $\|U\| \leq \|F\|$.

Since for any x, $a = |F(x)|/F(x)$ [$F(x) \neq 0$], we have

$$|F(x)| = F(ax) = U(ax) \leq \|U\| \cdot \|ax\| \leq \|U\| \cdot \|x\|$$

and thus $\|F\| \leq \|U\|$.

Proof of Theorem 2.11.1 [The Bohnenblust-Sobczyk theorem (1938)] Let f be a linear functional on L_1; by Lemma 2.11.2 it is of the form

$$f(x) = u(x) - iu(ix)$$

where u is a real linear functional. By the Hahn-Banach theorem we find the extention U of u such that $\|u\| = \|U\|$. Clearly,

$$F(x) = U(x) - iU(ix)$$

satisfies the theorem.

2.11.3 REMARK Corollaries 2.10.2 and 2.10.3 remain true for the case of complex Banach spaces.

We show now that there exists a nice connection between L and the dual L**.

2.11.4 THEOREM If L is a Banach space, then there exists an application

$$\phi: L \to L^{**}$$

with the following properties:

1. $\|x\| = \|\phi(x)\|$.
2. ϕ is a linear operator.

Proof. For any $x \in L$ we define the linear functional $f_x \in L^*$ with the following properties:

1. $f_x(x) = \|x\|^2$
2. $\|f_x\| = \|x\|$

and the existence of f_x is a consequence of the Hahn-Banach theorem or of Theorem 2.11.1.

We define a linear functional on L^* by the formula

$$F_x(f) = f(x)$$

Using the properties of f_x, it follows that

$$\phi: x \to F_x$$

satisfies all the requirements.

2.11.5 REMARK Any operator with the property $\|Tx\| = \|x\|$ is called an *isometry*. From the theorem we have that ϕ is an isometric operator.

2.11.6 REMARK If $L^{**}/\phi(L)$ is of finite dimension, then L is called *quasi-reflexive*. For details about these spaces see the paper of Civin and Yood (1957).

2.11.7 DEFINITION A Banach space L is called *reflexive* if $\phi(L) = L^{**}$.

There exists a Banach space L for which L and L^{**} are isometric and L is not reflexive [Day (1958) and James (1957)].

2.12 THREE BASIC THEOREMS OF LINEAR ANALYSIS

Two of the most useful theorems of functional analysis are the so-called principle of uniform boundedness and the so-called closed graph theorem.

The purpose of the present section is to give a unified approach to these theorems as well as to the open mapping theorem.

For this we need the following fundamental lemma, whose proof has many common points with earlier proofs of the open mapping theorem.

2.12.1 LEMMA [Zabreiko (1959)] Let L be a Banach space and $F: L \to R$ a function with the following properties:

1. $\lim_{t\to 0} F(tx) = 0$ for all $x \in L$.
2. For each convergent series, $\Sigma_{n=1}^{\infty} x_n$, $F(\Sigma_{n=1}^{\infty}) \leq \Sigma_{n=1}^{\infty} F(x_n)$.

In this case G is a continuous function.

Proof. First we remark that from 2 it follows that F is *subadditive*, i.e., for any x,y in L,

$$F(x + y) \leq F(x) + F(y)$$

We prove that F is continuous at $x = 0$. Let $\varepsilon > 0$ and define

$$G_{\varepsilon} = \{x \mid x \in L,\ F(x) + F(-x) \leq \frac{\varepsilon}{2}\}$$

From 1 it follows that for any $x \in L$ there exists an integer n such that $x \in G_{n\varepsilon}$, and thus $L = \bigcup_{n=1}^{\infty} G_{n\varepsilon}$; clearly, $L = \bigcup_{n=1}^{\infty} \bar{G}_{n\varepsilon}$. By Baire's theorem $\bar{G}_{n_0\varepsilon}$ contains an open set, say

$$S(x_0, r) = \{x \mid \|x - x_0\| < r\}$$

Since $\bar{G}_{\varepsilon}$ is symmetric, it also contains $S(-x_0, r)$, and since F is subadditive, we can suppose, without loss of generality, that $\bar{G}_{\varepsilon}$ contains an open sphere of the form $S(0, d_{\varepsilon})$. We can assume that $d_{\varepsilon} < \varepsilon$, and thus for all x in this sphere we have

$$F(x) \leq 2\varepsilon$$

Indeed, let $\tilde{G}_{\varepsilon} = G_{n_0\varepsilon} \cap S(0, d\varepsilon)$, which is dense in $S(0, d\varepsilon)$, and we choose $x_1 \in \tilde{G}_{\varepsilon}$ such that $x - x_1 \in S(0, d_{\varepsilon}/2)$. Suppose that $x_1, x_2, \ldots, x_n$ are chosen; we choose $x_{n+1} \in \tilde{G}_{\varepsilon/2^n}$ such that $(x - (x_1 + \cdots + x_n) \in S(0, \varepsilon/2^n)$.

It is clear that $x = \Sigma x_n$, and from 2 it follows that

$$F(x) \leq \sum F(x_n) \leq \sum \frac{\varepsilon}{2^n} = 2\varepsilon$$

Thus F is continuous at the point $x = 0$. The continuity at every point of L follows from the following lemma.

2.12.2 LEMMA If $h: L \to R$ has the following properties:

1. $h(x + y) \le h(x) + h(y)$
2. $h(x) \ge 0$ if $\|x\| \ge M$

then if $h(0) = 0$ and h is continuous at $x = 0$, then it is continuous at every point in L.

Proof. Let $y = \tilde{M}x$ and thus $0 \le h(y) \le \tilde{M}h(x)$, where M is an integer such that $|M - \tilde{M}| \le 1/2$. This gives that $h(x) \ge 0$. Now let $x_n \to x$, and thus $x_n - x \to 0$. We have

$$h(x) = h(x - x_n + x_n) \le h(x - x_n) + h(x_n)$$

and thus

$$h(x) - h(x_n) \le h(x - x_n)$$

Similarly

$$h(x_n) - h(x) \le h(x - x_n)$$

which gives that

$$|h(x_n) - h(x)| \le h(x - x_n)$$

Clearly this implies the continuity of h at x.

As a consequence of Lemma 2.12.1 we obtain three basic theorems of linear analysis. It is worth remarking that the proof is valid in more general spaces, namely, those in which we can apply Baire's theorem.

For the applications we need the following simple fact: that condition 2 is satisfied when F is lower semicontinuous and subadditive. Indeed, for each $x = \Sigma x_n$ we have

$$F(x) \le \liminf F\left(\sum_{n=1}^{n} x_n\right) \le \lim \sum_{m=1}^{m} F(x_n) = \sum F(x_n)$$

Also for applications it is useful to have the following two consequences of Lemma 2.12.1.

2.12.3 LEMMA Let I be an index set and for each $i \in I$

$$F_i: L \to R$$

such that for each i, F_i is subadditive and positive; also

$$\lim_{t \to 0} \sup F_i(tx) = 0$$

and each F_i is continuous. In this case,

$$\lim_{\|x\| \to 0} \sup F_i(x) = 0$$

Proof. Since each F_i is continuous, it is clear that

$$x \mapsto \sup F_i(x)$$

is subadditive and lower semicontinuous. Thus we can apply Lemma 2.12.1 and this gives the assertion of the lemma.

2.12.4 LEMMA [Gelfand's Lemma (1936)] If $f: L \to R$ is an arbitrary subadditive lower semicontinuous function such that

$$f(ax) = |a| f(x)$$

then there exists $M > 0$ such that

$$f(x) \leq M\|x\|$$

We now give three basic theorems of linear analysis.

2.12.5 THEOREM (The Principle of Uniform Boundedness) Let I be an index set and for each $i \in I$, T_i be a linear operator

$$T_i: L \to L_1 \qquad T_i \in L(L, L_1)$$

where L and L_1 are Banach spaces. Then

1. There exists $M < \infty$ such that $\|T_i\| \leq M$ for all $i \in I$.

or

2. $\sup_i \|T_i x\| = \infty$ for all $x \in G_\delta \subset L$.†

---

†G_δ means a set which is the intersection of a sequence of open sets.

Proof. For each $i \in I$ let

$$x \mapsto \|T_i x\|$$

which is obviously a continuous subadditive function and in this case

$$x \mapsto \sup_i \|T_i x\|$$

is lower semicontinuous. Now, if the first assertion is not true, it follows that

$$G_n = \{x \mid \sup_{i \in I} \|T_i x\| > n\}$$

is dense in L for each n and thus on

$$G_\delta = \cap G_n$$

we have

$$\sup \|T_i x\| = \infty$$

and the assertion is proved.

We give now a theorem known as "the open mapping theorem." For the formulation of the theorem we need the following definition.

2.12.6 DEFINITION A function $f: L \to L_1$ is called *surjective* if $f(L) = L_1$, i.e., for all $y \in L_1$ there exists at least one $x \in L$ such that $f(x) = y$.

2.12.7 THEOREM If $T: L \to L_1$, where L and L_1 are Banach spaces, is a linear continuous surjective function, then there exists $d > 0$ such that

$$T(S_L(0, 1)) \supset dS_{L_1}(0,1)$$

Proof. For each $x \in L$ we define

$$x \mapsto F(x) = \inf\{\|y\| \mid Tx = y\}$$

which satisfies the conditions of Lemma 2.12.1. Thus F is continuous, and this implies the theorem.

We give now the so-called closed graph theorem. First we give the following definition.

2.12.8 DEFINITION If $f: L \to L_1$, where L and L_1 are Banach spaces, is a linear operator, then f is called *closed* if for any sequence $\{x_n\} \subset L$ with $x_n \to x$ and $fx_n \to y$, we have $y = fx$.

2.12.9 THEOREM (The Closed Graph Theorem) If $T: L \to L_1$ is a closed linear operator, then $T \in L(L, L_1)$.

Proof. For each x we define

$$x \mapsto \|Tx\|$$

which obviously satisfies condition 1 of Lemma 2.12.1. For condition 2, let Σx_n be a convergent series in L and

$$\sum_{n=1}^{\infty} \|Tx_n\| < \infty$$

(in the contrary case there is nothing to prove). Obviously

$$y = \sum_{n=1}^{\infty} Tx_n = \lim_{m\to\infty} \sum_{n=1}^{m} Tx_n$$

and since T is closed, we have

$$T\left(\sum_{n=1}^{\infty} x_n\right) = \sum_{n=1}^{\infty} Tx_n$$

and

$$\left\|T\left(\sum_{n=1}^{\infty} x_n\right)\right\| = \left\|\sum_{n=1}^{\infty} Tx_n\right\| \le \sum_{n=1}^{\infty} \|Tx_n\|$$

and 2 is also satisfied. In this case T is continuous and the theorem is proved.

We now give some applications of the above theorems.

2.12.10 THEOREM If $\{T_n\}$ is a sequence of linear operators

$$T_n: L \to L_1$$

where L and L_1 are Banach spaces, and for each $x \in L$,

$$\sup \|T_n x\| < \infty$$

then $\sup \|T_n\| < \infty$.

Proof. Follows immediately from Theorem 2.12.5.

2.12.11 THEOREM If $T_n: L \to L_1$, L and L_1 as above, and for each x, $\lim T_n x$ exists, then $Tx = \lim T_n x$ is a linear operator in $L(L, L_1)$.

Proof. Follows from Theorem 2.12.10.

2.12.12 THEOREM [E. Landau (1907)] Let $p > 1$ and $\{a_n\}$ be a sequence of complex numbers such that for any sequence $b = \{b_n\} \in \ell^q$ $(1/p + 1/q = 1)$ and $\{a_n b_n\} \in \ell^1$, then $a = \{a_n\} \in \ell^p$.

Proof. For each n we define on the space ℓ^q the functional

$$f_n(b) = \sum_{i=1}^{n} a_i b_i$$

and it is easy to see that f_n is a bounded, linear functional and

$$\|f_n\| = \left(\sum_{i=1}^{n} |a_i|^p \right)^{1/p}$$

Since $f_n \to f$, where $f(b) = \Sigma_{i=1}^{\infty} a_i b_i$, we have that $\|f\| < \infty$, and since $\lim \|f_n\| = \|f\|$, the assertion follows.

2.12.13 THEOREM If $\{f_n\} \subseteq L^*$ is uniformly bounded, then

$$\{x \mid \lim f_n(x) \text{ exists}\}$$

is a closed subspace of L.

Proof. It is clear that for each x, $\{f_n(x)\} \in m$, the space of all bounded sequences and $d: x \to (f_n(x))$ is a continuous application of L into the closed subspace c, the set of all convergent sequences. The assertion follows since $d^{-1}(c)$ is a closed subspace of L.

Another interesting application refers to the theory of trigonometric series. As is well known, for any periodic continuous function $f\colon [-\pi, \pi] \to C$ we can define the Fourier coefficients $\{a_n(f)\}$ and $\{b_n(f)\}$ and let

$$U_n(f)(t) = \frac{a_0}{2} + \sum_{k=1}^{n} (a_k \cos kt + b_k \sin kt)$$

The problem is about approximations of f using $U_n(f)$.

The first example of a continuous function for which $\{U_n(f)\}$ does not converge at some point of $[-\pi, \pi]$ was constructed by DuBois-Reymond. Later a more simple example was constructed by L. Fejer. [See Zygmund (1959).] We prove now an existence of a continuous function with the above property as an application of the principle of uniform boundedness.

For terminology we remark that when $\{U_n(f)\}$ is not convergent at some point, we say that it is divergent at that point.

Consider the Banach space of all continuous periodic functions on $[-\pi, \pi]$ and for each point $t_0 \in [-\pi, \pi]$ the continuous linear functional

$$F_n(f) = U_n(f)(t_0)$$

If for each t_0, $\sup |F_n(f)| < \infty$, by the principle of uniform boundedness it follows that $\sup \|U_n\| < \infty$. We shall show that this is not the case.

Indeed, let

$$g_n(t) = \frac{\sin(n + 1/2)x}{2\pi} \sin \frac{1}{2x}$$

and we have for any function $f \in C_{[-\pi,\pi]}$, $f(-\pi) = f(\pi)$, the following relation:

$$U_n(f)(t) = \int_{-\pi}^{\pi} f(x) g_n(x - t)\ dt$$

and

$$\|U_n\| = \int_{-\pi}^{\pi} |g_n(t)|\, dt$$

Now,

$$\|U_n\| \geq \sum_{k=0}^{n-1} \int_{(4k+1)/(4k+2)}^{(4k+3)/(4k+2)} |g_n(t)|\, dt > \frac{1}{2\pi} 2^{1/2} \sum_{k=0}^{n-1} \int \frac{dx}{\sin(1/2)x}$$

$$> \frac{1}{2\pi} 2^{1/2} \sum_{k=0}^{n-1} \int 2\frac{dx}{x} > \frac{1}{\pi} 2^{1/2} \sum_{k=0}^{n-1} \log \frac{4k+3}{4k+1}$$

$$> \frac{1}{\pi} 2^{1/2} \sum_{k=0}^{n-1} \frac{2}{4k+3}$$

since $\log(1 + t) \geq t/(1 + t)$. This inequality proves that $\sup \|U_n\| = \infty$ and the existence of a function as in the Du Bois-Reymond example is proved.

Another useful application, which will be used later, in the theory of operators is concerned with the so-called Banach limit, or "glim."

2.12.14 THEOREM On the space of all bounded sequences m there exists a continuous linear functional denoted by "glim" having the following properties:

1. If $a_n \geq 0$, then $\text{glim}\,(a_n) \geq 0$.
2. $\text{glim}\,(a_n) = \text{glim}\,(a_{n+1})$ (i.e., translation invariance).
3. If (a_n) is convergent, then $\text{glim}\,(a_n) = \lim\,(a_n)$.
4. If $a_n \in R$, then $\underline{\lim}\,(a_n) \leq \text{glim}\,(a_n) \leq \overline{\lim}\,(a_n)$.

Proof. First we remark that we need consider only real sequences since the extension to complex sequences is obvious.

On the space of all bounded real sequences m_r we define the function

$$a = (a_n) \mapsto \pi(a, n_1, n_2, \ldots, n_k) = \overline{\lim_{n\to\infty}} \frac{1}{k} \sum (a_{n+n_j})$$

and

$$a = (a_n) \mapsto p(a) = \inf_{(n_1,\ldots,n_k)} \pi(a, n_1, n_2, \ldots, n_k)$$

It is obvious that

$$p(sa) = sp(a) \qquad \text{for all } s \geq 0$$

and we show that

$$p(x + y \leq p(x) + p(y)$$

i.e., p is subadditive.

Indeed, for any $\varepsilon > 0$ let $(n_1, n_2, \ldots, n_p)$ and $(n_1, n_2, \ldots, n_k)$ such that

$$\pi(a, n_1, n_2, \ldots, n_p) \leq p(a) + \varepsilon$$
$$\pi(b, n_1, n_2, \ldots, n_k) \leq p(b) + \varepsilon$$

Defining

$$n_{j,l} = n_j + n_l$$

we have that

$$\pi(a + b, n_{1,1}, n_{1,2}, \ldots, n_{p,k}) \geq p(a + b)$$

and since

$$\begin{aligned}
&\pi(a + b, n, \ldots, n) \\
&\quad = \frac{1}{pk} \overline{\lim} \sum_{j,l} (a_{n+n_{j,l}} + b_{n+n_{j,l}}) \\
&\quad \leq \frac{1}{pk} \overline{\lim} \sum_{j,l} (a_{n+n_{j,l}}) + \frac{1}{pk} \overline{\lim} \sum_{j,l} (b_{n+n_{j,l}}) \\
&\quad \leq \frac{1}{k} \sum \sup \frac{1}{p} \sum (a_{n+n_j+n_l}) + \frac{1}{p} \sum \frac{1}{k} \sum (b_{n+n_j+n_l}) \\
&\quad \leq \pi(a, n_1, n_2, \ldots, n_p) + (b, n_1, n_2, \ldots, n_k) \\
&\quad \leq p(a) + p(b) + 2\varepsilon
\end{aligned}$$

and since ε is arbitrary, property 2 follows. Since m_r contains the closed subspace of all convergent sequences c and the functional

$$f(x) = \lim x_n \qquad x = (x_n)$$

is continuous on c and

$$|f(x)| \leq p(\tilde{x}) \qquad \tilde{x} = (|x_n|) \qquad x = (x_n)$$

by the extension theorem, we find an extension of this functional to the entire space m_r; we denote this extension by glim. Clearly this functional has the property

$$\text{glim } (a_n) = \text{glim } (a_{n+1})$$

since

$$-p(-a) \leq \text{glim } a \leq p(a)$$

and taking the sequence $\{a_{n+1} - a_n\}$ we obtain the assertion. Other properties stated in the theorem are obvious from the definition.

Another very useful application refers to the so-called semi-inner product first considered by Lumer (1961a).

2.12.15 DEFINITION An application $[,]: L \times L \to C$, where L is a Banach space, is called a *semi-inner product* if the following conditions are satisfied:

1. $[x + y, z] = [x, z] + [y, z]$ and $[sx, y] = s[x, y]$
2. $[x, x] > 0$ if $x \neq 0$
3. $|[x, y]|^2 \leq \|x\| \cdot \|y\|$
4. $[x, sy] = s^*[x, y]$

for all $x,y,z \in L$ and $s \in C$.

2.12.16 THEOREM On every Banach space there exists a semi-inner product with the property that $[x, x] = \|x\|^2$ (i.e., it is compatible with the norm).

Proof. For any $x \in L$ we define the functional $f_x \in L^*$ with the following properties:

1. $\|f_x\| = \|x\|$
2. $f_x(x) = \|x\|^2$

(the existence of this functional follows from the extension theorem) and we define

$$[x, y] = f_y(x)$$

for each $s \in C$ and $x \in L$,

$$f_{sx} = s^* f_x$$

and it is easy to see that properties 1 through 4 are satisfied.

2.12.17 REMARK The existence of a semi-inner product with property 4 was noted by J. R. Giles (1967).

Other applications are given in this book and also in the references.

2.13 CONVERGENCE IN BANACH SPACES

Let L be a Banach space and $\{x_n\}$ be a sequence of elements of L. There exist several types of convergence for this sequence and we give some of those which are important for what follows.

Of course, the first type of convergence refers to the convergence derived from the metric induced by the norm.

2.13.1 DEFINITION A sequence $\{x_n\}$ of elements of a Banach space L is strongly convergent to x if

$$\lim \|x_n - x\| = 0$$

2.13.2 REMARK It is easy to see that every strong (or convergent) sequence in a Banach space is bounded.

The following example shows that there exist sequences in Banach spaces which are bounded and without any convergent subsequence.

Indeed, let $L = \ell^2$ and consider the sequence $\{e_n\}$, where

$$e_n = (0, \ldots, 0, \underset{n}{1}, 0, \ldots)$$

and it is clear that for $n \neq m$ we have

$$\|e_n - e_m\| = 2^{1/2}$$

and this sequence is without convergent subsequences.

On the space $L(L, L_1)$, where L and L_1 are Banach spaces, there exist several types of convergence.

2.13.3 DEFINITION A sequence $\{T_n\} \subset L(L, L_1)$ is called *uniformly convergent* to $T \in L(L, L_1)$ if $\lim \|T_n - T\| = 0$.

2.13.4 DEFINITION A sequence $\{T_n\} \subset L(L, L_1)$ is called *strongly convergent* to T if for all $x \in L$

$$\lim \|T_n x - Tx\| = 0 \qquad (\text{or } T_n \xrightarrow{s} T)$$

2.13.5 DEFINITION A sequence $\{T_n\} \subset L(L, L_1)$ is called *weakly convergent* to T if for any $x \in L$ and $x_1^* \in L_1^*$

$$\lim x_1^*(T_n x) = x_1^*(Tx) \qquad (\text{or } T_n \xrightarrow{w} T)$$

Since the relations between these types of convergence are obvious, we give some facts about the dual space.

Let L be a Banach space and L* be its dual. The strong topology $\beta(L^*, L)$ on L* is defined by the norm

$$x^* \mapsto \|x^*\| = \sup \{|x^*(x)| \mid \|x\| \le 1\}$$

and the weak topology, which we denote by $\sigma(L^*, L)$, has as a fundamental system of neighborhoods of zero the family of sets

$$\{x^* \mid |x^*(x_j)| < \varepsilon, \quad j = 1, 2, \ldots, n\}$$

and where $\varepsilon > 0$ and $\{x_j\}$ is a finite family of elements in L.

Also, on the space L we can define a useful topology, which we denote by $\sigma(L, L^*)$, which has as a fundamental system of neighborhoods of zero the family of sets

$$\{x \mid |x_j^*(x)| < \varepsilon, \quad j = 1, 2, \ldots, n\}$$

where $\varepsilon > 0$ and $\{x_j^*\}$ is a finite family of elements of L*.

From the definitions it is obvious that

$$\beta(L^*, L) \supset \sigma(L^*, L)$$
$$\beta(L, L^*) \supset \sigma(L, L^*)$$

In practice it is useful to have some criteria for the convergence.

In what follows we give such criteria and we apply them to some concrete Banach spaces. Also we give some useful facts about the connections between strong and weak convergence in some classical Banach spaces.

2.13.6 *DEFINITION* Let L be a Banach space and $\{x_n\}$ a sequence in L. The sequence is called *weak convergent* to x if for each $x^* \in L^*$ $\lim x^*(x_n) = x^*(x)$. A sequence $\{x_n\}$ is called *weak convergent* if for each $x^* \in L^*$, $x^*(x_n)$ is convergent.

2.13.7 *DEFINITION* A Banach space is called *sequentially weak complete* if every weak convergent sequence is weakly convergent to an element of L.

2.13.8 *REMARK* If $\{x_n\}$ is a weakly convergent sequence, then there exists at most one element x such that $\{x_n\}$ is weakly convergent to x.

2.13.9 *EXAMPLE* Let $L = \ell^2$ and $\{x_n\}$ be the sequence $x_n = e_n$. In this case $\{x_n\}$ converges weakly to zero and clearly is not strongly convergent.

The following theorem gives a useful criterion for the weak convergence of a sequence in a Banach space.

2.13.10 *THEOREM* A sequence $\{x_n\}$ in a Banach space L converges weakly to an element $x \in L$ if and only if the following conditions are satisfied:

1. $\sup_n \|x_n\| < \infty$
2. $\lim x(x_n) = x^*(x)$ for x^* in an arbitrary dense set in L^*

Proof. The conditions are necessary. Indeed, we consider the sequence $\{F_n\}$ of functionals on L^* defined by

$$F_n(x^*) = x^*(x_n)$$

and from the principle of uniform boundedness we have that

$$\sup_n \sup_{x^*} |F_n(x^*)| < \infty$$

and since

$$\|x_n\| = \sup \{|F_n(x^*)|\}$$

the assertion that $\{x_n\}$ is a bounded sequence is proved. Condition 2 is obviously necessary. We prove now that these are also sufficient. Let $y^* \in L^*$ be an arbitrary element and $\varepsilon > 0$.

We find an element x^* in the dense set of the theorem such that $\|x^* - y^*\| < \varepsilon$. But for $\varepsilon > 0$ we can find an N_ε such that

$$|x^*(x_n) - x^*(x)| < \varepsilon$$

for all $n > N_\varepsilon$, and thus we obtain

$$\begin{aligned}|y^*(x_n) - y^*(x)| &= |y^*(x_n) - x^*(x_n) + x^*(x_n) - x^*(x) \\ &\quad + x^*(x) - y^*(x)| \\ &< \varepsilon\|x_n\| + \varepsilon + \varepsilon\|x\| = \varepsilon(\|x_n\| + 1 + \|x\|)\end{aligned}$$

This proves that $y^*(x_n) \to x^*(x)$.

The proof of the following theorem is very simple and thus we omit it.

2.13.11 THEOREM If $\{x_n\}$ is a strongly convergent sequence, then $\{x_n\}$ is a weakly convergent sequence.

The following theorem of Mazur gives a useful connection between the weak and strong convergence.

2.13.12 THEOREM If $\{x_n\}$ is a weak convergent sequence in L and converges weakly to x, then there exists a sequence in conv $\{x_n\}$ such that x is the strong limit of this sequence.

Proof. Let $M = \text{conv}\{x_n\}$ and suppose that x is not in M.

We can suppose, without loss of generality, that $x \neq 0$. In this case we find an $\varepsilon > 0$ such that $\|x - m\| > \varepsilon$ for all $m \in M$.

Let $\tilde{M} = \{x \mid \|x - m\| \leq \varepsilon/2$, for some $m \in M\}$ which is a convex neighborhood of the origin.

Consider the Minkowski functional of $\tilde{M}$, say p. Since $x = a^{-1}m_0$, where m_0 has the property $p(m_0) = 1$, for the linear space

$$L_1 = \{x \mid x = am_0,\ a \in R\}$$

the linear functional on L_1,

$$x_1^*(x) = a \qquad x = am_0$$

satisfies the relation

$$x_1^*(x) \le p(x)$$

By the extension theorem we can consider x_1^* defined on L and the inequality $x_1^*(x) \le p(x)$ is preserved. This shows that x_1^* is a continuous linear functional defined on the real normed space L, and since

$$\sup_{x \in M} x_1^*(x) \le \sup_{x \in \tilde{M}} x_1^*(x) \le \sup_{x \in \tilde{M}} x_1^*(x) = 1 < a^{-1} = x_1^*(a^{-1}m_0) = x_1^*(x)$$

we have a contradiction, and the assertion is proved.

Another type of convergence is the so-called weak *-convergence.

2.13.13 DEFINITION A sequence $\{x_n^*\}$ of L* is called *weak *-convergent* if for any $x \in L$ the sequence $\{x_n^*(x)\}$ is convergent; the sequence is said to be weak *-convergent to x* if the sequence $\{x_n^*(x) - x^*(x)\}$ converges to zero for all $x \in L$.

The proof of the following theorem is similar to the proof of Theorem 2.13.10 and thus we omit it.

2.13.14 THEOREM A sequence $\{x_n^*\}$ is weak *-convergent to an element $x^* \in L^*$ if and only if the following conditions are satisfied:

1. $\sup \|x_n^*\| < \infty$
2. $\lim x_n^*(x) = x^*(x)$ for any x in a dense subset of L.

The following theorem gives a large class of sequentially complete spaces.

2.13.15 THEOREM If L is a reflexive Banach space, then L is sequentially weakly complete.

Proof. Let $\{x_n\}$ be a sequence which is weak convergent. We define on L* the following functional:

$$F(x^*) = \lim F_n(x) = \lim x^*(x_n)$$

which is continuous. Since L is reflexive, we find $\tilde{x} \in L$ such that

$$\lim F_n(x^*) = x^*(\tilde{x})$$

and it is clear that $\tilde{x}$ is the weak limit of $\{x_n\}$.

In this connection with the weak and strong convergence we prove now a theorem of Dunford (1967) which asserts that the weak and strong analytic (holomorphic) functions are the same.

First we give the following definition.

2.13.16 DEFINITION Let L be a Banach space and D be a domain in the complex plane. A function f: D → L is called *weakly holomorphic* if for any $x^* \in L^*$, $x^*(f(\cdot))$ is an analytic function.

2.13.17 DEFINITION A function f: D → L is called *strongly analytic* if for any $s_0 \in D$ there exists $d > 0$ such that for $|z - s_0| < d$

$$f(z) = \sum_{i=0}^{\infty} a_i (z - s_0)^i$$

where $a_i \in L$.

It is easy to extend many concepts of the classical theory; in particular, the radius of convergence is given by the Hadamard formula

$$\frac{1}{R} = \lim\sup \|a_n\|^{1/n}$$

2.13.18 THEOREM The function f: D → L is weakly holomorphic if and only if it is strongly holomorphic.

Proof. Let C be a rectifiable Jordan curve such that the closed, bounded domain $\bar{C}$ enclosed by C is in D, and let $s_0 \in \bar{C}\backslash C$. Let D_0 be any domain in the complex plane containing s_0 and such that $\bar{D}_0 \subset \bar{C}$.

By the Cauchy formula we have,

$$x^*(f(x_0)) = \frac{1}{2\pi i} \int_C \frac{x^*(f(z)}{z - s_0}\, dz$$

and since $s_0 + h$ and $s_0 + k$ are in D_0, we obtain that

$$(k - h)^{-1}\left\{\frac{x^*(f(s + h)) - x^*(f(s_0))}{h} - \frac{x^*(f(s_0 + k)) - x^*(f(s_0))}{k}\right\}$$

$$= \frac{1}{2\pi i}\int_C x^*(f(z))\left[\frac{1}{z - s_0 - h}(h - k - s_0)(z - s_0)\right] dz$$

Since the distance between D_0 and C is strictly positive, for each $x^* \in L^*$, the absolute value of the right-hand side is uniformly bounded when s_0, $s_0 + h$, $s_0 + k$ are in D_0. By the principle of uniform boundedness we have that

$$\sup \frac{1}{|h - k|}\left\|\frac{f(s + h) - f(s)}{h} - \frac{f(s + k) - f(s)}{k}\right\|$$

is finite, and since L is a Banach space, it follows that f is strongly holomorphic.

2.13.19 COLOLLARY

1. If f is as in Theorem 2.13.18, then

$$f(s_0) = \frac{1}{2\pi i}\int_C \frac{f(z)}{z - s_0}\, dz$$

2. For any point s_0 at which f is holomorphic,

$$f(s) = \sum_{i=0}^{\infty} a_i (s - s_0)^i = \sum \frac{f^i(s_0)}{i!}(s - s_0)^i$$

where s is in some neighborhood of s_0 and where

$$f^i(s_0) = \frac{1}{2\pi i}\int_C \frac{f(z)\, dz}{(z - s_0)^{n+1}} \qquad i = 1, 2, 3, \ldots$$

3. (Liouville's Theorem) If f is strongly holomorphic in the complex plane and $\sup \|f(z)\| < \infty$, then f = constant.

2.13.20 REMARK Another related result refers to measurability; for results in this direction see, for example, the monographs of Hille and Phillips (1957) and Dunford and Schwartz (1967).

2.14 THE ADJOINT OF AN OPERATOR

Suppose that L and L_1 are two Banach spaces over the same K. Let L* and L_1^* be their duals as Banach spaces. Consider the Banach space $L(L, L_1)$ and T an element of this set. For each $g_1 \in L_1^*$ we can define the linear functional g on the space L by the relation

$$g(x) = g_1(Tx)$$

Since

$$|g(x)| \leq \|g_1\| \cdot \|T\| \cdot \|x\|$$

and thus $g \in L^*$. In this way for any $g_1 \in L_1^*$ we have an element $g \in L^*$ and denote the function defined in this way as follows:

$$T^*g_1 = g$$

and thus $T^*: L_1^* \to L^*$.

2.14.1 DEFINITION The function T* is called the *adjoint* of T.

The following theorem gives the basic properties of the adjoint.

2.14.2 THEOREM The function T* is a linear operator and

$$\|T^*\| = \|T\|$$

Proof. Let $g_1 = h_1 + \ell_1$; thus for any $x \in L$ we have

$$g(x) = g_1(Tx) = h_1(Tx) + \ell_1(Tx) = (T^*h_1)(x) + (T^*\ell_1)(x)$$

and for any scalar z,

$$g(zx) = T^*(zg)(x) = (z^*T^*)(g(x))$$

and thus T is a lincar operator. Also, since for any $g_1 \in L_1^*$

$$\|g\| \leq \|T\| \cdot \|g_1\|$$

it follows that

$$\|T^*\| \leq \|T\|$$

We show now that we have in fact an equality. Indeed, for any $x \in L$ there exists $g_1 \in L_1^*$ such that

1. $g_1(Tx) = \|Tx\|$
2. $\|g_1\| = 1$

and this gives

$$\|Tx\| = g(Tx) = (T^*g_1)(x) \le \|T + g\| \cdot \|x\| \le \|T^*\| \cdot \|x\|$$

or $\|T^*\| \ge \|T\|$, and the equality is proved.

We recall that a function $f: X \to Y$ is called *injective* if and only if $x \ne y$ implies that $f(x) \ne f(y)$.

2.14.3 THEOREM The operator $T^* \in L(L_1^*, L^*)$ is injective if and only if $\{Tx \mid x \in L\}$ is dense in L_1.

Proof. Suppose that T^* is injective and $TL = \{Tx \mid x \in L\}$ is a subspace of L_1. To prove that it is dense, it is sufficient to show that a linear functional which is zero on this subspace is identically zero. Indeed, for any $x \in L$ we have

$$(T^*g_1)(x) = g(Tx) = 0$$

and by the injectivity of T^* we obtain that $g_1 = 0$, and thus TL is a dense subspace.

For the converse we remark that if $g \in L_1^*$ and $g/TL = 0$, then

$$g(Tx) = (T^*g_1)(x) = 0$$

and so $g = 0$. The theorem is proved.

2.14.4 THEOREM If $T \in L(L, L_1)$ has the property that $\|Tx\| \ge \|x\|$, then T^* is injective if and only if T is surjective.

Proof. If T^* is injective, then TL is dense in L_1 and for any ℓ_1 in L_1 we find a sequence $g_n \in TL$ such that,

$$g_n \to \ell_1 \qquad g_n = Tx_n$$

Now since $\|g_n - g_m\| = \|T(x_n - x_m)\| \ge \|x_n - x_m\|$, it follows that $\{x_n\}$ is a Cauchy sequence. Let $\ell = \lim x_n$ and clearly $\ell_1 = T\ell$.

Thus T is a surjective operator. Other assertions of the theorem are obvious.

2.14.5 THEOREM If $T \in L(L, L_1)$ and $S \in L(L_1, L_2)$, then

$$(ST)^* = T^*S^*$$

Proof. Since the proof is simple, we omit it.

2.14.6 EXAMPLE Let $L = \ell^2$ and $T: L \to L$ defined as follows: if $x = (x_i)$, then $Tx = (0, x_1, x_2, \ldots)$, and it is easy to see that $T^*(y_1, y_2, y_3, \ldots) = (y_2, y_3, \ldots)$.

2.15 THE SPECTRUM OF AN OPERATOR

In what follows we suppose that L is a Banach space and $T \in L(L)$; we consider two basic notions connected with T: the resolvent and the spectrum.

2.15.1 DEFINITION The *resolvent* set of T, which is denoted by $\rho(T)$, is the set of all complex numbers z such that $R(z, T) = R_z(T) = (z - T)^{-1}$ exists as a bounded operator defined on L. The *spectrum* of T, denoted by $\sigma(T)$, is the complement of $\rho(T)$. The function defined on $\rho(T)$ by

$$z \to T(z, T)$$

is called the *resolvent function* (or simply, the *resolvent* of T).

The following theorem gives some basic properties of $\sigma(T)$ and $\rho(T)$.

2.15.2 THEOREM For any operator $T \in L(L)$ the set $\rho(T)$ is open and $R(z, T)$ is an analytic function on $\rho(T)$.

Proof. First we show that $\rho(T)$ is an open set. Indeed let z_0 be a point in $\rho(T)$ such that $(z_0 - T)^{-1}$ exists, and take z such that

$$|z_0 - z| < \frac{1}{\|(z_0 - T)^{-1}\|}$$

In this case it is easy to see that $[I + \Sigma (z - z_0)^n (z_0 - T)^{-n}] (z_0 - T)^{-1}$ is an element of L(L), and a simple computation shows that this is the inverse of $(z - T)^{-1}$. Thus for any $z_0 \in \rho(T)$ we find an open disk

$$D_{z_0} = \left\{ z \mid |z - z_0| < \frac{1}{\|(z_0 - T)^{-1}\|} \right\}$$

which is in $\rho(T)$. This proves that $\rho(T)$ is an open set.

For the second assertion we remark that the form of the inverse shows that R(z, T) is an analytic function on $\rho(T)$.

The following theorem gives information about the spectrum of an operator $T \in L(L)$ and the iterates of the operator.

2.15.3 THEOREM For each operator $T \in L(L)$,

$$r_T = \sup \{|z| \mid z \in \sigma(T)\} = \lim \|T^n\|^{1/n}$$

and for z, $|z| > r_T$,

$$R_z(T) = \frac{I}{z} + \sum \frac{T^n}{z^{n+1}}$$

Proof. The right member of $R_z(T)$ is a convergent series for $|z| > r_T$ and this element is exactly the inverse of (z - T).

Also, $\{z \mid |z| = r_T\} \cap \sigma(T)$ is a nonempty set; in the contrary case, the inverse $(z - T)^{-1}$ exists for $|z| > r_T - \varepsilon$, which is a contradiction since this gives $r_T < r_T - \varepsilon$.

2.15.4 COROLLARY For each $T \in L(L)$,

1. $r_T \leq \|T\|$
2. $\sigma(T) \subseteq \{z \mid |z| \leq \|T\|\}$

2.15.5 COROLLARY If d(z) is the distance from z to $\sigma(T)$, then

$$\|R_z(T)\| \geq \frac{1}{d(z)}$$

2.15.6 THEOREM For any $T \in L(L)$, $\sigma(T)$ is a nonempty compact set.

Proof. Suppose that $\sigma(T)$ is empty and in this case $R_z(T)$ is an entire function, and from the form of $R_z(T)$, this is an analytic function at ∞ and it is bounded. By Liouville's theorem it is a constant, and from the form at ∞, it follows that the constant is zero. This is a contradiction. Thus $\sigma(T)$ is a nonempty set which is clearly compact. See Theorem 2.15.3 for another proof of the nonemptiness of sets.

2.15.7 THEOREM (Hilbert Resolvent Identity) If $T \in L(L)$ and if z_1 and z_2 are points in $\rho(T)$, then

$$R_{z_2}(T) - R_{z_1}(T) = (z_1 - z_2)R_{z_2}(T)R_{z_1}(T)$$

Proof. Since we have the relation

$$\begin{aligned}(z_1 - T)(z_2 - T)[R_{z_2}(T) - R_{z_1}(T)] &= (z_2 - T) - (z_1 - T)\\ &= z_1 - z_2\end{aligned}$$

and multiplying by $R_{z_2}(T)R_{z_1}(T)$ we obtain the identity.

For bounded operators we can define a functional calculus using the Cauchy formula for analytic vector-valued functions. In what follows we give some results in this direction; for further details, we refer to Dunford and Schwartz (1967).

Let T be a bounded operator on a Banach space L and let $\mathcal{F}(T)$ be the set of all analytic functions on some neighborhood of $\sigma(T)$.

2.15.8 DEFINITION For $f \in \mathcal{F}(T)$ we define the operator $f(T) \in L(L)$ in the following manner: for U an open set with boundary ∂U consisting of a finite number of rectifiable Jordan curves oriented in the positive sense and $U \supset \sigma(T)$, $U \cup \partial U$ is in the domain of holomorphy of f, and

$$f(T) = \frac{1}{2\pi i}\int_{\partial U} f(z)R_z(T)\,dz$$

It is easy to see that the operator $f(T)$ depends only on T.

The following result is useful in theory as well as in applications.

2.15.9 THEOREM For any $f,g \in \mathcal{F}(T)$ and complex numbers a and b the following relations hold:

1. $af + bg \in \mathcal{F}(T)$ and $(af + bg)(T) = af(T) + bg(T)$.
2. $fg \in \mathcal{F}(T)$ and $(fg)(T) = f(T)g(T)$.
3. If f is of the form $f(z) = \Sigma_{n=0}^{\infty} a_n z^n$, then $f(T) = \Sigma\, a_n T^n$.
4. $f(T^*) = [f(T)]^*$.

Proof. Assertion 1 is clear from the definition of the operator f(T). We prove now assertion 2. Let $f,g \in \mathcal{F}(T)$ and U_1 and U_2 be two neighborhoods of $\mathcal{F}(T)$ whose boundaries are as in Definition 2.15.8, and $U_1 \cup \partial U_1 \subset U_2$. We suppose that $U_2 \cup \partial U_2$ is contained in the domain of holomorphy of f and g. In this case we have

$$f(T)g(T) = -\frac{1}{4\pi^2}\left(\int_{\partial U_1} f(z)R_z(T)\, dz\right) \cdot \left(\int_{\partial U_2} g(s)R_s(T)\, ds\right)$$

$$= -\frac{1}{4\pi^2}\int_{\partial U_1}\int_{\partial U_2} f(z)g(s)R_z(T)R_s(T)\, dz\, ds$$

$$= -\frac{1}{4\pi^2}\int_{\partial U_1}\int_{\partial U_2} f(z)g(s)\,\frac{R_z(T) - R_s(T)}{z - s}\, dz\, ds$$

$$= -\frac{1}{4\pi^2}\int_{\partial U_2} f(z)R_z(T)\left(\int_{\partial U_2} \frac{g(s)}{z - s}\, ds\right) dz$$

$$+ \frac{1}{4\pi^2}\int_{\partial U_2} g(s)R_s(T)\left(\int_{\partial U_1} \frac{f(z)}{s - z}\, ds\right) dz$$

$$= \frac{1}{2\pi i}\int_{\partial U_1} f(z)g(z)R_z(T)\, dz = (fg)(T)$$

To prove 3 we remark that if f(z) has the radius of convergence such that the domain of convergence contains $\{z \mid |z| \le r_T + \varepsilon\}$, then if C denotes the boundary, we have

$$f(T) = \frac{1}{2\pi i}\int_C \left(\sum a_n z^n\right) R_z(T)\, dz = \frac{1}{2\pi i}\sum a_n \int_C z^n R_z(T)\, dz$$

$$= \frac{1}{2\pi i}\sum a_n \int_C z^z \left(\sum \frac{T^1}{z^{n+1}}\right) dz = \sum a_n T^n$$

The last assertion follows from the definition of the operator f(T) and the definition of the adjoint.

The following theorem gives information about the spectrum of T and of f(T).

2.15.10 THEOREM (Spectral Mapping Theorem) For any $f \in \mathcal{F}(T)$ the following holds:

$$\sigma(f(T))) = f(\sigma(T)) = \{f(s) \mid s \in \sigma(T)\}$$

Proof. Let $s \in \sigma(T)$ and $f \in \mathcal{F}(T)$. We define the function

$$g(z) = \frac{f(s) - f(z)}{s - z}$$

of the domain of holomorphy of f. Then by Theorem 2.15.9 we have that

$$f(s) - f(T) = (s - T)g(T)$$

and thus if $f(s) - f(T)$ has a bounded inverse defined everywhere, say M, then $g(T)M$ is a bounded inverse for $s - T$ and this gives $f(s) \in \sigma(f(T))$.

For the converse, let $s \in \sigma(T)$ and suppose that $f(s) \in \sigma(f(T))$. Then $g(z) = (f(z) - s)^{-1}$ is in $\mathcal{F}(T)$ and

$$g(T)(f(T) - s) = I$$

which is a contradiction.

The following result can be proved in a similar manner.

2.15.11 THEOREM

1. If $f \in \mathcal{F}(T)$ and $g \in \mathcal{F}(f(T))$, then $F(z) = g(f(z))$ is in $\mathcal{F}(T)$ and $F(T) = g(f(T))$.

2. If $f_n \in \mathcal{F}(T)$ and all are holomorphic in a fixed neighborhood of $\mathcal{F}(T)$ and $f_n \to f$ uniformly on that neighborhood, then $f \in \mathcal{F}(T)$ and $f_n(T)$ converges uniformly to $f(T)$.

For later use, and also since they are important notations, we consider now certain classes of points in $\sigma(T)$.

2.15.12 DEFINITION

1. A point $z_0 \in \sigma(T)$ is an *isolated point* in $\sigma(T)$ if there exists a neighborhood V of z_0 such that $V \cap \sigma(T) = \{z_0\}$.

2. The point $z_0 \in \sigma(T)$ is called a *pole* of T if it is a pole of $R_z(T)$ and the order of z_0 is the order of z_0 as a pole of $R_z(T)$.

3. A subset M of $\sigma(T)$ is called a *spectral set* if it is a closed and open subset of $\sigma(T)$.

For further information concerning other kinds of spectral sets, see Lebow (1963).

2.15.13 REMARK The results in this section are valid in a more general setting, i.e., Banach algebras. See Chapter 4.

2.16 THE LOCAL SPECTRUM OF AN OPERATOR

Suppose that T is a bounded linear operator on a Banach space L and let $z \in \rho(T)$. The function

$$z \mapsto R_z(T)$$

is an analytic function on $\rho(T)$; it is possible that for some x the function

$$z \mapsto R_z(T)x$$

which is an analytic L-valued function, has analytic extension into $\sigma(T)$, the spectrum of T. This possibility leads to localization of the spectrum of T at x.

The notion of local spectrum and applications to spectral operators were considered by Dunford (1952).

2.16.1 DEFINITION Let $L \in L(L)$ and $x \in L$. The *local resolvent* $\rho(x, T)$ of T at x is the set of all complex numbers z for which there is a neighborhood V_z of z and an analytic function $U: V_z \to L$ such that

$$(z - T)U(z) = x$$

for all $z \in V_z$.

2.16.2 DEFINITION The *local spectrum* $\sigma(T, x)$ of T at x is the complement, in the complex plane, of the set $\rho(T, x)$.

2.16.3 DEFINITION The *local spectral radius* of T at x is the number $r_{T,x}$ defined by the relation

$$r_{T,x} = \sup \{|z| \mid z \in \sigma(T, x)\}$$

2.16.4 DEFINITION Any function U as in Definition 2.16.1 is called an *extended resolvent* of T at x on V_z.

2.16.5 THEOREM Let V be an open set such that $V \cap \rho(T) \neq \emptyset$. Then the function U is an extended resolvent of T at x on V if and only if U is an analytic continuation of $R_z(T)x$ into V.

Proof. Suppose that U is an analytic continuation of $R_z(T)x$ to V for $z \in \sigma(T)$. From the permanence of functional equations for analytic functions, we have that U has the property

$$[1 - (r - s)R_r(T)]U(s) = R_r(T)x$$

for $s \in V \cap \rho(T)$ and $r \in \sigma(T)$. Thus

$$(s - T)U(s) = x$$

and by Definition 2.16.1 U is an extended resolvent of T at x.

Conversely, if U is an extended resolvent of T at x, analytic on $V \cap \rho(T)$, then $T_zU(z) = x$ if $T_z = T - z$; also $T_zR_z(T)x = x$, at least for $z \in \rho(T)$. Thus we have

$$T_z(U(z) - R_z(T)x) = 0$$

and this gives that $U(z) = R_z(T)x$ if $z \in \rho(T)$. This clearly shows that U is an analytic continuation of the resolvent of T into V.

The following result relates $\sigma(T, x)$ to $\sigma(T)$.

2.16.6 THEOREM If $S \in L(L)$ and commutes with T, then for all $x \in L$,

$$\sigma(Sx, T) \subset \sigma(T, x)$$

Proof. Let $z \in \sigma(T, x)$; thus we can find a neighborhood V of z and a function $U: V \to L$ such that

$$(z - T)U(z) = x$$

for all $z \in V$. In this case we have also

$$Sx = (z - T)SU(z) = zSU(z) - TSU(z)$$

and thus $SU(\cdot)$ is analytic on V and is an extended resolvent of T at Sx. The assertion is proved.

The following theorem gives information about the local spectral radius of T at x.

2.16.7 THEOREM If $T \in L(L)$, then

$$r_{T,x} = \lim\sup \|T^n x\|^{1/n}$$

Proof. From Theorem 2.16.5 and the definition of extended resolvent, it follows that $r_{T,x} \le r_T$.

Let $s > r_{T,x}$. Thus there exists an analytic continuation of the resolvent U which is represented by an absolutely convergent Laurent series about $z = \infty$, $U(z) = \Sigma a_n z^{-n}$, where $a_n \in L$. From the Neumann expansion, for $|z| > r_T \ge r_{T,x}$ it follows that

$$z^{-1} \Sigma z^{-n} T^n x$$

converges absolutely and represents an extended resolvent for each x, and by the principle of uniqueness for vector-valued analytic functions, the two series are identical. From the Hadamard formula for radius of convergence, the assertion follows.

We give now an example to show that in general $\lim \|T^n x\|^{1/n}$ does not exist.

Let m be the space of all bounded sequences, $x = (x_i)$ with the norm

$$\|x\| = \sup |x_i|$$

and the operator $T: m \to m$ defined as follows: $T(x_1, x_2, \ldots) = (0, a_1x_1, a_2x_2, \ldots)$, where $a_i \in (0, 1)$. It is clear that

$$T^n x = (0, 0, \ldots, 0, a_1a_2, \cdots, a_nx_1, \ldots)$$

Since $\|T\| = 1$, $T \in L(L)$ and for $x = (1, 0, 0, \ldots)$ we obtain that for the case of the sequence

$$a_n = \begin{cases} 2^{-n} & \text{if } n = 10^N \text{ for some } N > 0 \\ 1 & \text{otherwise} \end{cases}$$

$\|T^n x\|^{1/2} \to 2^{-10/9}$ if $n = 10^p \to \infty$ and $\|T^n x\|^{1/2} \to 1$ in the contrary case.

As we know, for any $T \in L(L)$, $\sigma(T)$ is a compact nonempty set in the complex plane. The following theorem gives the same result for the local spectrum.

2.16.8 THEOREM If $T \in L(L)$ and $x \in L$, then $\sigma(T, x)$ is a closed subset in the complex plane and $\sigma(T, x)$ is void if and only if $x = 0$.

Proof. Since $\rho(T, x)$ is an open set, $\sigma(T, x)$ is closed and we have proved that $\sigma(T, x) \subset \sigma(T)$. Thus $\sigma(T, x)$ is a compact set in the complex plane. Suppose now that $\sigma(T, x)$ is empty. Since

$$z^{-1} \sum_{n=0}^{\infty} z^{-n}T^n x$$

converges to an extended resolvent for all nonzero points z, U is an entire function. By Liouville's theorem $U = 0$, and this gives $zU - TU = x$, which is a contradiction except in the case $x = 0$.

The definition of extended resolvent of T at x was a local definition; it is possible, if U_1 and U_2 are two extended resolvents of T at x, that $U_1 - U_2$ is not zero on $V_1 \cap V_2$. The following example illustrates this fact.

2.16.9 EXAMPLE Let H be a space ℓ^2 and $\{e_i\}_1^\infty$ be the orthonormal basis $e_i = (0, \ldots, 0, 1, 0, \ldots)$, and define

$$Ve_i = \begin{cases} 0 & n = 1 \\ e_{i-1} & n > 1 \end{cases}$$

We show that there exist two analytic extensions of $R_z(V)e_i$. First let $f_1(z) = -e_1/z$ for $z \neq 0$, and thus $(V - z)f_1(z) = (-1/z)(V - z)e_1 = e_1$. Since f_1 agrees with $R_z(V)$ on $\{z \mid |z| > 1\}$, it is clearly an analytic extension of $R_z(T)e_1$. Now let

$$f_2(z) = \begin{cases} -\dfrac{e_1}{z} & |z| > 1 \\ \sum z^n e_{n+1} & |z| < 1 \end{cases}$$

and it is clear that f_2 has the right form for $|z| > 1$. For $|z| < 1$ we have

$$(V - z)f_2(z) = \sum (z^n e_{n+1} - z^{n+1} e_{n+2}) = e_1$$

Thus f_2 is also an analytic extension of $R_z(V)$ and clearly does not coincide with f_1.

This leads to the following definition.

2.16.10 DEFINITION If $T \in L(L)$ and there exists only one extended resolvent of T at x, analytic on $\sigma(T, x)$, then T is called an *operator with the single-valued extension property*.

2.16.11 THEOREM If $\sigma_p(T)$ denotes the point spectrum of T, i.e., the set of all points s in $\sigma(T)$ such that there exists $x_0 \neq 0$ with the property that $Tx_0 = sx_0$, then a sufficient condition for T to have the single-valued extension property is that for each component k of $\sigma(T, x)$, the set $k \setminus \sigma(T)$ has a limit point in k.

Proof. Since $\rho(T, x)$ is open and the complex plane is a locally connected space, each component of $\rho(T, x)$ is open. Let U_1 and U_2 be two extensions of the resolvent of T at x, analytic at $\rho(T, x)$. If k is an arbitrary component of the local resolvent for all $z \in k$

$$(T - z)U_1 = x = (T - z)U_2$$

and thus

$$(T - z)(U_2(z) - U_1(z)) = 0$$

This gives that $U_1(z) = U_2(z)$ for all $z \in k \setminus \rho(T)$ and thus $U_2(z) = U_1(z) + \quad (z)$, where $\varepsilon(z)$ is analytic and is in the null space of $(T - z)$ for each z. From our hypothesis, in $k \setminus \sigma_p(T)$ we find a sequence (z_n) such that $z_n \to z \in k$, and since $\varepsilon(z_n) \to 0$, we obtain that $\varepsilon(z) = 0$ and thus $U_1 = U_2$ for all z.

2.16.12 COROLLARY If $T \in L(L)$ and no components of $\rho(T, x)$ are contained in $\sigma(T)$, then T has the single-valued extension property.

2.16.13 COROLLARY If $\sigma(T)$ is a nowhere dense set, then T has the single-valued extension property.

We mention now without proof the following results [Gray (1968)].

2.16.14 THEOREM If $T \in L(L)$, then

1. $T \to \sigma(T, x)$ is semicontinuous on $L(L)$.
2. If T has the single-valued extension property, U_x is an admissible set at x (see the definition below), and $f \in F(T)$,

$$\sigma(f(T), x) = f(\sigma(T, x))$$

2.16.15 DEFINITION An open set U_x is called T-*admissible* at x if

1. $U_x \supset \rho(T, x)$
2. U_x has only finite number of components.
3. The boundary ∂U_x of U_x consists of a finite number of closed, disjoint, rectifiable Jordan curves.

For any $f \in \mathcal{F}(T, x)$, the set of all holomorphic functions on a neighborhood of the local spectrum of T at x, we define

$$f(T)x = \frac{1}{2\pi i}\int_{\partial U_x} U(z, x, T)f(z)\,dz$$

where $z \mapsto U(z, x, T)$ is the extended resolvent of T at x in $\sigma(T, x)$. It is easy to see that the element $f(T)x$ is independent of the particular T-admissible U_x which we use.

2.17 ANALYTIC REPRESENTATION OF THE DUAL OF SOME BANACH SPACES

We give now some examples of duals of some Banach spaces which present some interest and are useful in what follows.

2.17.1 THEOREM If $L = C^n$ is the space of all sequences of complex numbers $z = (z_1, z_2, \ldots, z_n)$, then $C^{n*} = C^n$.

This result is well known from linear algebra and thus we omit the proof.

2.17.2 THEOREM If $L = \ell^p = \{x = (x_i) \mid \Sigma |x_i|^p < \infty,\ p \in (1, \infty)\}$, then $(\ell^p)^* = \ell^q$, where $1/p + 1/q = 1$.

Proof. If we define $e_i = (0, 0, \ldots, 0, 1, 0, \ldots)$, then it is clear that $e_i \in \ell^p$ and every $x \in \ell^p$ has the form

$$x = \sum x_i e_i$$

(in other words, $\{e_i\}$ is a basis for ℓ^p). Let $x^* \in L^*$ and thus we have

$$x^*(x) = \sum x_i x^*(e_i)$$

We define $x^*(e_i) = a_i$, $i = 1, 2, 3, \ldots$, and we show that $(a_i) \in \ell^q$.

Indeed for any integer n, the linear functional

$$x_n^*(x) = x^*(\Sigma_{i=1}^n x_i e_i)$$

is continuous and has the norm equal to

$$\|x_n^*\| = (\Sigma_{i=1}^n |a|^q)^{1/q}$$

and since $x_n^* \to x^*$, it follows that

$$\|x^*\| = (\Sigma |a_i|^q)^{1/q}$$

Thus for each $x^* \in L^*$ we can define an element $\{a_i\} \in \ell^q$ such that

$$x^*(x) = \Sigma a_i x_i$$

where $x = \Sigma x_i e_i$ and where

$$\|x^*\| = (\Sigma |a_i|^q)^{1/q}$$

and thus it follows that $L^* = \ell^q$ and these spaces are reflexive.

2.17.3 REMARK The above argument works also in the case $p = 1$ and we obtain that $(\ell^1)^* = m = \ell^\infty$. These spaces are not reflexive.

2.17.4 THEOREM If $L = c_0$ the space of all convergent sequences to zero, then $(c_0)^* = \ell^1$.

Proof. Let e_i be as above and thus a_i is as above; then it is easy see that $\Sigma |a_i| < \infty$. This clearly implies the assertion of the theorem.

2.17.5 THEOREM If $L = C_{[0,1]}$, or more generally, $C(X)$, where X is a compact Hausdorff space, then every functional x^* in the dual is of the form

$$x^*(f) = \int_X f(t)\, d\mu(t)$$

where μ is a Radon measure.

For a proof see, for example, N. Dunford and J. T. Schwartz (1967).

2.17.6 EXAMPLE Let L be a Banach space. We consider the following sequence spaces:

$$c_0(L) = \{x \mid x = (x_n) \ \lim \|x_n\| = 0\}$$

$$\ell^p(L) = \{x \mid x = (x_n) \ \Sigma \|x_n\|^p < \infty\}$$

$$\ell^\infty(L) = \{x \mid x = (x_n), \quad \{\|x_n\|\} \text{ is a bounded sequence}\}$$

As is clear for all n we have supposed that $x_n \in L$ and the operations are defined in a natural manner. Also these are Banach spaces and the norm is clear in each case.

Concerning the duals of these spaces, we have the following results:

1. $(c_0(L))^*$ has elements of the form

$$f(x) = \Sigma\, x_n^*(x_n) \qquad x = (x_n)$$

where $x_n^* \in L^*$ and $\|f\| = \Sigma \|x_n^*\|$. The proof is not essentially different from the case L = C, the complex numbers.

2. $[\ell^p(L)]^*$ has the elements of the form

$$f(x) = \Sigma\, n_n^*(x_n) \qquad x = (x_n)$$

where $x_n^* \in L^*$ and $\|f\| = (\Sigma \|x_n^*\|^q)^{1/q}$. Also the proof is similar to the case L = C.

3. $[\ell^\infty(L)]^*$ has the elements of the form

$$f(x) = \int_N x_n \, d\mu(n)$$

where μ is a finitely additive measure on the set of positive numbers (here $(N, \Sigma = 2^W, \mu)$ is a measure space and is an L*-valued measure. For a proof see G. Köthe (1966).

We present now a result connected with the famous result of Schur (1921) (in ℓ^1 the set of weakly convergent sequences and the strong convergent sequences are the same).

2.17.7 PROPOSITION The Banach space $L^1(L)$ has the same weakly convergent sequences and strong convergent sequences if and only if L has the same weakly convergent and strong convergent sequences.

Proof. Suppose first that L has the same weakly convergent and strong convergent sequences. For the proof of the property for $\ell^1(L)$, it will be sufficient to prove that $\ell^1(L)$ has the same weakly convergent and strong convergent null sequences.

Let $\{a_n\}$ be a sequence in $\ell^1(L)$ which converges weakly to zero; this implies that for each m, $a_m^n \to 0$ and thus for each m, $a_m^n \to 0$ because the assertion is true for L.

It remains to show that

$$\|a_n\| = \Sigma \|a_m^n\| \to 0$$

Suppose that this is not true, and so we find an $\varepsilon > 0$ such that

$$\limsup_{n\to\infty} \Sigma \|a_m^n\| > \varepsilon$$

From this we obtain that there exist two sequences $\{n_m\}$ and $\{r_m\}$ such that

1. $n_m \to \infty$, and $r_m \to \infty$ for $m \to \infty$.
2. n_1 is the smallest n such that $\Sigma_{i=1}^{\infty} \|a_i^n\| > \varepsilon$.
3. r_1 is the smallest r such that $\Sigma_{i=1}^{r} \|a_i^n\| > \varepsilon/2$ and $\Sigma_{i=r}^{\infty} \|a_i^n\| < \varepsilon/5$.
4. n_m is the smallest integer greater than n_{m-1} such that $\Sigma_{i=1}^{\infty} \|a_i^n\| > \varepsilon$ and $\Sigma_{i=1}^{r_{m-1}} \|a_i^n\| < \varepsilon/5$.
5. r_m is the smallest integer greater than r_{m-1} such that

$$\Sigma_{i=r_{m-1}}^{r_m} \|a_i^n\| > \varepsilon/2 \quad \text{and} \quad \Sigma_{r_{m+1}}^{\infty} \|a_i^n\| < \varepsilon/5$$

From the extension theorems, for each $n \geq 1$ and $m \geq 1$ there exists $b_m^n \in L^*$ such that $\|b_m^n\| = 1$ and $b_m^n(a_m^n) = \|a_m^n\|$.

For each $i \geq 1$ we define

$$b_i = \begin{cases} b_j^n & 1 \leq i \leq r_1 \\ b_i^{n_{m+1}} & r_m < i < r_{m+1} \end{cases} \qquad m = 1, 2, \ldots$$

and it is clear that $\|b_i\| = 1$ and $f = \{b_i\}$ is an element of $\ell^\infty(L^*)$.

Now since $a_n \to 0$ in $\ell^1(L)$, we have that

$$\lim_{m\to\infty} \sum_{i=1}^{m} b_i(a_i^n) = 0$$

Since the sequences $\{n_m\}$ and $\{r_m\}$ have the above property, we obtain

$$\left|\sum_{i=1}^{\infty} b_i(a_i^{n_m})\right| \geq \left|\sum_{i=r_{m-1}+1}^{r_m} b_i(a_i^{n_m})\right| - \sum_{i=1}^{r_{m-1}} |b_i(a_i^{n_m})|$$

$$- \sum_{i=r_{m+1}}^{\infty} |b_i(a_i^{n_m})|$$

$$> \left|\sum_{i=r_{m-1}+1}^{r_m} b_i(a_i^{n_m})\right| - \sum_{i=1}^{r_{m-1}} \|b_i\| \cdot \|a_i^{n_m}\|$$

$$- \sum_{i=r_{m+1}}^{\infty} \|b_i\| \cdot \|a_i^{n_m}\|$$

and since $\|b_i\| = 1$ and for $r_m < i < r_{m-1}$, $b_i(a_i^n) = b_i^n(a_i^n) = \|a_i^n\|$, we conclude that

$$\left|\sum_{i=1}^{\infty} b_i(a_i^{n_m})\right| > \sum_{i=r_{m-1}}^{r_m} \|a_i^{n_m}\| - \sum_{i=1}^{r_{m-1}} \|a_i^{n_m}\| - \sum_{i=r_m+1}^{\infty} \|a_i^{n}\|$$

$$> \frac{\varepsilon}{2} - \frac{\varepsilon}{5} - \frac{\varepsilon}{5} = \frac{\varepsilon}{10}$$

This clearly is a contradiction and thus $\|a_n\| \to 0$.

Suppose now that L has the same property, i.e., the two sets of sequences are the same. This obviously implies that L has the same property.

An interesting property of L^p spaces is the following one: if $\{x_n\}$ is a sequence of elements in L^p $(1 < p < \infty)$ and

1. $\|x_n\| \to \|x\|$
2. $x_n \rightharpoonup x$

then $x_n \to x$.

A Banach space having this property is called a space with the *Radon-Riesz property*.

2.17.8 THEOREM If L is a locally uniformly convex space, then L has the Radon-Riesz property.

Proof. Suppose that L is a locally uniformly convex space and let $\{x_n\}$ be a sequence with the above properties. Suppose that $\{x_n\}$ is not convergent to x. We can assume, without loss of generality, that $x \neq 0$. We can find a subsequence $\{x_{n_k}\}$ such that

$$\|\tfrac{1}{2}(x_{n_k} + x)\| < a < 1 \qquad \text{for } k = 1, 2, \ldots$$

By the Hahn-Banach theorem, we can find $x^* \in L^*$ such that

$$x^*(x) = 1 \qquad \|x^*\| = 1$$

and since $x_1 \rightharpoonup x$, we can obtain

$$\frac{1}{2}\,|x^*(x_{n_k} + x)| < a < 1$$

which is a contradiction since the left part is a sequence converging to 1.

This contradiction proves the assertion of the theorem.

2.17.9 THEOREM Let L be a Banach space. Then L has the Radon-Riesz property if and only if ℓ^p has the Radon-Riesz property, $1 < p < \infty$.

Proof. Suppose that L has the Radon-Riesz property and let $\{a_n\}$ be a sequence in ℓ^p with the properties stated in the Radon-Riesz property. In this case we have that for each m, $a_m^n \to a_m$, and since L has the Radon-Riesz property, in order to have $\|a_n - a\| \to 0$, it is sufficient to show that $\|a_m^n\| \to \|a_m\|$.

We consider the sequence of functions $\{f_n\}$ defined by

$$f_n(m) = \|a_m^n\|^p$$

and since a_n is in $\ell^p(L)$, we can find $M > 0$ such that

$$\sum_{m=1}^{\infty} \|a_m^n\|^p < M$$

for all $n \geq 1$. This gives that $0 \leq f_n(m) \leq M$, i.e., $\{f_n\} \in [0, M]^w$ for all $m \geq 1$, and since $[0, M]^w$ is a compact metric space, the sequence $\{f_n\}$ has a convergent subsequence $\{f_{n_i}\}$. Let $f_{n_i} \to f$. Since the product topology is equivalent with the coordinatewise convergence, we obtain that

$$f_{n_i}(m) \to f(m)$$

for all m when $i \to \infty$. Thus $\|a_m^{n_i}\| \to f(m)$. As is easy to see, the norm on any Banach space is weakly lower semicontinuous, and thus since $\|a_n\| \to \|a\|$, we obtain

$$\|a_m\|^p < \lim_{n_i \to \infty} \|a_m^{n_i}\|^p$$

Suppose now that there exists m_0 such that

$$\|a_{m_0}\| \neq \lim_{n_i \to \infty} \|a_m^{n_i}\|$$

This implies that for some $\varepsilon > 0$,

$$\|a_{m_0}\|^p < \|a_{m_0}\|^p + \varepsilon < \lim \|a_{m_0}^{n_i}\|$$

Let N such that for $m_0 < N$ and

$$\sum_{m=N+1}^{\infty} \|a_m\|^p < \frac{\varepsilon}{2}$$

and thus

$$\sum_{m=1}^{N} \|a_m\|^p = \sum_{m=1}^{\infty} \|a_m\|^p - \sum_{m=N+1}^{\infty} \|a_m\|^p > \sum_{m=1}^{\infty} \|a_m\|^p - \frac{\varepsilon}{2}$$

Since $m_0 < N$, we obtain that

$$\lim_{n_i \to \infty} \|a_m^{n_i}\|^p > \sum_{m=1}^{\infty} \|a_m\|^p + \varepsilon > \sum_{m=1}^{\infty} \|a_m\|^p + \varepsilon - \frac{\varepsilon}{2} = \sum_{m=1}^{\infty} \|a_m\| + \frac{\varepsilon}{2}$$

and thus,

$$\lim_{n_i \to \infty} \|a_m^{n_i}\| > \sum_{m=1}^{\infty} \|a_m\|^p + \frac{\varepsilon}{2}$$

which contradicts the fact that $\|a_n\| \to \|a\|$. Thus we have that $\|a_m^{n_i}\| \to \|a_m\|$, and since $f_{n_i}(m) \to f(m)$, we obtain $f(m) = \|a_m\|^p$ for all $m \geq 1$. In fact, the above proof shows that every subsequence of $\{f_n\}$ contains a convergent subsequence and the limit is independent of the subsequence. This proves that $\{f_n\}$ is convergent to f, and thus $\|a_m^n\| \to \|a_m\|$ for all m.

Thus the assertion in one direction is proved.

Suppose now that $\ell^p(L)$ has the Radon-Riesz property and let $\{x_n\}$ in L such that

1. $\|x_n\| \to \|x\|$
2. $x_n \rightharpoonup x$

and we define

$$a_n = (x_n, 0, 0, \ldots) \qquad a = (x, 0, 0, \ldots)$$

Clearly $\|a_n\| \to \|a\|$, and we show that $a_n \rightharpoonup a$ in $\ell^p(L)$. Indeed, since for every $f^* \in \ell^q(L) = (\ell^p(L))^*$ is of the form

$$f^*(x) = \Sigma\, bm(x)$$

where $x = (x_m)$, $b_m \in L^*$, $\{\|b_m\|\} \in \ell^q$, we have that

$$f^*(x - x) \to 0$$

and thus $a_n \rightharpoonup a$. This forces $\|a_n - a\| = \|x_n - x\| \to 0$.

2.18 MEASURES OF NONCOMPACTNESS AND CLASSES OF MAPPINGS ON BANACH SPACES

A. Measures of Noncompactness

2.18.1 DEFINITION A closed set F in a complete metric space (X, ρ) is called *compact* if whenever $\{x_n\}$ is a sequence of elements of F, there exists a convergent subsequence $\{x_{n_k}\}$.

2.18.2 DEFINITION A set M in a complete metric space is called *relatively compact* if the closure $\bar{M}$ of M is compact.

The family of compact sets plays a fundamental role in analysis and thus properties and criteria for a set to be compact are of great importance. For what follows we need the following characterizations; these characterizations suggest the introduction of some numbers for a bounded set which in some sense measure how noncompact is a set.

For proofs of these assertions we refer the reader to books on general topology.

2.18.3 THEOREM The closed set F is a compact set if and only if for each $\varepsilon > 0$ there exists a finitely many covering of F with sets having diameter less than ε.

2.18.4 THEOREM The closed set F is compact if and only if for each $\varepsilon > 0$ there exists a finite set $F_\varepsilon \subset F$ such that for all $x \in F$ there exists $x_\varepsilon \in F_\varepsilon$ with $\rho(x, x_\varepsilon) < \varepsilon$.

These characterizations suggest the introduction of the following measures of noncompactness.

2.18.5 DEFINITION [K. Kuratowski (1929)] For each bounded set A in a complete metric space $\alpha(A)$ denotes the infimum of all $\varepsilon > 0$ such that A can be covered by finitely many subsets of diameter less than ε.

2.18.6 DEFINITIONS For each bounded set A, $\chi(A)$ denotes the infimum of all $\varepsilon > 0$ such that there exists a finite set $F_\varepsilon \subset F$ with the property that for any $x \in M$ there exists $x_\varepsilon \in F_\varepsilon$ such that $\rho(x, x_\varepsilon) < \varepsilon$.

From the definitions and from the above characterizations of compact sets we have the following corollary.

2.18.7 COROLLARY A bounded set A is relatively compact if and only if $\alpha(A) = \chi(A) = 0$.

The following theorem gives the basic properties of these numbers. Since the proofs for $\chi(A)$ are not essentially different from those for $\alpha(A)$, we present in what follows only the case of Kuratowski's measure of noncompactness. For other measures of noncompactness as well as for some comparison theorems, we refer to Danes (1972).

2.18.8 THEOREM If X is a complete metric space, then

1. $0 \le \alpha(A) \le \operatorname{diam} A = \sup \{\rho(x, y) \mid x,y \in A\}$.
2. $0 = \alpha(A)$ if and only if A is relatively compact.
3. $\alpha(A) = \alpha(\bar{A})$.
4. $\alpha(A \cup B) = \max \{\alpha(A), \alpha(B)\}$.
5. $\alpha(A \cup \{x\}) = (A)$.
6. If $A \subset B$, then $\alpha(A) \le \alpha(B)$.

Proof. Since the proof is easy, we omit it.

Interesting properties of Kuratowski's measure of noncompactness are shown in the case of subsets of a Banach space.

2.18.9 THEOREM If A is a bounded set in a Banach space, then $\alpha(A) = \alpha(\operatorname{conv} A)$.

Proof. First we prove that if A_1 and A_2 are two bounded convex sets in a Banach space, then

$$\alpha(\operatorname{conv}(A_1 \cup A_2)) \le \max \{\alpha(A_1, \alpha(A_2)\}$$

Indeed, since A_1 and A_2 are bounded convex sets, it follows that

$$\operatorname{conv}(A_1 \cup A_2) = \bigcup\{tA_1 + (1 - t)A_2 \mid 0 \le t \le 1\}$$

Let $\varepsilon > 0$ and since the sets A_1 and A_2 are bounded, we find $M > 0$ such that $\|x\| \le M$ for all $x \in A_1 + A_2$. Also, if

$a_1, a_2, \ldots, a_m$ is an $\varepsilon/2M$ net for $[0, 1]$, we have that

$$\operatorname{conv}(A_1 \cup A_2) \subset \bigcup_{i=1}^{m} \{a_i A_1 + (1 - a_i)A_2 + \varepsilon S(0, 1)\}$$

and thus from the properties of $\alpha(A)$, we find

$$(\operatorname{conv}(A_1 \cup A_2)) \leq \max \{a_i \alpha(A_1) + (1 - a_i)\alpha(A_2) + 2\varepsilon\}$$
$$= \max \{\alpha(A_1), \alpha(A_2)\} + 2\varepsilon$$

Since ε is arbitrary, the assertion follows.

To prove the theorem it is sufficient to prove the inequality

$$\alpha(\operatorname{conv} A) \leq \alpha(A)$$

Let $a > \alpha(A)$ and thus there exists a covering of A with sets of diameter less than a. Let $A_1, A_2, \ldots, A_m$ be these sets. Since diam A = diam (conv (A)), we may assume that A_i are convex sets. Define

$$A^{\sim} = \operatorname{conv}(\cdots \operatorname{conv}(\operatorname{conv}(A_1 \cup A_2) \cup A_3) \cdots \cup A_m)$$

and by induction, the inequality just proved gives

$$\alpha(A^{\sim}) \leq \max \{\alpha(A_i)\}$$

and since

$$\operatorname{conv} A \subset \operatorname{conv}\left(\bigcup_{1}^{m} A_i\right) \subseteq A^{\sim}$$

the assertion follows.

B. Finite Dimensional Operators

We consider now an important class of mappings on Banach spaces, the so-called finite dimensional operators.

2.18.10 DEFINITION An operator $T \in L(L_1, L_2)$ is called *finite dimensional* if the space $\{y \mid Tx = y \text{ for some } x \in L_1\}$ is a finite dimensional.

2.18.11 THEOREM The set of all finite dimensional operators between two spaces is a linear space.

Proof. Since the proof is easy, we omit it.

Also the following result is obvious.

2.18.12 THEOREM If $L_1 = L_2$ and if T and S are finite dimensional operators, then TS and ST are finite dimensional operators.

C. Completely Continuous Operators

Using the class of finite dimensional operators, we can introduce the following useful class of operators.

2.18.13 DEFINITION An operator $T \in L(L_1, L_2)$ is called *completely continuous* if there exists a sequence $\{T_n\}$ of finite dimensional operators such that

$$\lim \|T_n - T\| = 0$$

The set of all completely continuous operators is denoted by $K_{c,c}(L_1, L_2)$.

2.18.14 THEOREM The set $K_{c,c}(L_1, L_2)$ is a Banach space, and if $L_1 = L_2$, this set is a two-sided ideal (closed).

Proof. All the properties stated in the theorem are obvious except possibly the completeness. We prove only that property. Let $\{T_n\}$ be a Cauchy sequence in $K_{c,c}(L_1, L_2)$ and let $\varepsilon > 0$ such that for $n > N_\varepsilon$, $\|T_n - T\| < \varepsilon/2$. Since T_n is in $K_{c,c}(L_1, L_2)$, we find a finite dimensional operator, say $\tilde{T}_n$, such that $\|T_n - \tilde{T}_n\| < \varepsilon/2$ and thus

$$\|T - \tilde{T}_n\| = \|T - T_n + T_n - \tilde{T}_n\| < \frac{\varepsilon}{2} + \frac{\varepsilon}{2} = \varepsilon$$

and the assertion is proved.

D. Compact Operators

An operator $T \in L(L_1, L_2)$ is called *compact* if for each bounded set M the set $\overline{TM}$ is compact.

The following result is obvious.

2.18.15 THEOREM Every completely continuous operator is compact.

2.18.16 THEOREM If $K_c(L_1, L_2)$ denotes the set of all compact operators, then

1. K_c is a Banach space.
2. If $L_1 = L_2$, then K_c is a closed two-sided ideal.

Proof. The fact that K_c is a linear space is easy to see. We prove only the completeness. Let $\{T_n\}$ be a sequence of compact operators and T such that $\lim \|T_n - T\| = 0$. In this case for any bounded set M we have that T_nM is a compact set. Thus we find a finite set F_1 such that for any $\varepsilon > 0$ every element of T_nM is in the sphere $S(\tilde{x}, \varepsilon/2)$ for some $\tilde{x} \in F_1$. Also for $n > N_\varepsilon$ such that $\|T_n - T\| < \varepsilon/2$ we have, for all $x \in M$,

$$\|T_nx - Tx\| \le \frac{\varepsilon}{2}$$

where $R = \sup \{\|x\| \mid x \in M\}$. Thus F_1 is an ε-net for TM, and by the Hausdorff characterization of compact sets, TM is a compact set.

2.18.17 REMARK By the famous result of Enflo (1973), the inclusion $K_{c,c} \subset K_c$ is strict for some Banach spaces. However, in every Banach space with basis it is easy to see that we have the equality between these classes of operators.

The following result is very useful in applications in the theory of linear equations.

2.18.18 THEOREM If $T \in L(L_1, L_2)$ is a compact operator, then $T^* \in L(L_2^*, L_1^*)$ is also a compact operator.

Proof. Let $\varepsilon > 0$ and consider a covering of the set $TS(0, 1)$, with sets of the form $S(s, \varepsilon/2)$. By compactness of this set, we find a finite number of elements, say $y_1, y_2, \ldots, y_n$, such that

$\{S(y, \varepsilon/2)\}$ is a covering of $TS(0, 1)$. Let $S^*(0, 1) = \{x^*, \|x^*\| \le 1\}$ and we define the application

$$U^*(y^*) = \{y^*(y)\}_1^2$$

of the space L_2^* in the complex space C^n. Since this is a finite dimensional operator, it is obviously compact and thus $U^*S^*(0, 1)$ can be covered by open spheres with radius $\varepsilon/4$ and the centers in $U^*(y_i^*)$. In this case, for each $y^* \in S^*(0, 1)$, we find y_i such that

$$|y^*(y_h) - y_j^*(y_h)| < \frac{\varepsilon}{4} \qquad h = 1, 2, \ldots, n$$

Since for any $x \in S(0, 1)$ we can find y_h such that $\|Tx - y_h\| < \varepsilon/4$, we obtain

$$|y^*(Tx) - y_h^*Tx)| = |y^*(Tx) - y^*(y_h)| + |y^*(y_h) - y_h^*(y_k)| + |y_h^*(y_k) - y_h^*(Tx)| < \frac{3}{4}\varepsilon$$

and this gives us that

$$\|T^*y^* - TI(y_h^*)\| = \sup_{\|x\|\le 1} |y^*(Tx) - y_h^*(Tx)|$$

and thus

$$T^*S^*(0, 1) \subset \bigcup_1^k S^*(T^*(y_h^*), \varepsilon)$$

Since $\varepsilon > 0$ is arbitrary, the assertion follows.

The following results are basic in the theory of compact operators and are given essentially by Fr. Riesz [see Riesz-Nagy (1952)].

2.18.19 THEOREM Let $L \in L(L)$ be a compact operator and let a be a complex number. If $T - aI = T - a$ is an injective operator, then

$$R(T_a) = R(T - a) = \{y \mid y = T_a x, x \in L\}$$

is a closed subspace.

Proof. The fact that this is a subspace is obvious. We prove now that this is a closed space. Let $\{y_n\}$ be a sequence such that $y_n \to y$. We show that $y \in R(T_a)$. Indeed, let $y_n = T_a x_n$.

We have two cases:

1. $\{x_n\}$ is a bounded sequence. In this case, by compactness of T and the fact $a \neq 0$, we obtain a convergent subsequence of $\{x_n\}$. It is obvious that $\tilde{x} = \lim x_{n_k}$ satisfies the relation $y = T\tilde{x}$.

2. $\{x_n\}$ does not contain any bounded sequence, and we define

$$x_n' = \frac{x_n}{\|x_n\|}$$

and it is clear that we have the relations

(a) $\|x_n'\| = 1$

(b) $Tx_n' - ax_n' \to 0$

Since T is compact, we obtain a convergent subsequence of $\{x_n'\}$, $\lim x_{n_k}' = x'$ and $Tx' = ax'$, $\|x'\| = 1$, which is a contradiction to the injectivity of T and the theorem is proved.

From this theorem we obtain the following corollary.

2.18.20 COROLLARY Suppose that $T \in L(L)$ is a compact operator and a is such that $T - a$ has no inverse $(a \neq 0)$. In this case there exists x such that $Tx = ax$ or there exists x^* such that $T^*x^* = ax^*$.

Proof. We consider the subspace $R(T_a)$. If $T - a$ is injective and $R(T_a)$ is a dense subspace, then necessarily, $R(T_a)$ is closed and in this case T_a has an inverse, which is a contradiction. If T_a is not injective, then the assertion of the corollary is clear. It remains to prove the corollary in the case when T_a is injective and $R(T_a)$ is not dense. Then by the Hahn-Banach theorem we find x^* such that

1. $\|x^*\| = 1$
2. $x^*(R(T_a)) = 0$

and this is equivalent with the last assertion of the theorem.

For the structure of the set $\{a \mid T_a \text{ is not invertible}\}$ in the case of certain classes of operators including the class of compact operators, we need the following beautiful result of Riesz.

2.18.21 LEMMA (Riesz's Lemma) If L_1 and L_2 are two Banach spaces and $L_1 \subset L_2$, $L_1 \neq L_2$, then for any $\varepsilon > 0$ there exists $y \in L_2$ such that

1. $\|y\| = 1$
2. $\|y - x\| \geq \varepsilon$

for all $x \in L_1$.

Proof. Let $b \in L_2$ and $b \notin L_1$. Then $d = \inf \{\|b - x\| \mid x \in L_1\} > 0$ and thus we find $a_0 \in L_1$ such that $\|b - a_0\| < d(1 + \varepsilon)$. We define $b_1 = b - a_0$ and obtain

$$(1 + \varepsilon) \inf_{a \in L_1} \|b_1 - a\| = (1 + \varepsilon)d > \|b_1\|$$

Let $y = b/\|b\|$ and thus

$$\|y - a\| = \left\|\frac{b_1}{\|b_1\|} - a\right\| = \frac{1}{\|b_1\|} \|b_1 - a'\| > \frac{1}{d(1 + \varepsilon)} \|b_1 - a'\|$$

$$> \frac{1}{1 + \varepsilon} > 1 - \varepsilon$$

and the lemma is proved.

2.18.22 THEOREM Let $L \in (L)$ be a compact operator and $Tx_n - a_n x_n = 0$, $a_n \neq a_m$ for $n \neq m$, $\|x_n\| \neq 0$. Then $a_n \to 0$.

Proof. We suppose that $\{a_n\}$ is not convergent to zero. Thus there exists $\varepsilon > 0$ such that for some subsequence $\{a_{n_k}\}$, $|a_{n_k}| \geq \varepsilon$.

We define $b_i = a_{n_i}$ and $x_{n_i} = y_i$, and we set E_n equal to the space generated by $y_1, y_2, \ldots, y_n$ for all $n = 1, 2, 3, \ldots$. First we show that $E_n \subset E_{n+1}$. Since the inclusion is obvious, we show that it is a strict inclusion. Suppose, on the contrary, that for some n we have $E_n = E_{n+1}$. Then we have

$$y_{n+1} = a_1'y_1 + a_2'y_2 + \cdots + a_n'y_n$$

and thus applying T, we obtain

$$0 = (T - a_{n+1})y = a_1'(a_1 - a_{n+1})y_1 + \cdots + a_n'(a_n - a_{n+1})y_n$$

Since $a_n \neq a_{n+1}$, we obtain that $y_1, \ldots, y_n$ are linearly independent elements. But we can suppose, without loss of generality, that this set of elements is linearly independent. Thus $a_i' = 0$; this is a contradiction, and the assertion is proved.

Thus $E_n \subset E_{n+1}$ is strict and by Riesz's lemma we can find $z_{n+1} \in E_{n+1}$ such that

1. $\|z_{n+1}\| = 1$

2. $\|z_{n+1} - x\| \geq 1/2$ for all $x \in E_n$

But $z_{n+1} = \Sigma\, a_i' y_i$ and this gives $(T - a_{n+1})z_{n+1} \in E_n$. Indeed,

$$(T - a_{n+1})z_{n+1} = \sum_{i=1}^{n+1} a_i'(T - a_{n+1})y_i = \sum_{i=1}^{n} a_i'(a_i - a_{n+1})y_i$$

which is clearly an element of E_n. Also for $n > m$ the element

$$z_{n,m} = (z_{n+1} - a_{n+1}^{-1}Tz_{n+1}) + a_m^{-1}Tz_m$$

is in E_n, and by Riesz's lemma we have

$$\left\|T\left(\frac{1}{a_{n+1}} z_{n+1}\right) - T\left(\frac{1}{a_m} z_m\right)\right\| = \|z_{n+1} - z_{n,m}\| \geq \frac{1}{2}$$

From this it follows that the sequence $\{T((1/a_n)z_n)\}$ has no convergent subsequences. Since $\{(1/a_n)z_n\}$ is a bounded sequence,

$$\left\|\frac{1}{a_n} z_n\right\| = \frac{1}{|a_n|} \|z_n\| \leq \frac{1}{\varepsilon}$$

and T is compact. Clearly we have a contradiction and the assertion is proved.

E. Riesz's Operators

An interesting class of operators containing properly the class of compact operators is the class of Riesz's operators. To define this class we need some notions which are of independent interest.

2.18.23 DEFINITION Let L be a complex Banach space and M and N be two closed subspaces of L. The pair (M, N) is said to *reduce* T if the following conditions hold:

1. $M \oplus N = L$ (i.e., $x = x_m + x_n$, $x_m \in M$, $x_n \in N$, and this decomposition is unique).
2. M and N are invariant subspaces for T, i.e., $TN \subset N$ and $TM \subset M$.

For any operator we set

$$N(T) = \{x \mid Tx = 0\} \qquad R(T) = \{y \mid y = Tx \text{ for some } x \in L\}$$

2.18.24 DEFINITION A point $s \in \sigma(T)$ is called a *Riesz point* of T if $(N(T - s), R(T - s))$ reduces T and the following conditions hold:

1. $\dim N(T - s) < \infty$.
2. $T/N(T - s)$ is nilpotent [i.e., $\sigma(\cdot) = \{0\}$].
3. $T/R(T - s)$ is a homeomorphism.

The notion of a Riesz operator is now introduced.

2.18.25 DEFINITION An operator $T \in L(L)$ is called a *Riesz operator* if every point $s \in \sigma(T)$, $s \neq 0$, is a Riesz point.

It is clear that every compact operator is a Riesz operator.

We give now some properties of Riesz operators which show that this class is from the class of compact operators.

2.18.26 THEOREM If $T \in L(L)$ is a Riesz operator, then every s in $\sigma(T)$, $s \neq 0$, is an isolated point in $\sigma(T)$.

Proof. Since $L = N(T - s) \oplus R(T - s)$, then $T = T_1 \oplus T_2$, where $T_1 = T/N(T - s)$ and $T_2 = T/R(T - s)$. We find a neighborhood V_s of s such that $(T_2) \cap V_s = \emptyset$ and thus $\sigma(T) = \sigma(T_2) \cup \{s\}$.

The assertion is proved.

2.18.27 COROLLARY If $T \in L(L)$ is a Riesz operator, then $\sigma(T)$ is at most countable and has zero as an accumulation point (possible) and if $s \in \sigma(T)$, $s \neq 0$, then s is an eigenvalue.

Proof. The first assertion follows from the above theorem in an obvious way. We prove only the last assertion. Let $s \neq 0$ in $\sigma(T)$

and C_ε be a circle around s and disjoint from $\sigma(T) \cap \{s\}$.

We define

$$P_s(T) = \frac{1}{2\pi i} \int_{C_\varepsilon} R_z(T)\, dz$$

which is a projection ($P_s^2 = P_s$) and $R(P_s(T)) = N(T - s)$.

F. Fredholm Operators

Let L be a Banach space and $K \in L(L)$ be a compact operator. The operator $T = I + K$, where I is the identity operator, has the following properties:

1. $N(T)$ is a finite dimensional subspace, $N(T) = \ker T$.
2. The space $L/R(T)$ is also finite dimensional, $L/R(T) = \operatorname{coker} T$.

The class which we define now is considered using these two properties. We mention that the class of Fredholm operators is of great importance in the recent theories in algebraic geometry and global analysis.

2.18.28 DEFINITION An operator $T \in L(L)$ is called a *Fredholm operator* if ker T and coker T are finite dimensional spaces.

If T is a Fredholm operator, the index ind T is defined by the relation

$$\operatorname{ind} T = \dim \ker T - \dim \operatorname{coker} T$$

For the study of Fredholm operators we need some results which are also of independent interest.

2.18.29 PROPOSITION Let L be a complex Banach space and S be a closed subspace of E. Suppose that the following is true:

$$\dim S < \infty \qquad \text{or} \qquad \dim L/S < \infty$$

then there exists a closed subspace $S^\sim$ such that $L = S \oplus S^\sim$ and every $x \in L$ has a unique decomposition $x = s + s^\sim$ with $s \in S$ and $s^\sim \in S^\sim$; the functions

$$x \mapsto s = P_S x$$

$$x \mapsto s^{\sim} = P_{S^{\sim}} x$$

are continuous.

2.18.30 REMARK In the case where for a closed subspace S there exists $S^{\sim}$ with the above properties, we say that S has a complement or that S is a complemented subspace.

Proof. If S is finite dimensional, then there exists a basis for S, say $\{s_1, s_2, \ldots, s_n\}$, and $\{s_1^*, s_2^*, \ldots, s_n^*\}$ be in S^*, such that

$$s_i^*(s_j) = s_i s_j \qquad i,j = 1, 2, \ldots, n$$

We can consider s_i^* defined on L (by the extension theorems) and we define the operator

$$P_S(x) = \Sigma\, s_i^*(x) s_i$$

which is bounded and ker $P_S = P_S^{-1}(0)$ is S. Obviously, $P_{S^{\sim}} = I - P_S$ and in this case the assertion is proved.

Suppose now that dim $L/S < \infty$. First we give the following lemma.

2.18.31 LEMMA If $L_1 \subset L$ is a linear subspace, then there exists $L_2 \subset L$ such that

1. $L_1 \cap L_2 = \{0\}$.
2. Every $x \in L$ has the form $x = x_1 + x_2$ with $x_i \in L_i$.

Proof. Indeed, we consider the quotient space $L/L_1 = \{\tilde{x}\}$ and let $\{\tilde{x}_i\}$ be a Hammel basis for this space. For each i we choose $x_i \in \tilde{x}_i$; it is easy to see that this is also an independent set. We denote by L_2 the space generated by x_i. Clearly the first assertion is true. We prove only the second. Let $\tilde{x} = \Sigma\, a_i \tilde{x}_i$ and set $x_2 = x - \Sigma\, a_i x_i$, which is in L_1, and the lemma is proved.

The following lemma gives more information about these spaces.

2.18.32 LEMMA The subspaces L/L_1 and L_2 are isomorphic.

Proof. For any $x \in L$ we define

$$\varphi(x) = \tilde{x}$$

which has the following properties:

1. $\varphi(x + y) = \varphi(x) + \varphi(y)$
2. $\varphi(ax) = a\varphi(x)$

For the proof of the lemma it will be sufficient to prove the following assertions:

1. φ is surjective.
2. φ is injective.

The first assertion follows from the properties of the spaces L_1 and L_2. The surjectivity is proved as follows: Let $\tilde{x}$ be a equivalence class and choose an arbitrary $x \in \tilde{x}$. Since $x = x_1 + x_2$, then $\varphi(x_2) = \tilde{x}$.

Now the proof of Proposition 2.18.29 can be continued as follows: We consider the subspace constructed by the above lemma, and since it is finite dimensional, we can apply the first part.

Let L and L' be two Banach spaces, L_1 a closed subspace of L, and L* the dual of L. Let

$$L_1^{\perp} = \{x^* \mid x^* \in L^*, x^*(X) = 0 \text{ for all } x \in L_1\}$$

We can prove the following result.

2.18.33 THEOREM For any $T \in L(L, L')$ the following equalities hold:

1. $\ker T^* = (TL)^{\perp}$
2. If TL is closed in L' then $T^*L'^* = (\ker T)^{\perp}$.

Proof. Since we have the relations

$$T^*(x^*) = 0$$

$$T^*x^*(x) = 0 \qquad \text{for } x \in L$$

$$x^*(Tx) = 0 \qquad \text{for all } x \in L$$

$$x^* \in (TL)^{\perp}$$

and thus the first equality is proved.

For the second assertion we remark that

$$T^{\sim}\colon\ L/\ker T \to TL$$

is an isomorphism. Indeed this is obviously a linear operator. We prove that this is surjective since the injectivity is also obvious. For each $y = Tx$ we take the equivalence class of x, $\tilde{x}$, and this gives the surjectivity. From the open mapping theorem it follows that $T^{\sim}$ is an isomorphism. Let S be the inverse of $T^{\sim}$. In this case we have

$$S\colon\ TL \to\ L/\ker T$$

Let us consider the canonical map

$$p\colon\ L \to\ L/\ker T$$

and let $f^{\sim} \in (\ker T)^{\perp}$ such that $f^{\sim}_0 p = 1$ and $S^*f^{\sim} \in (TL)^*$.

By the extension theorems, this can be extended to an element of L'^*, and we denote this element by f'. Since $T^{\sim}_0 p = T$ we have $p = ST^{\sim}$, $p = ST$, and thus

$$T^*F'(x) = f'(Tx) = S^*F^{\sim}(Tx) = f^{\sim}(STx) = f^{\sim}(px) = 1(x)$$

i.e., $Tf' = 1$, which gives

$$(\ker T)^{\perp} \subseteq T^*(F^*)$$

But $\ell \in T^*(L'^*)$ and $\ell = T^*f'$ which implies that

$$\ell(x) = T^*f'(x) = f'(Tx) = 0 \qquad \text{if } x \in \ker T$$

and thus $\ell \in \ker T^{\perp}$. This gives us that $T^*(L'^*) = (\ker T)^{\perp}$ and the theorem is proved.

2.18.34 COROLLARY Let L be a Banach space and L be a closed subspace of L. In this case

$$p_1\colon\ L \to L/L_1$$

has the property that p_1^* is an isomorphism between $(L/L_1)^*$ and $L_1^{\perp}$.

Proof. Since p_1 is surjective, p_1^* is also surjective, and since $(\ker p_1)^\perp = L_1^\perp$ is closed in L_1^*, the open mapping theorem gives the assertion.

Similarly, we have the following corollary.

2.18.35 COROLLARY If $T \in L(L, L')$ is such that TL is a closed subspace, then there exist isomorphisms

1. $\ker T^* \to (L'/TL)^* = (\text{coker } T)^*$
2. $\ker T^* \to (L^*/T^*L'^*) = (\text{coker } T)^*$

In the following theorem we give some properties of Fredholm operators.

2.18.36 THEOREM Let $T \in L(L, L')$ be a Fredholm operator. Then

1. TL is a closed subspace of L'.
2. T^* is a Fredholm operator.
3. $\text{ind } T = - \text{ind } T^*$.

Proof. If we consider the operators

$$h: L \to L/\ker T \qquad s: E/\ker T \to L'$$

then $T = s \circ h$. Since TL has a complement W, $\dim W < \infty$, we can write $L \oplus W$. Considering the norm

$$\|(\ell, w)\| = \|\ell\| + \|w\|$$

and

$$T^\sim(\ell, w) = T\ell + w$$

we have a continuous operator, and this gives the fact that $T^\sim$ is an isomorphism of L' onto L'. Since L is closed in $L \oplus W$, TL is closed in L' and the first assertion is proved.

Now

$$\dim \ker T^* = \dim L/TL \qquad \dim T^* = \dim L^*/T^*L'^*$$

and thus T^* is also a Fredholm operator.

G. Regularizable Operators

In several problems, the consideration of the following quotient algebra, called the *Calkin algebra* of the Banach space L, $\mathcal{C} = L(L)/K_c(L)$, appears to be of some interest.

Connected with this algebra are the classes of regularizable operators: left and right regularizable operators.

2.18.37 DEFINITION An operator $T \in L(L, L')$ is called *left regularizable* if $QT = I + K$, where K is a compact operator on L, I is the identity operator, and $Q \in L(L', L)$.

2.18.38 DEFINITION An operator $T \in L(L', L)$ has the operator $Q \in L(L, L')$ as *right regularizable* if $TQ = I + K^{\sim}$ where $K^{\sim}$ is a compact operator on L'.

We give now a characterization of regularizable operators.

2.18.39 THEOREM Let $T \in L(L, L')$. Then T has a left regulator if and only if

1. $\dim \ker T < \infty$.
2. R(T) is a complemented subspace.

For the proof we need the following lemma, whose proof can be adapted from the proof of Theorem 2.18.36.

2.18.40 LEMMA If E is a Banach space and M and N two closed subspaces such that $M \cap N = \{0\}$, then $M \oplus N$ is closed in E.

Proof of Theorem 2.18.29. We prove first that these conditions are sufficient. Indeed, let $L = \ker T \oplus M$, $L' = R(T) \oplus W$, where M and W are closed subspaces. Since the restriction of T at M, T/M is a closed and injective operator and $R(T) = R(T/M)$, it follows that $(T/M)^{-1}$ exists and $Q = (T/M)^{-1} \circ P$, where P is the projection of L' onto R(T), has the property that $QT = I$ on the space M.

Now since $R(QT - I) = \ker T$, $QT = I$ is a compact operator and thus $QT = I + K$.

We prove now the necessity of these conditions. Since $\ker T \subseteq \ker QT$, $\dim \ker QT < \infty$, it follows that $\dim \ker T < \infty$.

Now let $L_2 = R(T) \cup \ker Q$, which is a finite dimensional subspace, and suppose T has the property that $T(\ker QT) = L_2$. We set further,

$$Y_0 = R(T) + \ker Q$$

and it is clear that $QY_0 = RQTL'$ is closed since $\dim L/(QT)L' < \infty$.

But Q is a bounded operator and Y_0 is closed, and if X_2 is a closed subspace in ker Q such that $\ker Q = X_1 \oplus X_2$, we have $Y_0 = R(T) \oplus X_2$.

Now if X_3 is the complement of Y_0 in L', we have

$$\ker Q = Y_0, \qquad L' = Y_0 \oplus X_3$$

and thus Q is an injective application with the property

$$Q(L') = Q(Y_0) \oplus Q(X_3) = R(QT) \oplus Q(X_3)$$

and because $\dim L'/(QT)L < \infty$, $\dim Q(X_3) < \infty$. Thus X_3 is a complemented subspace. Also $X_3 \oplus X_3$ is closed and $L' = R(T) \oplus X_3 \oplus X_3$.

This implies that R(T) is a closed and complemented subspace.

The theorem is proved.

In connection with the above result we mention the following conjecture of Mikhlin: If $T \in L(L, L')$ has a left regulator, then $\dim \ker T < \infty$ and R(T) is closed, and these conditions are necessary and sufficient. The following example shows that Mikhlin's conjecture is false.

Let X be a Banach space and X_1 be a subspace which is closed and uncomplemented. Define

$$T: X \to X_1$$

by the relation $Tx = x$. It is clear that the following assertions hold:

1. $\dim \ker T = 0$.
2. $R(T) = X_1$ is closed.

and by the above results T has no left regulator.

An example of a pair of Banach spaces satisfying the above properties is the following:

$$X = \left\{ f \mid \int_{-\pi}^{\pi} |f(t)|\, dt < \infty \right\}$$

$$X_1 = \{ f \mid f \in X,\ c_{-n}(f) = 0,\ n = 1, 2, 3, \ldots \}$$

and a famous result of D. J. Newmann (1960) shows that X_1 is not complemented.

2.18.41 REMARK W. Rudin (1962) has generalized the result of Newmann to certain classes of locally compact groups (abelian groups).

2.18.42 REMARK We mention that Mikhlin's conjecture is true for the case of Hilbert spaces.

The following theorem can be proved in a similar way as above and we omit the proof.

2.18.43 THEOREM An operator $T \in L(L, L')$ is with right regulator if and only if

1. $\dim \operatorname{coker} T < \infty$.
2. $\ker T$ is a complemented subspace.

From the results just proved, we have the following characterization of Fredholm elements in $L(L)$ in connection with the Calkin algebra.

2.18.44 THEOREM An operator $T \in L(L)$ is Fredholm if and only if it is invertible modulo the compact operators; in other words, T is invertible in the Calkin algebra.

This result was proved by F. V. Atkinson (1951). The next theorem also follows from the above results.

2.18.45 THEOREM If $T \in L(L, L')$ and $S \in L(L', L'')$ are Fredholm operators, then ST is a Fredholm operator and $\operatorname{ind} ST = \operatorname{ind} S + \operatorname{ind} T$.

H. α-Contractions and Generalizations

Let L be a Banach space and $T \in L(L)$. The following class of mappings was first considered by G. Darbo (1955), and since they can be also considered as nonlinear mappings, they play a fundamental role in the recent extensions of the Schauder fixed point theorem.

2.18.46 DEFINITION A continuous mapping (possible nonlinear) $T: L \to L$ is called *α-contraction* if for any bounded noncompact set ML, $\alpha(TM) < k\alpha(M)$, where k is a constant and $k \in [0, 1)$.

We give now an example of a mapping which is not compact and is not a strict contraction, but is an α-contraction.

2.18.47 EXAMPLE Let T be an operator in L(L) such that $\|T\| < 1$ and T_1 be a compact operator. Clearly $T + T_1$ is not compact and is not a strict contraction. Since for any bounded noncompact set M we have

$$\alpha((T + T_1)M) \le \|T\|\alpha(M)$$

$T + T_1$ is an α-contraction.

A more general class of operators (mappings) can be introduced as follows.

2.18.48 DEFINITION A continuous mapping $T: L \to L$ is called *densifying* if for any noncompact set M, $\alpha(TM) < \alpha(M)$.

2.18.49 DEFINITION A continuous mapping $T: L \to L$ is called a *locally power α-contraction* if for each bounded noncompact set M there exists an integer $n = n(M)$ such that $\alpha(T^nM) \le k\alpha(M)$, where $k \in [0, 1)$.

2.18.50 DEFINITION A continuous mapping $T: L \to L$ is called *locally power densifying* if for each bounded noncompact set M there exists an integer $n = n(M)$ such that $\alpha(T^nM) < \alpha(M)$.

In connection with the class of mappings considered in Definition 2.18.49, it is worth noting that this class contains the class of quasi-compact operators, important for applications in probability theory.

We recall the definition of quasi-compact operators.

2.18.51 DEFINITION An operator $T \in L(L)$ is called quasi-compact if the following properties hold:

1. $\sup \|T^n\| = M < \infty$.
2. There exists a compact operator $Q \in L(L)$ and an integer m such that $\|T^m - Q\| < 1$.

It is obvious that the following assertion holds.

2.18.52 THEOREM Every quasi-compact operator T is a locally power α-contraction.

2.18.53 REMARK It is obvious that we can consider other measures of noncompactness, and in this case almost all results are valid for these classes, except for those in which several quantities involving the measure of noncompactness appears. For a correct version it is enough to see the exact relations between the measure of noncompactness considered and the Kuratowski's measure.

2.18.54 DEFINITION Let T be a continuous mapping which is also an α-contraction or a locally power α-contraction. The infimum of all k for which

$$\alpha(TM) \le k\alpha(M) \qquad \text{or} \qquad \alpha(T^n M) \le k\alpha(M)$$

holds for all noncompact sets M is called the *modulus of the α-contraction*, respectively, the *modulus of the locally power α-contraction*. These are denoted by h_T and h_T^{loc}, respectively.

The results in the following theorem are simple consequences of the definitions.

2.18.55 THEOREM

1. If T and S are α-contractions, then TS and ST are α-contractions, and

$$h_{TS} \le h_T \circ h_S \qquad h_{ST} \le h_T \circ h_S$$

2. If T and S are commuting locally power α-contractions, then

$$h_{TS}^{loc} \le h_T^{loc} \circ h_S^{loc}$$

3. If T is a locally power α-contraction and S a compact operator, then T + S is a locally power α-contraction.

4. If $\{T_i\}$ are α-contractions and $\{a_i\}$ is a sequence with the property that $\Sigma\ |a_i| h_{T_i} \le 1$, then $\Sigma\ a_i T_i$ is an α-contraction.

Now we give some results about spectral properties of operators in the classes introduced above. These properties are in some sense, extensions of properties known for compact operators.

2.18.56 THEOREM If $T \in L(L)$ is a locally power α-contraction then $\{s \mid s$ eigenvalue of T and $|s| \ge 1\}$ is a finite set.

Proof. Suppose on the contrary that this set is infinite. In this case we find an infinite sequence $\{s_n\}$ of eigenvalues of T with $Tx_n = s_n x_n$, $|s_n| \ge 1$, and $s_n \ne s_m$. We can suppose that the set $\{x_n\}$ is linearly independent, and if we denote by A_n the space generated by the elements $x_1, x_2, \ldots, x_n$, then the inclusions

$$A_n \subset A_{n+1}$$

are strict. By the Riesz lemma, for each n we can find y_n, $\|y_n\| = 1$, with the properties

1. $y_n \in A_n$, $n = 1, 2, 3, \ldots$
2. $\|y_n - x\| \ge 1/2$ for all $x \in A_{n-1}$, $n = 2, 3, \ldots$

Since for every $p \ge 1$, $T^p y_n - s_n^p y_n \in A_{n-1}$, $n = 2, 3, 4, \ldots$, we obtain that for $n > m$,

$$z_{n,m} = y_n - \frac{1}{s_n^p} T^p y_n + \frac{1}{s_m^p} T^p y_m \in A_{n-1}$$

and thus,

$$\left\| T^p\left(\frac{y_n}{s_n^p}\right) - T^p\left(\frac{y_m}{s_n^p}\right) \right\| = \|z_{n,m} - y_n\| \ge \frac{1}{2} \qquad (*)$$

Let S(0, 1) be the unit ball of L and $B = \{y_n/s_n^p\}$ and since it is clear that $B \subseteq S(0, 1)$, $\alpha(B) < \alpha(S(0, 1))$. Now since T is a locally power α-contraction, we find a sequence of integers $\{p_i\}$ such that

$$p_{i+1} = n(T^{p_i+p_{i-1}+\cdots+p_1}B)$$

and where $p_1 = n(B)$. In this case we obtain that

$$\alpha(T^{p_q+\cdots+p_1}B) \quad k^q\alpha(B) \quad k^q\alpha(S(0, 1))$$

Since $k < 1$, we can choose q such that $k^q\alpha(S(0, 1)) < 1/2$, and this is obviously a contradiction.

2.18.57 THEOREM Let $T \in L(L)$ be a locally power α-contraction and z be a complex number such that $|z| \geq 1$. If $\{x_n\}$ and $\{y_n\}$ are two sequences with the property that

$$Tx_n - zx_n = y_n \qquad n = 1, 2, 3, \ldots$$

and $\{y_n\}$ is a convergent sequence $y_n \to y$ and $\{x_n\}$ is bounded, then $\{x_n\}$ is compact.

Proof. Let $A = \{x_n\}$. Since for every n we have $zx_n = Tx_n - y_n$, we obtain the following relations:

$$T^2x_n = T(Tx_n) = T(zx_n - y_n) = zTx_n - Ty_n = z^2x_n - zx_n - Ty_n$$

$$T^3x_n = T(T^2x_n) = z^3x_n - z^2x_n - zTy_n - T^2y_n$$

$$\vdots$$

$$T^px_n = z^px_n - z^{p-1}y_n - zTy_n - \cdots - zT^{p-2}y_n - T^{p-1}y_n$$

and therefore

$$A \; \frac{1}{a^p}\, T^pA + \frac{1}{z}\, B + \cdots + \frac{1}{z^{p-1}}\, T^{p-1}B$$

Now we take $p = n(A)$ and we get

$$\alpha(A) \leq \frac{1}{|z|^p}\, \alpha(T^pA) \leq \frac{k}{|z|^p}\, \alpha(A) \leq k\alpha(A)$$

which forces $\alpha(A) = 0$, and thus the assertion of the theorem is proved.

2.18.58 COROLLARY

1. If $T \in L(L)$ is a locally power α-contraction and $z \notin \sigma_p(T)$, $|z| \geq 1$, then the range of $T - z$ is closed.

2. If $|z| \geq 1$, $z \notin \sigma_p(T)$ and T is in $L(L)$ and is also a locally power α-contraction, then $(T - z)^{-1}$ exists as a bounded operator defined on the range of $T - z$.

The following theorem gives information about the structure of the spectrum of the locally power linear α-contractions.

2.18.59 THEOREM If $T \in L(L)$ is a locally power α-contraction, then

$$\sigma(T) \cap \{z \mid |z| \geq 1\} = \sigma_p(T) \cap \{z \mid |z| \geq 1\}$$

Proof. In $\{z \mid |z| \geq 1\}$ let Σ be the complement of the eigenvalues and let also $\Sigma' = \Sigma \cap \rho(T)$ and Σ'' the rest of the set. Obviously, we have $\Sigma = \Sigma' \cup \Sigma''$. The assertion is proved if we show that Σ'' is an empty set.

We consider now the relative topology on the set Σ; thus Σ is connected and open. Also Σ' is open. Now if Σ'' is nonempty, then we can find $s \in \Sigma''$ which is an accumulation point for Σ'. Let $\{z_n\}$ such that $z_n \to s$ and $z \in \Sigma'$. Since by the above corollary, $\|(T - z_n)^{-1}\| < M$, $n = 1, 2, 3, \ldots$. Let $d = M^{-1}$ and thus $|z - s| < d \leq \|R_{z_n}(T)\|^{-1}$ and this implies that s is a regular point.

This contradiction proves the assertion.

In the case of α-contractions a more precise result is valid.

2.18.60 THEOREM If $T \in L(L)$ is an α-contraction, then

$$\sigma(T) \cap \{z \mid |z| \geq h_T + \varepsilon\}$$

is a finite set for any $\varepsilon > 0$.

Proof. This follows from the fact that $T^{\sim} = 1/(h_T + \varepsilon)$. T is again an α-contraction.

2.18.61 THEOREM If $T \in L(L)$ is a locally power α-contraction, then ker $(T - I)$ is finite dimensional.

Proof. Let $A = S(0, 1)$ of the space ker $(T - I)$, and thus we have

$$\alpha(A) = \alpha(T^n A) \leq k\alpha(A)$$

which implies that A is a compact set, and the assertion is proved.

Suppose now that $T \in L(L)$ is locally power α-contraction and let $z_1, z_2, \ldots, z_n$ be the isolated points in $\sigma(T) \cap \{z \mid |z| \geq 1\}$; these are eigenvalues of T. Consider λ_k as a circle with the radius r_k and the center z_k.

We define the projections P_k by the usual formula

$$-P_k = -P(z_k, T) = T \frac{1}{2\pi i} \int_{\lambda_k} \frac{(T - z)^{-1}\, dz}{z}$$

and also we define the operator $-E_k$ by the formula

$$E_k = -\frac{1}{2\pi i} \int_{\lambda_k} \frac{(T - z)^{-1}\, dz}{z}$$

It is clear that the following relations hold:

$$P_k = TE_k = P_k^m = T^m E_k^m$$

for all integers $m \geq 1$.

The following theorem gives the basic property of the operators P_k.

2.18.62 THEOREM The operators P_k are locally power α-contractions.

Proof. From the relation given above, we obtain that

$$\|E_k^m\| \leq r_k \frac{1}{(|z_k| - r_k)^m} \sup \|(T - z)^{-1}\|$$

which implies the existence of an integer m such that $\|E_k^m\| < 1$.

Since T is a locally power α-contraction, for any noncompact set A,

$$\alpha(T^q E k^q(A)) = \alpha(T^q A) \le k^q \alpha(A)$$

where $m < q$, $q = p_q + p_{q-1} + \cdots + p_1$ with $p_{i+1} = n(T^{p_1+\cdots+p_i}A)$. The assertion is clear from these formulas.

The following theorem gives the structure of locally α-contractions.

2.18.63 THEOREM If $T \in L(L)$ is a locally power α-contraction, then $T = T_1 + K$, where T_1 is an operator with the spectrum in $\{z \mid |z| < 1\}$ and K is a finite dimensional operator.

Proof. Let $\sigma' = \{z_k \mid z_k \in \sigma(T), |z_k| \ge 1\}$ and define the projections P_k. Let $\tilde{K} = \Sigma P_k$, $z_k \in \sigma'$. Clearly this is a finite dimensional operator and $T_1 = T(I - P)$, $K = T\tilde{K}$ satisfies the theorem.

Chapter 3

HILBERT SPACES

In this chapter we give some basic results in the theory of Hilbert spaces. Only facts which have similar proofs, as in the case of Banach spaces, will be mentioned.

3.1 INNER PRODUCTS ON LINEAR SPACES

3.1.1 DEFINITION Let E be a complex linear space. A function defined on $E \times E$ with values in C, denoted $\langle \, , \, \rangle$, is called an *inner product* on E if the following properties hold:

P1. $\langle ax + by, z\rangle = a\langle x, z\rangle + b\langle y, z\rangle$

P2. $\langle x, y\rangle = \overline{\langle y, x\rangle}$

P3. $\langle x, x\rangle > 0$ if $x \neq 0$

for all $x,y,z \in E$, $a,b \in C$.

It is clear from P1 and P2 that

1. $\langle 0, 0\rangle = 0$
2. $\langle 0, x\rangle = 0$ for all $x \in E$

Also from P1 and P2 we have

$$\langle x, ay + bz\rangle = a^*\langle x, y\rangle + b^*\langle x, z\rangle$$

3.1.2 DEFINITION For each $x \in E$ let

$$x \mapsto \|x\| = \langle x, x\rangle^{1/2}$$

and we call the number $\|x\|$ the *norm* of the element x.

The theorems which follow give a justification of this definition.

First we prove an inequality which is called the *Cauchy-Schwarz inequality*.

3.1.3 THEOREM If E is a linear space with the inner product $< \, , \, >$, then for any x and y in E,

$$|<x, y>| \leq \|x\| \cdot \|y\|$$

and the equality holds if and only if $x = ay$ for some $a \in C$.

Proof. Let x and y be arbitrary elements in E and ϑ be a real number such that $|<x, y>| = e^{i\vartheta}<x, y>$. In this case we have,

$$0 \leq \|e^{i\vartheta}x - y\|^2 = <e^{i\vartheta}x - y, e^{i\vartheta}x - y> = \|x\|^2 + \|y\|^2 - 2|<x, y>|$$

and this gives that

$$|<x, y>| \leq \frac{1}{2}(\|x\|^2 + \|y\|^2)$$

The Cauchy-Schwarz inequality is obvious if $\|x\| \cdot \|y\| = 0$ and thus we suppose that $\|x\| \neq 0 \neq \|y\|$. In this case, the above inequality applied to $x_1 = x/\|x\|$ and $y_1 = y/\|y\|$ gives

$$|<x_1, y_1>| \leq 1$$

and this is obviously equivalent with the Cauchy-Schwarz inequality.

Also from the computation given above it is clear the assertion for the case of equality, $|<x, y>| = \|x\| \cdot \|y\|$.

3.1.4 THEOREM The function

$$x \mapsto \|x\|$$

is a norm.

Proof. We recall that the properties of a norm are

1. $\|x + y\| \leq \|x\| + \|y\|$
2. $\|x\| \geq 0$ and $\|x\| = 0$ if and only if $x = 0$
3. $\|ax\| = |a| \cdot \|x\|$

for all x,y in E and $a \in C$. In our case properties 2 and 3 are obvious and thus it remains to prove 1. For x,y in E, we have

$$\|x + y\|^2 = \langle x + y, x + y\rangle = \|x\|^2 + \|y\|^2 + 2 \operatorname{Re} \langle x, y\rangle$$
$$\le \|x\|^2 + \|y\|^2 + 2|\langle x, y\rangle| \le \|x\|^2 + \|y\|^2 + 2\|x\| \cdot \|y\|$$
$$\le (\|x\| + \|y\|)^2$$

and this clearly implies property 1.

In the following we give some useful identities; some of these will be used to give an intrinsic characterization of Hilbert spaces.

3.1.5 THEOREM Let E be a linear space with an inner product $\langle\ ,\ \rangle$ and the norm as in Definition 3.1.2. Then for any x,y in E,

I. $\|x + y\|^2 + \|x - y\|^2 = 2[\|x\|^2 + \|y\|^2]$

II. $\|x + y\|^2 - \|x - y\|^2 = 2(\langle x, y\rangle + \langle x, y\rangle^*)$

III. $\|x + y\|^2 - \|x - y\|^2 + i\|x + iy\|^2 - \|x - iy\|^2 = 4\langle x, y\rangle$

Proof. Since these are easy consequences of P1 to P3, we omit the proof.

We are now ready to introduce the notion of a Hilbert space.

3.1.6 DEFINITION A *Hilbert space* is a Banach space such that there exists an inner product $\langle\ ,\ \rangle$ with

$$\|x\| = \langle x, x\rangle^{1/2}$$

3.1.7 THEOREM Every Hilbert space is uniformly convex.

Proof. From Theorem 3.1.5, we have for $\|x\| = 1$, $\|y\| = 1$

$$\left\|\frac{x + y}{2}\right\|^2 + \left\|\frac{x - y}{2}\right\|^2 = 1$$

and thus if $\|x - y\| \ge \varepsilon$, we obtain

$$\left\|\frac{x + y}{2}\right\|^2 \le [1 - (\tfrac{\varepsilon}{2})^2]^{1/2}$$

and if we put $d(\varepsilon) = 1 - [1 - (\varepsilon/2)]^{1/2}$, then

$$\|x + y\| \le 2[1 - d(\varepsilon)]$$

The following theorem gives information about the intrinsic properties of Hilbert spaces.

3.1.8 THEOREM Define on $L \times L \setminus (0, 0)$, where L is a Banach space, the function

$$r(x, y) = \frac{1}{2} \frac{\|x + y\|^2 + \|x - y\|^2}{\|x\|^2 + \|y\|^2}$$

and set

$$m = \inf r(x, y) \qquad M = \sup r(x, y)$$

The following assertions hold:

1. $(1/2)m \leq 1 \leq M \leq 2$, $mM = 1$.
2. L is a Hilbert space if and only if $r(x, y) = 1$.

Proof. Since

$$r(x, y) \leq \frac{(\|x\| + \|y\|)^2}{\|x\|^2 + \|y\|^2} = 1 + 2 \frac{\|x\| \cdot \|y\|}{\|x\|^2 + \|y\|^2} \leq 2$$

and we obtain that $M \leq 2$. Now since $r(x, y) > 0$ and

$$r(x + y, x - y) = \frac{1}{r(x, y)}$$

we obtain the fact that if r takes the value s, then it takes also the value 1/r. In this way it is clear that $mM = 1$.

Since

$$M \geq \frac{1}{2} \frac{\|x\|^2 + \|x\|^2}{\|x\|^2} = 1$$

it follows that $m \leq 1 \leq M$ and $m \geq 1/2$. Thus assertion 1 is proved.

We now prove assertion 2. First we remark that from Theorem 3.1.5 the condition is necessary. We show now that it is also sufficient.

Indeed, we define the function $\langle\ ,\ \rangle$ on $L \times L$ by the relation

$$4\langle x, y\rangle = \|x + y\|^2 - \|x - y\|^2 + i\|x + iy\|^2 - i\|x - iy\|^2$$

and we show that this is in fact an inner product.

First we show that

$$\langle x, y\rangle = \overline{\langle y, x\rangle}$$

Indeed,

$$\begin{aligned}
\overline{\langle y, x\rangle} &= \frac{1}{4}\,\overline{\{\|y + x\|^2 - \|y - x\|^2 + i\|y + ix\|^2 - i\|y - ix\|^2\}} \\
&= \frac{1}{4}\,\{\|x + y\|^2 - \|x - y\|^2 - i\|y + ix\|^2 + i\|y - ix\|^2\} \\
&= \frac{1}{4}\,\{\|x + y\|^2 - \|x - y\|^2 - i\|i(y + ix)\|^2 + i\|i(y - ix)\|^2 \\
&= \langle x, y\rangle
\end{aligned}$$

and thus the property is proved.

To prove the other properties we remark that

$$\mathrm{Re}\,(\langle x, y\rangle) = \frac{1}{4}\,(\|x + y\|^2 - \|x - y\|^2)$$

and

$$\langle x, y\rangle = \mathrm{Re}\,\langle x, y\rangle + i\mathrm{Re}\,\langle ix, y\rangle$$

We prove now the relation

$$\mathrm{Re}\,\langle x_1 + x_2, y\rangle + \mathrm{Re}\,\langle x_1 - x_2, y\rangle = 2\mathrm{Re}\,\langle x_1, y\rangle$$

Indeed,

$$\begin{aligned}
&\mathrm{Re}\,\langle x_1 + x_2, y\rangle + \mathrm{Re}\,\langle x_1 - x_2, y\rangle \\
&\quad = \frac{1}{4}\,(\|x_1 + x_2 + y\|^2 - \|x_1 + x_2 - y\|^2 + \|x_1 - x_2 + y\|^2 \\
&\qquad - \|x_1 - x_2 - y\|^2) \\
&\quad = \frac{1}{4}\,((\|x_1 + y + x_2\|^2 + \|x_1 + y - x_2\|^2) - (\|x_1 - y + x_2\|^2 \\
&\qquad - \|x_1 - y - x_2\|^2)) \\
&\quad = \frac{1}{4}((2\|x_1 + y\|^2 + 2\|x_2\|^2) - (2\|x_1 - y\|^2 + 2\|x_2\|^2)) \\
&\quad = \frac{1}{2}\,(\|x_1 + y\|^2 - \|x_1 - y\|^2) \\
&\quad = 2\mathrm{Re}\,\langle x_1, y\rangle
\end{aligned}$$

From this relation we obtain for $x_1 = x_2 = x$ that,

$$\mathrm{Re}\,\langle 2x, y\rangle = 2\mathrm{Re}\,\langle x, y\rangle$$

for all $x,y \in L$. This gives

$$\mathrm{Re}\,\langle x_1 + x_2, y\rangle + \mathrm{Re}\,\langle x_1 - x_2, y\rangle = \mathrm{Re}\,\langle 2x_1, y\rangle$$

and if we set

$$2x_1 = x_1' + x_2' \qquad 2x_2 = x_1' - x_2'$$

we obtain

$$\mathrm{Re}\,\langle x_1', y\rangle + \mathrm{Re}\,\langle x_2', y\rangle = \mathrm{Re}\,\langle x_1' + x_2', y\rangle$$

and if we set

$$x_1 = ix_1' \qquad x_2 = ix_2'$$

we obtain

$$\mathrm{Re}\,\langle ix_1', y\rangle + \mathrm{Re}\,\langle ix_2', y\rangle = \mathrm{Re}\,\langle i(x_1' + x_2'), y\rangle$$

In this way,

$$\langle x_1 + x_2, y\rangle = \langle x_1, y\rangle + \langle x_2, y\rangle$$

and the function

$$x \mapsto \langle x, y\rangle$$

is additive and obviously continuous. This gives that

$$\langle sx, y\rangle = s\langle x, y\rangle$$

for all real numbers since it holds for all rational numbers.

To prove that $\langle\ ,\ \rangle$ is an inner product, it remains to show that

$$\langle ix, y\rangle = i\langle x, y\rangle$$

Let $x,y \in L$, and thus

$$\begin{aligned}\langle ix, y\rangle &= \frac{1}{4}\left(\|ix + y\|^2 - \|ix - y\|^2 + i\|ix + iy\|^2 - i\|ix - iy\|^2\right)\\ &= \frac{1}{4}\left(\|ix + y\|^2 - \|ix - y\|^2 + i\|x + y\|^2 - i\|x - y\|^2\right)\\ &= \frac{i}{4}\left(\|x + y\|^2 - \|x - y\|^2 + i\|x + iy\|^2 - i\|x - iy\|^2\right)\\ &= i\langle x, y\rangle\end{aligned}$$

and the theorem is proved.

3.2 ORTHONORMAL BASES AND THE BESSEL INEQUALITY

Let H be a Hilbert space.

3.2.1 DEFINITION Two elements of H are called *orthogonal* if

$$\langle x, y\rangle = 0$$

and we write this as $x \perp y$.

3.2.2 DEFINITION A family $\mathcal{T}$ of elements of H is called *orthonormal* if $\mathcal{T} = \{x_\alpha\}_{\alpha\in I}$, then $\langle x_\alpha, x_\beta\rangle = \delta_{\alpha\beta}$, where $\delta_{\alpha\beta}$ is the Kronecker symbol.

Let $\mathcal{F}$ be the set of all families of orthonormal elements in H. We define a partial order by $\mathcal{T}_1 < \mathcal{T}_2$ if $\mathcal{T}_1 \subset \mathcal{T}_2$.

3.2.3 THEOREM For $\mathcal{F}$ we can apply Zorn's lemma.

Proof. Let $\{\mathcal{T}_i\}_{i\in I}$ be a set of families which is linearly ordered and define

$$\mathcal{T} = \bigcup_i \mathcal{T}_i$$

and it is easy to see that $\mathcal{T}$ is in $\mathcal{F}$ and $\mathcal{T} \geq \mathcal{T}_i$.

3.2.4 COROLLARY In any Hilbert space there exists a maximal family of elements which is an orthonormal family.

The result in the corollary gives us the possibility to define for Hilbert spaces the very important notion of a base for a Hilbert space H.

3.2.5 DEFINITION A *basis* for the Hilbert space H is any maximal orthonormal set.

The following theorem gives a useful characterization of bases.

3.2.6 THEOREM An orthonormal family $\{e_\alpha\}_{\alpha\in I}$ is a basis for H if and only if $\langle e_\alpha, x\rangle = 0$ for all $\alpha \in I$ implies $x = 0$.

Proof. Suppose that $\{e_\alpha\}_{\alpha\in I}$ is a maximal orthonormal set. In this case if $x \neq 0$, we can add the element $x/\|x\|$ to the family and we obtain a new orthonormal family, and this contradicts the maximality of $\{e_\alpha\}_{\alpha\in I}$.

Suppose now that $\{e_\alpha\}$ is an orthonormal family with the property that $<e_\alpha, x> = 0$ implies that $x = 0$. We show that $\{e_\alpha\}$ is maximal set. Indeed in the contrary case, we find a new family F_1 with the property that $F_1 > \{e_\alpha\}$. We find an element $\tilde{x} \in F_1$ which is not in $\{e_\alpha\}$ and thus since $<e_\alpha, \tilde{x}> = 0$ and the property of $\{e_\alpha\}$, we obtain that $\tilde{x} = 0$. This contradiction proves the theorem.

3.2.7 THEOREM If $\{e_\alpha\}_{\alpha \in J}$ is a finite orthonormal set and x an arbitrary element of H, then

$$\inf_{c_\alpha} \|x - \sum_{\alpha \in J} c_\alpha e_\alpha\| = \|x - \sum_{\alpha \in J} <x, e_\alpha> e_\alpha\|$$

Proof. Since for any complex numbers $\{c_\alpha\}$ we have

$$\begin{aligned} 0 \le \|x - \Sigma c_\alpha e_\alpha\|^2 &= <x - \Sigma c_\alpha e_\alpha, x - \Sigma c_\alpha e_\alpha> \\ &= \|x\|^2 - \Sigma c_\alpha^* <x, e_\alpha> - \Sigma c_\alpha <e_\alpha, x> + \Sigma |c_\alpha|^2 \\ &= \|x\|^2 + \Sigma (|c_\alpha|^2 - 2\mathrm{Re}\, c_\alpha <x, e_\alpha> + |<x, e_\alpha>|^2) - \Sigma |<x, e_\alpha>|^2 \\ &= \|x\|^2 - \Sigma |<x, e_\alpha>|^2 + \Sigma |c_\alpha - <x, e_\alpha>|^2 \end{aligned}$$

and thus the norm $\|x - \Sigma c_\alpha e_\alpha\|$ has the minimum value if and only if

$$\Sigma |c_\alpha - <x, e_\alpha>|^2 = 0$$

which gives that

$$c_\alpha = <x, e_\alpha>$$

and the assertion is proved.

3.2.8 COROLLARY If $\{e_\alpha\}$ is an orthonormal family, then

$$\|x\|^2 \ge \Sigma |<x, e_\alpha>|^2$$

Proof. Since for any finite part $J \subset I$, we have, by the above theorem,

$$\|x\|^2 \ge \Sigma |< x, e_\alpha>|^2$$

and the collary follows.

Let $x \in H$ and consider the element $\{\langle x, e_\alpha\rangle^2\}$ and it is clear that this family of complex numbers is summable. In what follows we show that any Hilbert space is essentially a space of sequences of the type described above.

3.2.9 COROLLARY If $\{e_\alpha\}$ is an orthonormal family and $x \in H$, then $\{\langle x, e_\alpha\rangle^2\}_\alpha$ contains at most a countable nonzero member.

Proof. This follows from Corollary 3.2.8.

3.2.10 THEOREM A necessary and sufficient condition for a family $F = \{e_\alpha\}_{\alpha\ \alpha\ I}$ to be maximal is the following: Every $x \in H$ has the representation

$$x = \Sigma\, c_\alpha e_\alpha \qquad c_\alpha = \langle x, e_\alpha\rangle$$

Proof. If any $x \in H$ has the above representation, then necessarily

$$\langle x, e_\alpha\rangle = 0 \qquad \alpha \in I$$

implies that $x = 0$, i.e., $\{e_\alpha\}$ is maximal.

Conversely, if $\{e_\alpha\}$ is maximal, i.e., $\{e_\alpha\}$ is a basis for H, we consider

$$\tilde{x} = \Sigma\, \langle x, e_\alpha\rangle e_\alpha$$

and thus

$$\langle x - \tilde{x}, e_\alpha\rangle = 0$$

and thus $x = \tilde{x}$. The theorem is proved.

The following theorem gives the form of the inner product of two elements as a function of the numbers c_α, with respect to a basis of H.

3.2.11 THEOREM Let H be a Hilbert space and $B = \{e_\alpha\}$ any orthonormal basis of H. If $x,y \in H$ are of the form

$$x = \Sigma\, x_\alpha e_\alpha \qquad y = \Sigma\, y_\alpha e_\alpha$$

then

$$\langle x, y\rangle = \Sigma\, x_\alpha y_\alpha^*$$

Proof. Since for any finite part $J \subset I$,

$$\left\langle \sum_{\alpha \in J} x_\alpha e_\alpha, \sum_{\alpha \in J} y_\alpha e_\alpha \right\rangle = \sum_{\alpha \in J} x_\alpha y_\alpha^*$$

and the assertion follows.

3.2.12 COROLLARY If T is an orthonormal family, $T = \{e_\alpha\}_{\alpha \in I}$ and $x = \Sigma\, a_\alpha e_\alpha$, then necessarily

$$\|x\|^2 = \sum_\alpha |\langle x, e_\alpha \rangle|^2$$

3.2.13 THEOREM Let H be a Hilbert space and $B = \{e_\alpha\}_{\alpha \in J}$ an orthonormal family of elements of H. A necessary and sufficient condition for B to be a basis is that for any $x \in H$,

$$\|x\|^2 = \sum_\alpha |\langle x, e_\alpha \rangle|^2$$

Proof. The fact that the condition is necessary follows from the above results. Suppose now that for any $x \in H$ the above relation holds. Let J be any finite part of I. In this case.

$$\left\| x - \sum_{\alpha \in J} \langle x, e_\alpha \rangle \right\|^2 = \|x\|^2 - \sum_{\alpha \in J} |\langle x, e_\alpha \rangle|^2$$

and thus

$$J \to \left\| x - \sum_{\alpha \in J} \langle x, e_\alpha \rangle e_\alpha \right\|$$

tends to zero. Thus

$$x = \Sigma \langle x, e_\alpha \rangle e_\alpha$$

and by Theorem 3.2.10, $\{e_\alpha\}_{\alpha \in I}$ is a basis.

Let H be a Hilbert space and $F = \{e_i\}_{i \in I}$ be an orthonormal family of elements of H. Define M_F as the closed space generated by the family F, i.e., the intersection of all closed subspaces containing F. The existence of such a space is obvious from Zorn's lemma.

3.2.14 THEOREM An element $x \in H$ is in M_F if and only if

$$x = \Sigma \langle x, e_i \rangle e_i$$

Proof. If $x = \Sigma \langle x, e_i\rangle e_i$, then clearly $x \in M_F$. Now if $x \in M_F$ and $\langle x, e_i\rangle = 0$, then $x = 0$. Since the set $\{z \mid \langle z, x\rangle = 0\}$ is a closed subspace which contains $\{e_i\}$ and it is equal to M_F, then $\langle x, x\rangle = 0$ and thus $x = 0$. The theorem is proved.

Let H be a Hilbert space and S_c be the set of all cardinal numbers α such that there exists $A \subset H$, card $A = \alpha$, and $\bar{A} = H$. We define

$$\alpha_H = \inf \{\alpha \mid \alpha \in S_c\}$$

and since $\alpha_H \in S_c$, we can define the following important notion.

3.2.15 DEFINITION The cardinal number α_H is called the *dimension* of H; if $\chi_0 = \alpha_H$, we call H a *separable* Hilbert space.

For any set S, card S is its cardinal number.

3.2.16 THEOREM If H is a Hilbert space and $\{e_i\}_{i\in I}$ an orthonormal family of elements of H, then

$$\alpha_H \geq \text{card } I$$

Proof. For any $i \in I$ let $x_i \in H$ such that

$$\|x_i - e_i\| \leq \frac{1}{2}$$

and we consider the function

$$i \mapsto x_i$$

which is one-to-one. This gives the assertion.

3.2.17 COROLLARY If H is a Hilbert space and $F = \{e_i\}_{i\in I}$ is a basis of H, then card $I = \alpha_H$.

3.3 SEPARABLE HILBERT SPACES: GRAM-SCHMIDT ORTHOGONALIZATION METHOD

As we know separable Hilbert spaces have the property that there exists a countable orthonormal basis $\{e_n\}_{n=1}^{\infty}$.

The Gram-Schmidt method gives a useful tool for constructing orthonormal bases in separable Hilbert spaces when we have a linearly independent set of elements of H.

3.3.1 THEOREM If H is a Hilbert space and $\{x_n\}_{n=1}^{\infty}$ is a linearly independent set in H (also we can suppose that this is finite), then there exists an orthonormal family $\{e_i\}_{i=1}^{\infty}$ which generates the same closed linear subspace.

Proof. Let $\{x_n\}$ be the linearly independent set in H. We define

$$e_i = \frac{x_i}{\|x_i\|}$$

$$y_2 = x_2 - \langle x_2, e_1\rangle \qquad e_2 = \frac{y_2}{\|y_2\|}$$

$$\vdots$$

$$y_{n+1} = y_n - \langle x_n, e_{n-1}\rangle - \cdots - \langle x_2, e_1\rangle \qquad e_{n+1} = \frac{y_{n+1}}{\|y_{n+1}\|}$$

$$\vdots$$

It is clear that if we have a finite family $\tilde{F} = \{x_i\}_{i=1}^{N}$, then we find $\{e_1, \ldots, e_N\}$ which is orthonormal, and clearly the spaces generated by these sets are the same. If the family is infinite, then it is also clear that the spaces are equal.

3.3.2 EXAMPLE Suppose that $H = L^2[0, 1]$, the space of all (classes) functions f, such that

$$\|f\| = \int_0^1 |f(t)|^2\, dt < \infty$$

Then $\{1, x, x^2, x^3, \ldots\}$ is a linearly independent set in L^2, and using the Gram-Schmidt orthogonalization method we obtain the system of Cebishev polynomials. Also other classes of orthogonal polynomials can be obtained using the same method and an appropriate Hilbert space.

3.3.3 EXAMPLE (The Trigonometric Orthogonal System) We consider the Hilbert space $L^2_{[-\pi,\pi]}$ and the following system of functions:

$$\frac{1}{\sqrt{2}}, \ \ldots, \ \sin\frac{nx}{\sqrt{\pi}}, \ \cos\frac{nx}{\sqrt{\pi}}, \ \ldots$$

It is easy to see that this is an orthonormal family of elements in $L^2_{[-\pi,\pi]}$. We prove that in fact this is a basis.

Let us define F by the relation

$$F(x) = c + \int_{-\pi}^{x} f(t)\,dt$$

where

$$c = -\frac{1}{2\pi}\int_{-\pi}^{\pi} dx \int_{-\pi}^{\pi} f(t)\,dt$$

and since

$$F(x) - F(x') = \int_{x}^{x'} f(t)\,dt$$

we have, if $x < x'$,

$$|F(x) - F(x')| \le \|f\|(x' - x)^{1/2}$$

and thus F is continuous. Clearly, $F(-\pi) = c$ and $F(\pi) = c$, $\int_{-\pi}^{\pi} F(s)\,ds = 0$.

We show now that if the Fourier coefficients of f are zero, then the Fourier coefficients of F are zero. Indeed,

$$0 = \int_{-\pi}^{\pi} f(t)\cos nt\,dt = F(t)\cos nt\Big|_{-\pi}^{\pi} + n\int_{-\pi}^{\pi} F(t)\sin nt\,dt$$

$$= n\int_{-\pi}^{\pi} F(t)\sin nt\,dt$$

and similarly

$$0 = \int_{-\pi}^{\pi} f(t) \sin nt\, dt = F(t) \sin nt \Big|_{-\pi}^{\pi} - n \int_{-\pi}^{\pi} F(t) \cos nt\, dt$$

$$= -n \int_{-\pi}^{\pi} F(t) \cos nt\, dt$$

The assertion follows from the following very important theorem of Féjer. See Zygmund (1959).

3.3.4 THEOREM If $f[-\pi, \pi] \to C$ is a continuous function, $f(-\pi) = f(\pi)$, and

$$S_k(x) = \frac{a_0}{2} + \sum_{n=1}^{k} (a_n \cos nx + b_n \sin nx)$$

$$\sigma_k(x) = \frac{1}{k+1} \sum_{1}^{k} S_j(x)$$

then $\sigma_k \to f$ uniformly on $[-\pi, \pi]$.

Proof. Here, a_n and b_n are the Fourier coefficients, i.e.,

$$a_n = \frac{1}{\pi} \int_{-\pi}^{\pi} f(t) \cos nt\, dt \qquad b_n = \frac{1}{\pi} \int_{-\pi}^{\pi} f(t) \sin nt\, dt$$

$$a_0 = \frac{1}{\pi} \int_{-\pi}^{\pi} f(t)\, dt$$

We remark that

$$S_n(x) = \frac{1}{2\pi} \int_{-\pi}^{\pi} f(t)\, dt + \frac{1}{\pi} \sum_{j=1}^{n} \left(\cos jx \int_{-\pi}^{\pi} f(t) \cos jt\, dt \right.$$

$$\left. + \sin jx \int_{-\pi}^{\pi} f(t) \sin jt\, dt \right)$$

$$= \frac{1}{\pi} \int_{-\pi}^{\pi} f(t) \left[\frac{1}{2} + \sum_{1}^{n} \cos jx \cos jt + \sin jx \sin jt \right] dt$$

$$= \frac{1}{\pi} \int_{-\pi}^{\pi} f(t) \left[\frac{1}{2} + \sum_{1}^{n} \cos j(x - t) \right] dt$$

$$= \frac{1}{\pi} \int_{-\pi}^{\pi} f(t) D_n(x - t)\, dt$$

where

$$D_n(s) = \frac{1}{2} + \sum_{j=1}^{n} \cos js$$

But it is well known and easy to prove that

$$D_n(s) = \sin \frac{(1/2 + n)s}{2} \sin \frac{s}{2}$$

and if we take $f \equiv 1$, then clearly $S_n = 1$, and thus

$$\frac{1}{\pi} \int_{-\pi}^{\pi} D_n(s)\, ds = 1$$

Since $D_n(s) = D_n(-s)$, we obtain that

$$\frac{1}{\pi} \int_{0}^{\pi} D_n(s)\, ds = \frac{1}{2}$$

Now we obtain

$$S_n(x) = \frac{1}{\pi} \int_0^{\pi} f(x + t) D_n(t)\, dt + \frac{1}{\pi} \int_0^{\pi} f(x - t) D_n(t)\, dt$$

$$= \frac{1}{\pi} \int_0^{\pi} [f(x + t) + f(x - t)] D_n(t)\, dt$$

which gives that

$$S_n(x) - f(x) = \frac{1}{\pi} \int_0^{\pi} [f(x + t) + f(x - t) - 2f(x)] D_n(t)\, dt$$

$$= \frac{1}{\pi} \int_0^{\pi} \phi_x(t) D_n(t)\, dt$$

where $\phi_x(t) = f(x + t) + f(x - t) - 2f(x)$.

We obtain further

$$\sigma_n(x) - f(x) = \frac{1}{n + 1} \sum_{0}^{n} S_n(x) - f(x) = \frac{1}{\pi} \int_0^{\pi} x(t) K_n(t)\, dt$$

where

$$K_n(t) = \frac{1}{n+1}\sum_0^n D_n(t) = \frac{1}{2(n+1)}\left(\frac{\sin (n+1)(1/2)t}{\sin (1/2)t}\right)^2$$

Since

$$K_n(t) \leq \frac{1}{2(n+1)} \frac{1}{\sin (1/2)}$$

if $0 < \delta < \pi$, $t \in [\delta, \pi]$, then $M_n(\delta) = \sup K_n(t)$ has the property

$$M_n(\delta) \leq \frac{1}{2(n+1)} \frac{1}{\sin (1/2)\delta}$$

and thus $\lim M_n(\delta) = 0$.

The properties of the Féjer kernels $\{K_n\}$ are as follows:

1. $K_n(t) \geq 0$, $n = 1, 2, 3, \ldots$
2. $1/\pi \int_0^\pi K_n(t)\, dt = 1/2$, $n = 1, 2, 3, \ldots$
3. $\lim M_n(\delta) = 0$

From these properties we obtain the assertion of the theorem as follows: Let $\varepsilon > 0$ and $\delta(\varepsilon)$ such that

$$|\phi_x(t)| < \varepsilon$$

which is well defined since f is uniformly continuous,

$$|\sigma_n(x) - f(x)| \leq \frac{1}{\pi}\int_0^\pi |\phi_x(t)|K_n(t)\, dt = \frac{1}{\pi}\int_0^\delta |\phi_x(t)|K_n(T)\, dt$$

$$+ \frac{1}{\pi}\int_\delta^\pi |\phi_x(t)|K_n(t)\, dt$$

$$\leq \frac{\varepsilon}{\pi}\int_0^\delta K_n(t)\, dt + \frac{M_n(\delta)}{\pi}\int_\delta^\pi |\phi_x(t)|\, dt$$

and from the above results, it is clear that we have $\sigma_n(x) - f(x) \to 0$ uniformly.

The proof that the sequence of trigonometric functions is a basis can be continued as follows: The function F is identically zero by Theorem 3.3.4, and since $F'(x) = f(x)$ a.e., we obtain that $f = 0$ a.e. The assertion is thus proved.

3.4 ORTHOGONAL SUBSPACES OF A HILBERT SPACE

Let H be a Hilbert space and m be a subspace of H.

3.4.1 DEFINITION The element $y \in H$ is called *orthogonal* on m if $\langle x, y\rangle = 0$ for all $x \in m$. The set of all elements orthogonal on m is denoted by $m^{\perp}$ and is obviously a subspace; it is called the *orthogonal subspace* of m.

3.4.2 REMARK For any m, $m^{\perp}$ is a closed subspace.

3.4.3 THEOREM For any m, $(m^{\perp})^{\perp} = m$.

3.4.4 THEOREM Let H be a Hilbert space and m be a closed subspace of H. Then every $x \in H$ has a unique decomposition

$$x = u + v$$

where $u \in m$ and $v \in m^{\perp}$.

Proof. The uniqueness is trivial and so we prove only the existence.

Let $\{e_i\}_{i \in I_m}$ be an orthonormal basis for m. Consider now all the orthonormal families of elements of H containing the basis of m.

Using Zorn's lemma, it is easy to see that there exists a maximal family which is also a basis for H. The index set for the basis of H is I and we set $I_2 = I \setminus I_m$. Denote by V the space generated by e_i with $i \in I_2$. It is easy to see that $m^{\perp} = V$.

Also since every element of H has the form

$$x = \sum \langle x, e_i\rangle e_i = \sum_{i \in I_m} \langle x, e_i\rangle e_i + \sum_{i \in I_2} \langle x, e_i\rangle e_i = u + v$$

the theorem is proved.

3.4.5 REMARK Using the above theorem we define the operators

$$x \mapsto u = P_m x$$
$$x \mapsto v = P_{m^{\perp}} x$$

which are with property $P^2 = P$ (i.e., are idempotents).

3.4.6 REMARK We can prove Theorem 3.4.4 in another way. Indeed we use the following lemma.

3.4.7 LEMMA If C is a convex closed set in a uniformly convex space, then there exists a unique element of minimal norm.

Proof of Theorem 3.4.4. By Theorem 3.1.7 every Hilbert space is uniformly convex, and if m is a closed subspace of H, let, for each fixed $x \in H$,

$$C_m = \{z \mid z = x - y,\ y \in m\}$$

which is a convex closed set. We find an element of minimal norm, say v, and let $u = x - v$. Since v is of minimal norm, we have for all complex numbers z,

$$\|v - zy\|^2 \geq \|v\|^2$$

which gives that

$$\|v\|^2 - 2\mathrm{Re}\, z \langle y, v\rangle + |z|^2\|y\|^2 \geq \|v\|^2$$

and for $z = (\langle v, y\rangle/\|y\|^2)$ we obtain that

$$-|\langle v, y\rangle| \geq 0$$

This clearly gives that $\langle v, y\rangle = 0$ for all $y \in m$. For $x - v = u$ we find that $\langle u, v\rangle = 0$. The existence is thus proved and the uniqueness of the decomposition is obvious.

3.5 THE DUAL OF A HILBERT SPACE

Let H be a Hilbert space and H^* be its dual. Let $x^* \in H^*$. The following famous result is the so-called Riesz representation theorem for functionals in H^*. This representation is a very useful tool in the study of Hilbert spaces as well as in applications.

3.5.1 THEOREM If $x^* \in H^*$, then there exists a unique element x_0 of H such that

1. $x^*(x) = \langle x, x_0\rangle$
2. $\|x^*\| = \|x_0\|$

Proof. We can assume, without loss of generality, that x^* is not identically zero and let

$$m = \{x \mid x^*(x) = 0\}$$

which is clearly a closed linear subspace of H. First we show that $m^{\perp}$ is one-dimensional. Indeed, suppose that there exist two linearly independent elements in $m^{\perp}$, say, y_1 and y_2. In this case the element

$$y = y_1 x^*(y_2) - y_2 x^*(y_1)$$

is in m and since y_1 and y_2 are in $m^{\perp}$, $y \in m^{\perp}$ and thus $\langle y, y\rangle = 0$, which gives $y = 0$. Thus $m^{\perp}$ is one-dimensional.

$$m^{\perp} = \{z\tilde{x}_0\} \qquad \|\tilde{x}_0\| = 1$$

We define $x_0 = x^*(\tilde{x}_0)^*\tilde{x}_0$ and we have for any $x \in H$,

$$\langle x, x_0\rangle = \langle (u + v), \tilde{x}_0\rangle = \langle v, \tilde{x}_0\rangle \qquad u \in m,\ v \in m$$

and since

$$x = \left(x - \frac{x^*(x)}{x^*(\tilde{x}_0)} \tilde{x}_0\right) + \frac{x^*(x)}{x^*(\tilde{x}_0)} \tilde{x}_0$$

the above relation gives

$$x^*(x) = \langle x, \tilde{x}_0\rangle$$

and the existence of the representation is proved. The uniqueness is obvious as well as the fact that $\|x^*\| = \|\tilde{x}_0\|$.

3.5.2 COROLLARY The dual of a Hilbert space can be organized as a Hilbert space.

Proof. For each x^* and y^* from H^* we define

$$\langle x^*, y^*\rangle = \langle x, y\rangle$$

where x and y are the elements given by the Riesz theorem.

Clearly the correspondence

$$g\colon x^* \to \tilde{x}_0 \qquad x^*(x) = \langle x, \tilde{x}_0\rangle$$

is conjugate linear.

3.5.3 COROLLARY Every Hilbert space is a reflexive space.

Proof. Let H^{**} be the second dual of H. Let $x^{**} \in H^{**}$ and since H^* is a Hilbert space, we find an element $a^* \in H^*$ such that

$$x^{**}(x^*) = \langle x^*, a^* \rangle$$

Now

$$x^{**}(x^*) = \langle g(x^*), g(a^*) \rangle = \langle g(a^*), g(x^*) \rangle$$

and since x^* is arbitrary

$$x^{**} = g(a^*)$$

and the assertion is proved.

Using the Riesz representation theorem, we can prove, following J. von Neumann, the Radon-Nicodym theorem. See Rudin (1972).

3.5.4 DEFINITION Let (S, M) be a pair, where S is a set and M a σ-algebra of subsets of S. If μ and ν are two measures such that $\mu(E) = 0$ implies $\nu(E) = 0$, we say that ν is absolutely continuous with respect to μ.

3.5.5 THEOREM (Radon-Nicodym Theorem) If ν is absolutely continuous with respect to μ, then there exists $f \in L^1(S, M, d\mu)$ such that for all $E \in M$,

$$\nu(E) = \int_S f(s)\, d\mu(s)$$

Proof. Let $\mu_1 = \mu + \nu$ and consider the Hilbert space $L^2(S, M, d\mu_1)$ and without loss of generality suppose that $\mu(S) < \infty$.

In this case the functional

$$f \to \int_S f(s)\, d\nu(s)$$

is in $(L^2)^*$. By the Riesz theorem, we can find $g \in L^2$ such that

$$\int_S fg\, d\mu_1 = \int_S f\, d\nu$$

Let $f = \chi_E$ be the indicator of an arbitrary set $E \in M$, $\mu(E) > 0$.

In this case we have

$$\mu(E)^{-1} \int_E g \, d\mu_1 \leq 1$$

and thus $g(s) \in [0, 1]$ a.e. (μ_1).

We can suppose without loss of generality that $g(x) \in [0, 1]$ for all $x \in S$. Let

$$S_0 = \{s \mid 0 \leq g(s) < 1\}$$

and

$$S_1 = \{s \mid g(s) = 1\}$$

Since $\int_S (1 - g)\chi_{S_1} \, d\nu = \int_S \chi_{S_1} \, d\mu$, we obtain that $\mu(S_1) = 0$. Let $E \in M$ and

$$\phi_n = (1 + g + g^2 + \cdots + g^n)\chi_E$$

and since

$$\int_E (1 - g^{n+1}) \, d\nu = \int_E \phi_n g \, d\mu_1$$

we obtain that $g\phi_n \to h$ since $g^{n+1} \downarrow 0$ and $g^{n+1} = 1$ on S_1.

Now h satisfies the property

$$\int_E h \, d\mu = \int_E d\nu = \nu(E)$$

The uniqueness is easy to prove and thus the Radon-Nicodym theorem is proved.

3.6 CLASSES OF BOUNDED LINEAR OPERATORS ON HILBERT SPACES

We present now several classes of bounded linear operators on Hilbert spaces; these classes are of fundamental importance in the study of arbitrary bounded operators on Hilbert spaces, as well as for applications.

A. Hermitian Operators

3.6.1 DEFINITION An operator $T \in L(H)$, H is a Hilbert space, is called *hermitian* if for all $x, y \in H$,

$$\langle Tx, y\rangle = \langle x, Ty\rangle$$

3.6.2 THEOREM An operator $T \in L(H)$ is hermitian if and only if $T = T^*$.

Proof. This follows immediately from the definition.

3.6.3 THEOREM An operator $T \in L(H)$ is hermitian if and only if

$$W(T) = \{\langle Tx, x\rangle \mid \|x\| = 1\}$$

is a subset of R.

Proof. It is obvious that the condition is necessary. The fact that it is sufficient follows from the identity

$$\langle T(x + x), (x + y)\rangle - \langle T(x - y), (x - y)\rangle = 2\langle Tx, y\rangle + 2\overline{\langle T^*x, y\rangle}$$

and thus,

$$\text{Im} \langle (T - T^*)x, y\rangle = 0$$

for all x, y which gives that $T = T^*$.

For another proof see Section 8.1.

The following theorem shows a nice connection between boundedness and being hermitian.

3.6.4 THEOREM (Hellinger-Toeplitz) [See F. Riesz and B. Nagy (1952).] If T is a linear operator on a Hilbert space and for all $x, y \in H$

$$\langle Tx, y\rangle = \langle x, Ty\rangle$$

then T is bounded.

Proof. Suppose that T is not bounded. We find a sequence $\{x_n\}$ such that $\|x_n\| = 1$, $\|Tx_n\| > n$. We consider for each n the functional

$$F_n(x) = \langle Tx, x_n\rangle = \langle x, Tx_n\rangle$$

and it is obvious that

$$|F_n(x)| \leq \|x\|\|Tx_n\|$$

and the function

$$x \mapsto \sup |F_n(x)|$$

is lower semicontinuous and satisfies all the conditions of the Gelfand lemma. Thus we find $M > 0$ such that

$$|F_n(x)| \leq M\|x\|$$

and for $x = Tx_n$ we obtain that

$$\|Tx_n\|^2 \leq M\|Tx_n\|$$

which gives $\|Tx_n\| \leq M$ for all n. This contradiction proves the assertion.

An important subclass of hermitian operators is the following.

3.6.5 DEFINITION An operator $T \in L(H)$ is called *positive* if

$$\langle Tx, x\rangle \geq 0$$

We write this simply as $T \geq 0$.

3.6.6 DEFINITION If T and S are two hermitian operators, we say that $T \leq S$ if $S - T$ is positive. Clearly this is an order relation on the set of all hermitian operators.

3.6.7 EXAMPLE For any $T \in L(H)$ the operators T^*T and TT^* are hermitian operators.

3.6.8 DEFINITION A hermitian projection P is any $P \in L(H)$ such that P is hermitian and $P^2 = P$.

3.6.9 EXAMPLE For any closed subspace m of a Hilbert space H, P_m the orthogonal projection of H onto m is a projection.

B. Normal Operators

3.6.10 DEFINITION An operator $T \in L(H)$ is called *normal* if

$$TT^* = T^*T$$

The following theorem gives the characterization of normal operators.

3.6.11 THEOREM An operator $T \in L(H)$ is normal if and only if for all $x \in H$

$$\|Tx\| = \|T^*x\|$$

Proof. The condition is obviously necessary. We show now that it is also sufficient. It is clear that it is equivalent to

$$<(T^*T - TT^*)x, x> = 0$$

and the assertion follows as a consequence of the following theorem, which is of interest itself.

3.6.12 THEOREM If $T \in L(H)$ and

$$w(T) = \sup \{|<Tx, x>| \mid \|x\| = 1\}$$

then

$$w(T) \le \|T\| \le 2w(T)$$

Proof. Let r and s be real numbers, $r \neq 0$. Then we have the following identity:

$$\|Tx\|^2 + e^{2is}<T^2x, x> = \frac{1}{2}<\{re^{2is}T^2x + r^{-1}e^{is}Tx\}, \{re^{is}Tx + r^{-1}x\}>$$
$$- \frac{1}{2}<\{re^{2is}T^2x - r^{-1}e^{is}Tx\}, \{re^{is}Tx - r^{-1}x\}>$$

and since for any y, $|<Ty, y>| \le w(T)\|y\|^2$, this gives the relation

$$\left|\|Tx\|^2 + e^{2is}<T^2x, x>\right| \le \frac{1}{2} w(T)\{\|re^{is}Tx + r^{-1}x\|^2 + \|re^{is}Tx - r^{-1}x\|^2\}$$
$$= w(T)(r^2\|Tx\|^2 + r^{-2}\|x\|^2)$$

If $Tx \neq 0$, we choose s such that $e^{2is}\langle T^2x, x\rangle = |\langle T^2x, x\rangle|$ and $r^2\|Tx\| = \|x\|$. In this case we obtain

$$\|Tx\|^2 + |\langle T^2x,x\rangle| \leq 2w(T)\|Tx\|\|x\|$$

and this gives further,

$$\|Tx\| \leq 2w(T)\|x\|$$

which proves the theorem.

3.6.13 COROLLARY For any $T \in L(H)$ we have the relation

$$w(T^2) \leq (w(T))^2$$

Proof. From the above inequality we obtain

$$[\|Tx\| - w(T)]^2 + |\langle T^2x, x\rangle| \leq (w(T))^2$$

and this implies our assertion.

3.6.14 REMARK From this follows that for all n,

$$w(T^{2^n}) \leq (w(T))^{2^n}$$

3.6.15 THEOREM If $T \in L(H)$ is normal, then

$$\|T\|^2 = \|T^2\|$$

Proof. Obviously we have the relation $\|T^2\| \leq \|T\|^2$, which is valid for any operator. Since

$$\|T^2x\|^2 = \langle T^2x, T^2x\rangle = \langle T^{*2}T^2x, x\rangle = \langle T^*Tx, T^*Tx\rangle = \|T^*Tx\|^2$$

we have

$$\|T^2\| = \|T^*T\|$$

and since for any operator

$$\|T^*T\| = \|T\|^2$$

we obtain the assertion.

3.6.16 THEOREM If $T \in L(H)$ is normal, then

1. $w(T) = \|T\|$
2. $\lim w(T^n)^{1/n} = \lim \|T^n\|^{1/n} = \|T\|$

Proof. Since for any p, T^p is normal, the assertions are consequences of the above results.

3.16.17 THEOREM If $T \in L(H)$, then T is normal if and only if T is of the form

$$T = S_1 + iS_2$$

where S_i are hermitian and $S_1S_2 = S_2S_1$.

Proof. It is easy to see that any operator is of the form

$$T = \frac{1}{2}(T + T^*) + i\left(\frac{1}{2i}(T - T^*)\right)$$
$$= \text{Re } T + i \text{ Im } T = S_1 + iS_2$$

If T is normal, then it is clear that our condition is necessary. Now, for the converse, we note that

$$T^*T = S_1^2 + i(S_1S_2 - S_2S_1) + S_2^2$$
$$TT^* = S_1^2 + i(S_2S_1 - S_1S_2) + S_2^2$$

and thus the condition is sufficient.

3.6.18 REMARK The above decomposition is suggested partly by the decomposition of complex numbers z, $z = x + iy = \text{Re } z + i \text{ Im } z$.

C. Unitary Operators

3.6.19 DEFINITION Let $T \in L(H)$. T is called a *unitary operator* if

1. $TH = H$
2. $\langle Tx, Ty\rangle = \langle x, y\rangle$

3.6.20 THEOREM A necessary and sufficient condition for T in L(H) to be unitary is that $T^*T = TT^* = I$.

Proof. It is clear that if T is unitary, the above conditions are satisfied. If these conditions are true, then

$$\langle Tx, Ty\rangle = \langle T^*Tx, y\rangle = \langle x, y\rangle$$

and since (as is easy to prove) T (and T*) are surjective operators, the assertion of the theorem is proved.

D. The Square Root of a Positive Hermitian Operator

Let H be a Hilbert space and $T \in L(H)$ be a positive operator. We are interested in the following problem: There exists an operator, which we denote by $T^{1/2}$, such that

1. $T^{1/2}$ is hermitian.
2. $(T^{1/2})^2 = T$.

We give a result answering these in the positive direction. We need for this some results which have their own interest.

3.6.21 THEOREM (Generalized Cauchy-Schwarz Inequality) Let $T \in L(H)$ be a positive operator. Then

$$|\langle Tx, y\rangle|^2 \le \langle Tx, x\rangle\langle Ty, y\rangle$$

Proof. Since the proof is essentially the same as for the case $T = I$, we omit it.

3.6.22 THEOREM If $\{T_n\}$ is a sequence of hermitian operators such that $T_n - T_m \ge 0$ for $n > m$, then $\{T_n\}$ converges strongly to a hermitian operator.

Proof. We can assume, without loss of generality, that $0 \le T_n \le I$ and thus applying the Cauchy-Schwarz inequality with $T = T_n - T_m$, we obtain

$$\begin{aligned}\|(T_n - T_m)x\|^4 &= \langle (T_n - T_m)x, (T_n - T_m)x\rangle^2 \\ &\le \langle (T_n - T_m)x, x\rangle\langle (T_n - T_m)^2x, (T_n - T_m)x\rangle \\ &\le \langle (T_n - T_m)x, x\rangle\|x\|^2\end{aligned}$$

Since the sequence $\{\langle T_n x, x\rangle\}$ is monotone (and bounded), it is convergent, and from the inequality just obtained, we conclude that $\{T_n x\}$ is a strong convergent sequence. Thus we can define

$$Tx = \lim T_n x$$

for each $x \in H$. Clearly this is a bounded hermitian operator.

3.6.23 LEMMA If T_1 and T_2 are two positive operators with $T_1T_2 = T_2T_1$, then T_1T_2 is positive.

Proof. First we remark that if T is positive and S is also positive, then T^2S is positive when TS = ST. Also if T and I - T are positive, then $T - T^2$ is positive.

Since the first assertion is obvious, we prove only the second assertion. Indeed, since

$$I - (T - T^2) \geq 0$$

we have that

$$I \geq (T - T^2)$$

and thus since

$$T - T^2 = T^2(I - T) + T(I - T)^2$$

the assertion follows.

We consider now the sequence

$$\tilde{T}_1 = \frac{T_1}{\|T_1\|} \qquad \tilde{T}_2 = \tilde{T}_1 - \tilde{T}_1^2 \quad \cdots \quad \tilde{T}_{n+1} = \tilde{T}_n - \tilde{T}_n^2 \quad \cdots$$

and by the above remarks we have

1. $0 \leq \tilde{T}_n \leq I$
2. $\Sigma_{i=1}^n \tilde{T}_i = \tilde{T}_1 - \tilde{T}_{n+1}$

From these we have that

$$\sum_{i=1}^{n} \|\tilde{T}_i x\|^2 < \langle T_1 x, x\rangle \qquad \text{for all } n$$

From this we obtain that

$$\lim \|\tilde{T}_i x\| = 0$$

and thus

$$\sum_{i=1}^{\infty} \tilde{T}_i x = \tilde{T}_1 x$$

Now

$$<T_2\tilde{T}_i x, \ x> = \sum_{i=1}^{\infty} <T_2\tilde{T}_i x, \ x> = \sum_{i=1}^{\infty} <T_2\tilde{T}_i x, \ \tilde{T}_i x> > 0$$

and thus

$$T_2\tilde{T}_1 \geq 0$$

and the assertion is proved.

3.6.24 THEOREM If $T \in L(H)$ is a positive operator, then there exists a unique hermitian operator $\tilde{T}$ such that $(\tilde{T})^2 = T$.

Proof. We can suppose, without loss of generality, that $0 \leq T \leq I$. We define the sequence of operators

1. $T_0 = 0$
2. $T_{n+1} = \frac{1}{2}(T - T_n^2) + T_n$ for $n = 0, 1, 2, \ldots$

which has the property that each T_n is hermitian. We show now that $\{T_n\}$ is monotone, i.e.,

$$T_n - T_m \geq 0 \qquad \text{for } n \geq m$$

Since T_n is an operator commuting with any operator commuting with T, we have

$$I - T_{n+1} = \frac{1}{2}(I - T_n)^2 + \frac{1}{2}(I - T)$$

For n = 0 we have that

$$T_1 - T_0 = \frac{1}{2} T \geq 0$$

and if the assertion is true for the operator

$$T_i - T_{i-1} \geq 0$$

for $i \leq n$, then from the above relation it holds for $n + 1$.

Thus the sequence $\{T_n\}$ is monotone and there exists a hermitian operator $\tilde{T}$ such that

$$\tilde{T} = \lim T_n x$$

From condition 2 of the sequence, we obtain for $n \to \infty$,

$$\tilde{T} = \tilde{T} + \frac{1}{2} (T - \tilde{T}^2)$$

and this gives that $\tilde{T}^2 = T$. The existence of the square root of a positive operator is proved.

Now we prove the uniqueness. Suppose that there exists $\tilde{T}_1$, another square root of T. Obviously,

$$T\tilde{T}_1 = \tilde{T}_1 T = \tilde{T}_1^3$$

and thus the operators T and $\tilde{T}_1$ are commuting. This gives that $\tilde{T}$ and $\tilde{T}_1$ are also commuting operators. Since $\tilde{T}$ and $\tilde{T}_1$ are positive operators, they have square roots. Let $\bar{T}$ and $\bar{T}_1$ be positive operators such that

$$\bar{T}^2 = \tilde{T} \qquad \bar{T}_1^2 = \tilde{T}_1$$

Let $x \in H$ be an arbitrary element and $y = (\tilde{T} - \tilde{T}_1)x$. We have,

$$\begin{aligned} \|\bar{T}y\|^2 + \|\bar{T}_1 y\|^2 &= \langle \bar{T}y, \bar{T}y\rangle + \langle \bar{T}_1 y, \bar{T}_1 y\rangle = \langle \bar{T}^2 y, y\rangle + \langle \bar{T}_1^2 y, y\rangle \\ &= \langle \tilde{T}y, y\rangle + \langle \tilde{T}_1 y, y\rangle = \langle (\tilde{T} + \tilde{T}_1)y, y\rangle \\ &= \langle (\tilde{T}^2 - \tilde{T}_1^2)x, y\rangle = 0 \end{aligned}$$

and thus

$$\tilde{T}y = \tilde{T}_1 y = 0$$

which gives that

$$\|(\tilde{T} - \tilde{T}_1)x\|^2 = \langle (\tilde{T} - \tilde{T}_1)x, x\rangle = \langle (\tilde{T} - \tilde{T}_1)y, x\rangle = 0$$

and thus

$$\tilde{T}x = \tilde{T}_1 x$$

which proves the uniqueness of the square root of a positive operator.

E. Operators and Matrices

Let I_1 and I_2 be two families of indices.

3.6.25 DEFINITION A matrix is a family of complex numbers

$$\{a_{\alpha\beta}\}_{(\alpha,\beta)\in I_1\times I_2}$$

The matrix with all $a_{\alpha\beta} = 0$ will be called the *zero matrix* and the matrix with $a_{\alpha\beta} = \delta_{\alpha\beta}$, $I_1 = I_2$ will be called the *identity matrix*.

If we define

$$\{a_{\alpha\beta}\} + \{b_{\alpha\beta}\} = \{a_{\alpha\beta} + b_{\alpha\beta}\}$$

$$z\{a_{\alpha\beta}\} = \{za_{\alpha\beta}\}$$

then the set of all matrices with the set of indices I_1 and I_2 is a linear space.

3.6.26 DEFINITION A matrix $\{a_{\alpha\beta}\}_{I_1\times I_2}$ is called *bounded* if there exists $M < \infty$ such that

$$\sum_{\alpha\in I_1} \left|\sum_{\beta\in\tilde{I}_2} z_\beta a_{\alpha\beta}\right|^2 \le M \sum_{\beta\in\tilde{I}_2} |z_\beta|^2$$

for all finite parts $\tilde{I}_2 \subset I_2$. The smallest constant M satisfying the inequality is called the *norm* of the matrix and is denoted $\|a_{\alpha\beta}\|$.

In what follows we consider only matrices with $I_1 = I_2$.

3.6.27 LEMMA If $\{a_{\alpha\beta}\}$ is a bounded matrix, then

$$\sum_{(\alpha,\beta)\in I\times I} |a_{\alpha\beta}|^2 < \infty$$

Proof. Let

$$c_\beta = \begin{cases} 0 & \beta \neq r \\ 1 & \beta = r \end{cases}$$

and since the matrix is bounded, we obtain

$$\sum_{\alpha\in I} |a_{\alpha r}|^2 < \infty$$

and for any finite part $J \subset I$,

$$\sum_{\alpha\in I} \sum_{\beta\in I} |a_{\alpha\beta}\bar{a}_{\beta r}|^2 \le M \sum_{\beta\in J} |a_{\beta r}|^2 \le M \sum_{\beta\in I} |a_{\beta r}|^2$$

which implies that

$$\sum_{(\alpha,\beta)\in I\times I} |a_{\alpha\beta}|^2 < \infty$$

and the assertion is proved.

Let H be a Hilbert space and $\{e_\alpha\}_{\alpha\ I}$ be a basis for H. If $T \in L(H)$, then we consider the matrix $\{a_{\alpha\beta}\}$, where

$$a_{\alpha\beta} = \langle Te_\alpha, e_\beta\rangle$$

and it is clear that for any $\beta \in I$

$$Te_\beta = \Sigma\, a_{\alpha\beta}e_\alpha$$

The mapping

$$m: T \to \{a_{\alpha\beta}\}_{\alpha,\beta\in I\times I}$$

has the following properties:

1. $m(T + S) = m(T) + m(S)$
2. $m(aT) = am(T)$
3. $m(0) = 0$
4. $m(I) = I = \{\delta_{\alpha\beta}\}$ $\qquad \delta_{\alpha\beta} = \begin{cases} 1 & \alpha = \beta \\ 0 & \alpha \neq \beta \end{cases}$
5. $\|T\| = \|a_{\alpha\beta}\|$

Since all the assertions are obvious except possibly 5, we give here the proof.

Let J be any finite part of I and

$$x = \sum_{\beta} c_\beta e_\beta$$

which gives that

$$\|Tx\|^2 = \sum_{\alpha\in I} |\sum_{\beta\in J} a_\beta c_\beta|^2$$

and thus

$$\|Tx\| \leq \|T\| \cdot \|x\|$$

if and only if

$$\sum_{\alpha\in I} |\sum_{\beta\in J} a_{\alpha\beta} c_\beta|^2 \leq \|T\|^2 \sum_{\beta\in J} |c_\beta|^2$$

Since the set of all elements like x is dense in H, we obtain that $\|m(T)\| = \|T\|$.

If $\{a_{\alpha\beta}\}$ and $\{b_{\alpha\beta}\}$ are two bounded matrices, then we can define their product as the matrix whose elements are

$$c_{\alpha\beta} = \sum_{r} a_{\alpha r} b_{r\beta}$$

and it is easy to see that

$$m(TS) = m(T)m(S)$$

If $m(T) = \{a_{\alpha\beta}\}$, then we have also $m(T^*) = \{a^1_{\alpha\beta}\}$, where

$$a^1_{\alpha\beta} = a^*_{\beta\alpha}$$

3.6.28 REMARK In terms of matrices, we can characterize the elements in L(H) which are hermitian, unitary, etc.

Chapter 4

BANACH ALGEBRAS

We give in this chapter a short introduction to Banach algebras; for further information we refer to Rickart (1960), Browder (1969), and Gelfand et al. (1964).

4.1 DEFINITIONS AND SOME EXAMPLES

4.1.1 DEFINITION A *Banach algebra* A is a complex Banach space with a product satisfying the following properties:

1. $(xy)z = x(yz)$
2. $x(y + z) = xy + xz$
3. $\alpha(xy) = (\alpha x)y$
4. $\|xy\| \le \|x\| \cdot \|y\| = \|x\|\|y\|$

for all $x,y,z \in A$ and $\alpha \in C$.

If $xy = yx$, then we say that A is a *commutative* Banach algebra; if there exists an element $e \in A$ such that $ex = xe = x$ for all $x \in A$, then we say that the Banach algebra is with *unit* or is a *unital Banach algebra* and e is called the *unit* of A.

4.1.2 DEFINITION If A is a Banach algebra with unit, an element x has an inverse if there exists an element $y \in A$ such that $xy = yx = e$.

We write $y = x^{-1}$ (since it is easy to see that the inverse is unique).

From axiom 4 of Definition 4.1.1, we have the following.

4.1.3 THEOREM If A is a Banach algebra, then the product is a continuous function in both variables.

Proof. Let x" and y" be in A. Then we have

$$\|xy - x'y'\| \le \|x - x'\|\|y\| + \|x'\|\|y - y'\|$$

and this proves the assertion.

We give now some examples of Banach algebras.

4.1.4 EXAMPLE Let S be a compact space and C(S) be the space of all continuous complex-valued functions with the product defined as follows:

$$(fg)(s) = f(s)g(s)$$

and with the norm

$$f \to \|f\|_S = \sup \{|f(s)| \mid s \in S\}$$

In this case it is easy to see that C(S) is a Banach algebra with unit. Clearly the unit is the function identical with 1.

4.1.5 EXAMPLE Let S be a locally compact, noncompact space and let $C_0(S)$ be the Banach space of all continuous complex-valued functions which vanishes at infinity. With the product defined as in the above example, $C_0(S)$ is a Banach algebra without unit.

4.1.6 EXAMPLE Let X be a complex Banach space and L(X) be the set of all linear and bounded operators on X. With the product defined as follows,

$$(TS)(x) = T(Sx)$$

L(X) is a Banach algebra with unit.

4.1.7 EXAMPLE Let A be the set of all continuous complex-valued functions defined on $[-\pi, \pi]$ such that

1. $f(-\pi) = f(\pi)$, $f \in A$.
2. The Fourier coefficients of f with negative indexes are zero.

Then A is a commutative Banach algebra with unit and where the product is defined as in Example 4.1.4. Another way to describe this algebra is as follows: A is isomorphic to the algebra of all continuous functions on $\{z \mid |z| \leq 1\}$ and analytic in $\{z \mid |z| < 1\}$.

4.1.8 EXAMPLE Let W be the set of all functions of the form

$$f(t) = \Sigma\, a_n e^{int} \qquad \Sigma\, |a_n| < \infty$$

This is a Banach algebra with the convolution as product. This Banach algebra was studied by Wiener (1922) and Gelfand (1938).

4.1.9 EXAMPLE Let $C_{[0,1]}$ be the space of all continuous complex-valued functions on [0, 1] with the product defined as follows:

$$(fg)(x) = \int_0^x f(t)g(x - t)\, dt$$

and in this case $C_{[0,1]}$ is a Banach algebra without unit.

In all of the above examples, when the Banach algebra has unit e, then $\|e\| = 1$. We show now that we can introduce a new norm $\|\cdot\|_1$ such that $\|e\|_1 = 1$.

4.1.10 THEOREM Let A be a Banach algebra with unit e. Then there exists an equivalent norm $\|\cdot\|_1$ such that $(A, \|\cdot\|_1)$ is a Banach algebra and $\|e\|_1 = 1$.

Proof. For this we consider the Banach algebra of all bounded linear operators on A. For any $a \in A$ we define

$$T_a(x) = ax$$

and it is clear that $\|T_a\| = \|a\|$. We denote by L_A the set of all operators T_a with $a \in A$. It is obvious that L_A is a Banach algebra. We define the norm $\|\cdot\|_1$ on A by the relation,

$$\|a\|_1 = \|T_a\|$$

and it is clear that $\|a\|_1 \leq \|a\|$. Now, since

$$\|T_a\| = \sup_{\|x\|\leq 1} \|ax\| = \sup_{\|x\|\leq\|e\|} \|ax\| \cdot \|e\|^{-1} \geq \|e\|^{-1}\|a\|$$

which yields

$$\|e\|^{-1}\|a\| \leq \|a\|_1 \leq \|a\|$$

and the assertion of the theorem is proved.

We prove now some elementary facts about elements in Banach algebras.

4.1.11 *THEOREM* Let A be a Banach algebra with unit and x in A. In this case,

1. If $\|x - e\| < 1$, then x^{-1} exists.
2. If s is such that $|s| > \|x\|$, then $(x - s)^{-1}$ exists.

Proof. We prove the first assertion using the well-known properties of geometric series.

Indeed, if

$$y = e + (e - x) + (e - x)^2 + (e - x)^3 + \cdots$$

and $x = e + (x - e)$, we obtain that

$$\begin{aligned} xy &= y - \sum_{n=1}^{\infty} (e - x)^{n+1} + (x - e) \\ &= e + \sum_{n=1}^{\infty} (e - x)^n - \sum_{n=2}^{\infty} (e - x)^n + (x - e) \\ &= e = yx \end{aligned}$$

and the assertion is proved.

To prove 2, we remark that

$$(x - s) = \frac{1}{s}\left(\frac{x}{s} - e\right)$$

and for x/s the first assertion applies and 2 is proved.

4.1.12 *DEFINITION* For any Banach algebra with unit

$$A^{-1} = \{a \in A \mid a^{-1} \text{ exists}\}$$

4.1.13 *THEOREM* A^{-1} is an open set in A.

Proof. From Theorem 4.1.11, it follows that

$$V_e = \{x \mid \|x - e\| < 1\}$$

is in A^{-1}. In this case for any z, $zx^{-1} \in V_e$, zx^{-1} has an inverse and this clearly implies that z has an inverse and the theorem is proved.

4.1.14 THEOREM If A is a Banach algebra with unit, then the function

$$f: A^{-1} \to A^{-1}$$

defined by the relation $f(x) = x^{-1}$ is continuous.

Proof. Let $\{x_n\}$ be from A^{-1} and $x_n \to x$. We show now that $x_n^{-1} \to x^{-1}$.

Since it is obvious that

$$x^{-1}x_n \to e$$

and since

$$(x^{-1}x_n)^{-1} = e + (e - x^{-1}x_n) + \cdots + (e - x^{-1}x_n)^m + \cdots$$

we show that the right member is a convergent series. Indeed we find an integer N such that $\|x^{-1}x_n - e\| < 1$ for all $n > N$.

Thus we obtain

$$e - (x^{-1}x_n) = -(e - x^{-1}x_n) - (e - x^{-1}x_n)^2 - \cdots$$

which gives

$$\|e - x_n^{-1}x\| < \sum_{n=1}^{\infty} \|e - x^{-1}x_n\|^m \leq \sum_{m=1}^{\infty} \|x^{-1}\|\|x - x_n\|^m$$

and thus for $n > N$ such that $\|x - x_n\| < (1/2)\|x^{-1}\|$, we obtain

$$\lim x_n^{-1}x = e$$

which is equivalent to $\lim x_n^{-1} = x^{-1}$.

4.2 THE SPECTRUM OF AN ELEMENT IN A BANACH ALGEBRA WITH UNIT

4.2.1 DEFINITION The *spectrum of an element* a in a Banach algebra with unit is the set of all complex numbers z such that a - ze has an inverse.

The inverse is denoted by $R_z(a)$ and the spectrum by $\sigma(a)$.

4.2.2 PROPOSITION For any z and z' the following relation holds:

$$R_z(a) - R_{z'}(a) = (z' - z)R_zR_{z'}$$

Proof. Since we have

$$\begin{aligned} R_z(a)^{-1}R_{z'}(a) &= (a - z)(a - z')^{-1} \\ &= [a - z' + (z' - z)](a - z')^{-1} \\ &= e + (z' - z)R_z(a)R_{z'}(a) \end{aligned}$$

This obviously implies the assertion.

4.2.3 DEFINITION If $a \in A$, A a Banach algebra with unit, the set of all complex numbers z for which a - z has an inverse is called the *set of regular points* and is denoted by $\rho(a)$.

Obviously, $\rho(a) = C\sigma(a)$.

4.2.4 PROPOSITION The set $\rho(a)$ is open.

Proof. Let $a \in A$ and $s_0 \in \rho(a)$. We take s such that

$$|s - s_0| \frac{1}{2} \|(a - s_0)^{-1}\| < 1$$

and consider the element y of the form

$$y = (s_0 - a)^{-1} e + \sum_{n=1}^{\infty} (s_0 - s)^n (s_0 - a)^{-n}$$

which exists and satisfies the relation

$$y(s - a) = (s - a)y = e$$

The proposition is proved.

4.2.5 PROPOSITION For any $a \in A$, the function

$$z \to (a - z)^{-1}$$

is an A-valued analytic function.

Proof. Since for any $z \neq z_0$

$$\frac{R_z(a) - R_{z_0}(a)}{z - z_0} = R_z(a)R_{z_0}(a)$$

it follows that when $z \to z_0$, we have

$$\lim \frac{R_z(a) - R_{z_0}(a)}{z - z_0} = R_{z_0}^2(a)$$

and the proposition is proved.

4.2.6 PROPOSITION For any $a \in A$, $\sigma(a)$ is a closed and nonempty set.

Proof. The fact that $\sigma(a)$ is a closed set follows from Proposition 4.2.4. The fact that it is a bounded set follows from Theorem 4.1.11. It remains to prove the fact that it is a nonempty set. Indeed, in the contrary case, we have that the function

$$z \mapsto (a - z)^{-1}$$

is analytic in the whole complex plane and it is a bounded function. By Liouville's theorem (which can be proved for vector-valued functions), it follows that it is a constant function. We prove now that this constant is zero. Indeed,

$$\lim_{z \to \infty} \left(\frac{e - 1}{z} a\right) = e$$

and thus the constant is zero. This forces $a = ze$, which is a contradiction.

4.2.7 PROPOSITION (*Mazur-Gelfand*) [See Gelfand et al. (1964).] If A is a Banach algebra with unit and $A^{-1} = A \setminus 0$, then $A = Ce$, i.e., A is isomorphic and isometric with the field of complex numbers.

Proof. Indeed, for any $a \in A$, $a \neq 0$, there exists z_a such that $a = z_a e$ and thus the application $a \mapsto z_a$ is an isomorphism and clearly $\|a\| = |z_a|$.

The following proposition gives information concerning the radius of the smallest circle which contains the spectrum of an element.

4.2.8 PROPOSITION The radius of the smallest circle with the origin as center and containing the spectrum $\sigma(a)$ of $a \in A$ (called the *spectral radius* of a, denoted r_a) can be computed with the formula

$$r_a = \lim \|a^n\|^{1/n}$$

Proof. From Theorem 4.1.11, we know that for s, $|s| > \|a\|$, $s \in \rho(a)$, and this shows that $r_a \leq \|a\|$. We suppose the following particular case of the spectral mapping theorem:

$$\sigma(a^n) = (\sigma(a))^n = \{s^n \mid s \in \sigma(a)\}$$

In this case we obtain

$$r_a n \leq \|a^n\|$$

and this gives that

$$r_a \leq \|a^n\|^{1/n}$$

or that

$$r_a \leq \lim\inf \|a^n\|^{1/n}$$

We prove now that

$$r_a \geq \sup \|a^n\|^{1/n}$$

For this we take an arbitrary functional $a^* \in A^*$ and define

$$F(s) = a^*((s - a)^{-1})$$

which is a holomorphic function in $\{z \mid |z| > r_a\}$ and since F has the form

$$F(s) = \sum \frac{a^*(a)}{s^{n+1}}$$

for s, $|s| > \|a\|$, then it is known from the complex function theory that $F(\cdot)$ has the same form for $\{s \mid |s| > r_a\}$ and thus $\{a^*(a^n)/s^n\}$ is a bounded sequence for any $a^* \in A^*$.

Then there exists a constant $k > 0$ such that

$$\left\|\frac{a^n}{s^n}\right\| < k$$

and this gives

$$\limsup \|a^n\|^{1/n} \leq |s| \qquad |s| > r_a$$

But the sequence $\{c_n\}$, where $c_n = \|a^n\|$ has the property

$$c_{n+m} \leq c_n c_m$$

and this gives that $\lim c_n^{1/n}$ exists. From these we have

$$r_a = \liminf c_n^{1/n} = \limsup c_n^{1/n} = \lim \|a^n\|^{1/n}$$

and the assertion is proved.

As in the case of linear and continuous operators on Banach spaces, we can consider the functions $f \in \mathcal{F}_a$, and as in that case, we can prove the following theorem.

4.2.9 THEOREM There exists a homomorphism of $\mathcal{F}_a$ in A such that for any $f \in \mathcal{F}_a$ the relation

$$\sigma(f(a)) = f(\sigma(a)) = \{f(s) \mid s \in \sigma(a)\}$$

holds.

The proof of this result is exactly the same as in the case of linear operators.

4.3 REPRESENTATION THEOREMS FOR COMMUTATIVE BANACH ALGEBRAS

First we give some results about ideals in Banach algebras.

4.3.1 DEFINITION Let A be a Banach algebra and I_ℓ be a subset of A. Then I_ℓ is called a *left ideal* of A if

1. For any $x \in I_\ell$ and any $y \in A$, $xy \in I_\ell$.
2. I_ℓ is a Banach algebra with the operations of A.

Similarly we say that a subset I_r of A is a *right ideal* if

1. For any $x \in I_r$ and any $y \in A$, $yx \in A$
2. I_r is a Banach algebra with the operations of A.

A subset I of A is called a *two-sided ideal* if it is a left and a right ideal.

Consider now a two-sided ideal I and we define the following equivalence relation: $x \tilde{} y$ if and only if $x - y \in I$.

For any $x \in A$ we set $\tilde{x} = \{y \mid x \tilde{} y\}$. The set of all equivalence classes $\{\tilde{x}\}$ can be organized as an algebra, and we call such an algebra the *quotient algebra* A/I.

4.3.2 DEFINITION An ideal (left, right, two-sided) is called *maximal* if there is no ideal of the same type different from A containing it.

4.3.3 PROPOSITION Any ideal is contained in a maximal ideal of the same type.

Proof. We prove the assertion only for left ideals. Let I_s be the left ideal and be the family of all left ideals which contains I_ℓ; we consider in this family an "order" defined by inclusion. Then it is easy to see that we can apply Zorn's lemma, obtaining a maximal left ideal.

4.3.4 PROPOSITION Every maximal ideal is closed.

The following theorem is concerned with the structure of certain classes of Banach algebras.

4.3.5 PROPOSITION If A is a Banach algebra with unit and has no left ideals different from {0} and A, then A is a field.

Proof. In fact, the assertion is valid for any algebra with unit. Indeed, let

$$I_x = \{xy \mid y \in A\}$$

which is obviously a left ideal. Now since in A there do not exist other ideals different from $\{0\}$ and A, we have the relation

$$I_x = A$$

for all $x \in A$, $x \neq 0$. Since A has unit we have that there exists an element y such that $xy = e$. For the element y we find an element z such that $yz = e$. We prove now that $y = x^{-1}$. Indeed,

$$yx = yx(e) = yx(yz) = y(xy)z = yz = e$$

The proposition is proved.

The following proposition is of fundamental importance in the theory of commutative Banach algebras.

4.3.6 PROPOSITION Let A be a Banach algebra with unit. A necessary and sufficient condition for A/I, I a closed two-sided ideal, to be a field is that I be a left maximal ideal.

Proof. Let $f: x \to \tilde{x}$ defined on A with values in A/I. Since f is a homomorphism, f(I) is an ideal in A/I. If $\tilde{I}$ is an ideal in A/I, then

$$f^{-1}(\tilde{I}) = \{x \mid \tilde{x} \in \tilde{I}\}$$

is a left ideal of A containing I. From this remark and Proposition 4.3.5, the assertion of the proposition follows.

The following proposition gives information about the connection between noninvertibility and maximal ideals.

4.3.7 PROPOSITION If A is a Banach algebra with unit, then every noninvertible element of A is contained in a nontrivial maximal ideal.

Proof. Suppose that $x \in A$ and noninvertible. We consider the set $I_x = \{xy \mid y \in A\}$ which is obviously a left ideal. By Proposition 4.3.3 there exists a left maximal ideal which contains I_x. This maximal ideal satisfies our assertion.

The following proposition gives a basic structure theorem for the quotient algebras A/I, when I is a closed two-sided ideal. An example of a two-sided closed ideal in a Banach algebra is the ideal of compact operators.

4.3.8 PROPOSITION Let A be a Banach algebra with unit and I be a closed two-sided ideal of A. In this case, $(A/I, \|,\|^{\sim})$ is a Banach algebra with unit and where

$$\|\tilde{x}\|^{\sim} = \inf \{\|x\| \mid x \in \tilde{x}\}$$

Proof. Clearly

$$\|x + \tilde{y}\| = \inf \{\|x + y\| \leq \inf \|x\| + \inf \|y\| = \|x\|^{\sim} + \|\tilde{y}\|^{\sim}\}$$

and

$$\|sx\|^{\sim} = \inf \|sx\| = |s| \inf \|x\| = |s| \cdot \|\tilde{x}\|^{\sim}$$

Now if $\|\tilde{x}\| = 0$, then $\tilde{x} = 0$ since $0 \in \tilde{x}$. Also, we find a sequence $\{x_n\}$ in I such that $\lim \|x_n\| = 0$. If $y \in \tilde{x}$, then $y - x_n \in I$ and since I is closed and $\lim (y - x_n) = y$, we obtain that $y \in I$.

For the next axiom of the norm we have

$$\|xy\| = \inf \|xy\| \leq \inf \|x\| \inf \|y\| = \|\tilde{x}\| \cdot \|\tilde{y}\|$$

which proves that $(A/I, \|,\|^{\sim})$ is a normed algebra. We prove now that in fact we have a Banach space.

Let $\{x_n\}$ be a Cauchy sequence in A/I. We can choose an element $\tilde{x}_{n_i}$ such that

$$\sum \|\tilde{x}_{n_{i+1}} - \tilde{x}_{n_i}\|^{\sim} < \infty$$

and for each n_i and element x_i such that

$$\sum \|x_{n_{i+1}} - x_{n_i}\| < \infty$$

In this case $\{x_{n_i}\}$ is a Cauchy sequence in A, and by letting

$$x = \lim x_{n_i}$$

we can show that

$$\tilde{x} = \lim \tilde{x}_n$$

Indeed, since

$$\|\tilde{x}_{n_i} - \tilde{x}\| \leq \| x_{n_i} - x\|$$

it follows that $\lim \tilde{x}_{n_i} = \tilde{x}$.

But we have

$$\limsup \|\tilde{x}_n - \tilde{x}\| \leq \limsup \|\tilde{x}_n - \tilde{x}_{n_i}\| + \limsup \|\tilde{x}_{n_i} - \tilde{x}\|$$

and the assertion follows.

4.3.9 REMARK The norm of $\tilde{e}$ is equal to 1.

4.4 STRUCTURE THEOREMS FOR COMMUTATIVE BANACH ALGEBRAS

Let A be a commutative Banach algebra with unit.

4.4.1 DEFINITION A linear functional on A is called a *multiplicative functional* on A if for any $x,y \in A$, $f(xy) = f(x)f(y)$.

The set of all multiplicative functionals on A is denoted by M_A. The following proposition gives some properties of M_A.

4.4.2 PROPOSITION If $f \in M_A$, then the following properties hold:

1. $f(e) = 1$.
2. If x^{-1} exists, then $f(x^{-1}) = 1/f(x)$.
3. $f \in A^*$ and more $\|f\| = 1$.

Proof. Properties 1 and 2 are obvious. We prove only 3. Let $s \in A$, $|s| > \|x\|$, and we know that $x - s \in A^{-1}$. In this case $f(x - s) \neq 0$ and this gives the fact that

$$|f(x)| \leq \|x\|$$

for all $x \in A$. This obviously shows that $\|f\| = 1$.

4.4.3 DEFINITION For any $x \in A$ we define the function

$$x\colon M_A \to C$$

by the relation

$$x(m) = m(x)$$

which is called the *Gelfand transform of* x.

We define now a topology on M_A which transforms M_A in a compact topological space.

4.4.4 DEFINITION The Gelfand topology on M_A is the weakest topology on M_A for which all the functions $\{x(m)\}_{x \in A}$ are continuous.

It is very important to note the connection between the Gelfand topology on M_A and the weak *-topology on A^*. Indeed, for this we remark that

$$\{f \in A^* \mid f(xy) = f(x)f(y)\}$$

is weak *-closed, and since

$$M_A = \{f \in A^* \mid f(xy) = f(x)f(y),\ \|f\| = 1\}$$

it follows that the Gelfand topology is the weak *-induced topology on M_A.

4.4.5 COROLLARY M_A is weak *-closed in $S_1^* = \{x^* \mid \|x^*\| \leq 1\}$.

From the properties of the multiplicative functionals, it follows that $\{x(m)\}$ is an algebra of continuous functions on a compact space M_A and we note this algebra by $\hat{A}$.

4.4.6 DEFINITION If S is a compact space and A_1 is a subalgebra of the algebra of all continuous functions on S, we say that A_1 *separates* the points of S if for any $s_1 \neq s_2$ there exists a function $g \in A_1$ such that $g(s_1) \neq g(s_2)$.

4.4.7 THEOREM $\hat{A}$ separates the points of M_A.

Proof. This follows from the definition of $\hat{A}$.

We show now a strong connection between multiplicative functionals and the maximal ideals of A.

Let m be a maximal ideal of A and consider the Banach algebra A/m; we know that this is isometric with the complex numbers. Thus for every maximal ideal we find a complex number x(m) such that $x - x(m)e \in m$. We define a linear functional on A by

$$\tilde{x}_m(x) = x(m)$$

and it is easy to see that this is a continuous multiplicative functional. Also it is clear that

$$\ker \tilde{x}_m \subset m$$

Conversely, if f is any multiplicative functional on A, then ker f is a maximal ideal of A.

Indeed, the fact that ker f is an ideal is obvious. We prove that it is maximal. Since f(e) = 1, ker f is a proper ideal.

Also, for any x we have the identity,

$$x = \left(x - \frac{f(x)}{f(y)} y\right) + \frac{f(x)}{f(y)} y$$

where y is an arbitrary element of the complement of ker f. This gives a contradiction if ker f is not a maximal ideal.

4.4.8 COROLLARY If A is as above and $a_1, a_2, \ldots, a_n \in A$, then

1. There exists $f \in A^*$ and multiplicative such that $f(a_i) = 0$ for all $0 \leq i \leq n$.

or

2. There exists $h_1, h_2, \ldots, h_n \in A$ such that $\Sigma_{i=1}^n a_i h_i = e$.

Proof. Let I be the ideal

$$\{\Sigma\, a_i y_i \mid y_i \in A\}$$

and m_I be the maximal ideal space which contains I. If m_I is not proper ideal, then 1 holds; in the contrary situation, case 2 holds.

4.4.9 COROLLARY For any $x \in A$,

$$\sup_{m \in M_A} |\tilde{x}(m)| = \lim \|x^n\|^{1/n}$$

For the subsequent discussion we need the following:

4.4.10 DEFINITION A Banach algebra A is called a *function algebra* if

1. S is a compact Hausdorff space.
2. $A \subset C(S)$ and for each pair $s_1 \neq s_2$ there exists $f \in A$ such that $f(s_1) \neq f(s_2)$.
3. A is a closed subspace of AC(S) and contains the constants.

In connection with the results about A, the following problem is of interest: What are the necessary and sufficient conditions such that $\hat{A}$ is a function algebra on M_A?

From Corollary 4.4.9, we have the following theorem.

4.4.11 THEOREM A necessary and sufficient condition to have the property that $\hat{A}$ is a function algebra on M_A is that for any $x \in A$, $\|x\|^2 = \|x^2\|$.

4.4.12 DEFINITION If A is a Banach algebra with unit, then the *radical* of A, rad A, is the intersection of all maximal ideals of A.

From Corollary 4.4.9, it follows that

$$\text{rad } A = \{x \mid \lim \|x^n\|^{1/n} = 0\}$$

4.4.13 DEFINITION A Banach algebra A is called *semisimple* if rad A = {0}.

In all the above definitions A is a commutative Banach algebra.

From results proved above, we obtain the following representation theorems.

4.4.14 THEOREM Let A be a commutative Banach algebra with unit. Then there exists a compact Hausdorff space M_A and an application

$$\phi: x \to \tilde{x}$$

defined on A with values in the Banach algebra of all continuous functions on M_A such that

1. ϕ is an homomorphism.
2. $\phi(e) = 1$.
3. $|\tilde{x}(m)| \leq \|x\|$.
4. For any $m_1 \neq m_2$ in M_A there exists $x \in A$ such that $\tilde{x}(m_1) \neq \tilde{x}(m_2)$.
5. If A is semisimple, then ϕ is an isomorphism.

We give now an important representation theorem for a class of Banach algebras, the so-called commutative Banach algebras with involution.

4.4.15 DEFINITION Suppose that A is an arbitrary Banach algebra (commutative or not). In this case a mapping $*: A \to A$ is called an *involution* of A, or A is an involutive Banach algebra, if

1. $(x + y)^* = x^* + y^*$
2. $(sx)^* = s^*x^*$
3. $(xy)^* = y^*x^*$
4. $(x^*)^* = x$

for all $x,y \in A$ and s a complex number.

4.4.16 EXAMPLES

1. Let A be the Banach algebra of all complex-valued functions on a compact Hausdorff space S. The involution is defined as the complex conjugation of the corresponding function.

2. Let A be the Banach algebra of all bounded linear operators on a Hilbert space. The involution here is the adjoint of the corresponding operator.

4.4.17 DEFINITION An involutive Banach algebra is called a *B*-algebra* if $\|x^*x\| = \|x\|^2$ for all $x \in A$.

If A_1 is a subalgebra of an involutive Banach algebra, then we say that it is *involutive* if it is invariant for the involution of $A \supset A_1$.

In the next theorem we give some simple facts about involutive Banach algebras.

4.4.18 PROPOSITION Let A be a B*-algebra with unit. Then the following properties hold:

1. $e^* = e$.
2. $\|x^*\| = \|x\|$.
3. $0^* = 0$.
4. For any $x \in A$ there exist x_1 and x_2 such that $x = x_1 + ix_2$ and $x_1^* = x_1$, $x_2^* = x_2$.

Proof. All the properties are obvious except possibly 2. For 2 we have

$$\|x\|^2 = \|x^*x\| = \|x^*(x^*)^*\| = \|x^*\|^2$$

and this proves our assertion.

In connection with the result of Proposition 4.4.18, part 4, we have the following definition.

4.4.19 DEFINITION If A is an involutive Banach algebra, then an element $x \in A$ is called *hermitian* if $x = x^*$.

This definition agrees with the usual definition in the case of algebra of operators on a Hilbert space.

2.4.20 THEOREM If A is a commutative B*-algebra, then

$$r_a = \|a\|$$

for all $a \in A$. In other words, any element of a commutative B*-algebra is normaloid.

Proof. Since

$$\|x\|^2 = \|x^*x\| = \|(x^*x)^2\|^{1/2} = \cdots = \|(x^*x)^{2^n}\|^{1/2^n}$$
$$= \|x^{2^n}x^{*2^n}\|^{1/2^n} = \|x^{2^n}\|^{1/2^n}\|x^{*2^n}\|^{1/2^n}$$

and thus we obtain

$$\|x\|^2 = r_x r_{x^*} = r_x^2$$

and the assertion is proved.

The following result was proved by Arens (1947).

4.4.21 THEOREM If A is a B*-algebra and x is a hermitian element, then the spectrum is a subset of real numbers.

Proof. Suppose, on the contrary, that there exists $u + iv \in \sigma(x)$, where u and v are real numbers, $v \neq 0$. We prove that this gives a contradiction.

Indeed, let t be any real number and consider the element $y_t = x + ite$, where e is the unit of A. In this case, we know that the number $u + iv + it \in \sigma(y_t)$ and that $u - iv - it \in \sigma(y_t^*)$. But y_t and y_t^* are polynomials in x and $zz^* \in \sigma(y_t y_t^*)$, where $z = u + iv + it$. Thus we obtain

$$u^2 + (v + t)^2 \leq \|x^2 + t^2 e\| \leq \|x\|^2 + t^2$$

which gives

$$u^2 + 2tv + v^2 \leq \|x\|^2$$

which is a contradiction if t is large.

The following result contains Theorem 4.4.21 as a special case.

4.4.22 THEOREM If A is involutive, commutative, and with unit element and $x = u + iv$, where u and v are hermitian elements, then for any $s \in \sigma(x)$

$$|\operatorname{Re} s \cos \alpha + \operatorname{Im} s \sin \alpha| \leq \|u \cos \alpha + v \sin \alpha\| \qquad \alpha \in [0, 2\pi]$$

Proof. Let $z = re^{i\theta}$ such that $r > \|x\|$ and in this case $z_1 - x$ has an inverse, for z_1 near to z, given by the formula

$$(z - x)^{-1}(e + \Sigma\ (z - z_1)^n(z - x)^{-n}$$

We compute now the radius of convergence for this series. It is known as the *Hadamard formula*:

$$\frac{1}{R} = \limsup \|(z - x)^{-n}\|^{1/n}$$

and since we have

$$\begin{aligned}\|(z - x)^{-n}\| \cdot \|(z^* - x^*)^{-n}\| &= \|(z - x)^{-n}(z^* - x^*)^{-n}\| \\ &= \|(z - x)^{-1}(z^* - x^*)^{-1}\|^n \\ &= \|y^{-1}\|^n\end{aligned}$$

where $y = (z - x)^{-1}(z^* - x^*)^{-1}$, in the formula for R, this gives

$$\begin{aligned}\frac{1}{R} &= \limsup \|(z - x)^{-n}\|^{1/2n}\|(z^* - x^*)^{-n}\|^{1/2n} \\ &= \limsup \|y^{-n}\|^{1/2n}\end{aligned}$$

Since

$$z - x = (er \cos \alpha - u) + i(er \sin \alpha - v)$$

we have

$$z^* = (er \cos \alpha - u) - i(er \sin \alpha - v)$$

and thus

$$y = r^2\left(e - \frac{2(u \cos \alpha + v \sin \alpha)}{r} + \frac{xx^*}{r^2}\right)$$

and

$$y = r^2(e - \tilde{z})$$

where

$$\tilde{z} = \frac{2(u \cos \alpha + v \sin \alpha)}{r} - \frac{xx^*}{r^2}$$

Let $s = y/r^2$ and for r large, $\|s - e\| < 1$, and

$$\|s^{-1}\| < \frac{1}{1 - \|e - s\|}$$

In this case we obtain

$$r^2\|y^{-1}\| < \frac{1}{1 - \|\tilde{z}\|}$$

But

$$\frac{1}{R} < \|y^{-1}\|^{1/2}$$

which gives that

$$\frac{1}{R} < \frac{1}{r(1 - \|\tilde{z}\|)^{1/2}}$$

or

$$R > r(1 - \|\tilde{z}\|)^{1/2}$$

Now for any s which is not in the spectrum of x

$$|s - z| > r(1 - \|\tilde{z}\|)^{1/2}$$

and thus

$$(\text{Re } s - r \cos \alpha)^2 + (\text{Im } s - r \sin \alpha)^2 > r^2(1 - \|\tilde{z}\|)$$

or

$$(\text{Re } s)^2 + (\text{Im } s)^2 + r^2 - 2r(\text{Re } s \cos \alpha + \text{Im } s \sin \alpha)$$
$$> r^2(1 - \|\tilde{z}\|)$$

and thus since

$$\|\tilde{z}\| < \frac{2}{r} \|u \cos \alpha + v \sin \alpha\| + \frac{\|xx^*\|}{r^2}$$

we obtain

$$\text{Re } s \cos \alpha + \text{Im } s \sin \alpha < \frac{|s|^2}{2r} + \|u \cos \alpha + v \sin \alpha\| + \frac{\|xx^*\|}{2r}$$

Let $r \to \infty$ and this gives

$$\text{Re } s \cos \alpha + \text{Im } s \sin \alpha \leq \|u \cos \alpha + v \sin \alpha\|$$

and replacing α by $\alpha + \pi$ we obtain the assertion of the theorem.

4.4.23 COROLLARY If $x = x^*$, then $\sigma(x) \subset R$.

Proof. In this case $v = 0$ and for any $s \in \sigma(x)$ we have

$$|\text{Im } s| \leq \|v\|$$

and the assertion follows.

The following theorem gives the structure of all involutive commutative B*-algebras.

4.4.24 THEOREM Let A be a commutative B*-algebra. Then A is isomorphic and isometric with the algebra of all continuous functions on some compact space.

Proof. Since for any $x \in A$, $r_x = \|x\|$, the homomorphism

$$\phi: x \to \hat{x}$$

is an isomorphism.

Also for any $x \in A$, $x = u + iv$, where u and v are hermitian elements, we have

$$x = u + iv$$

and thus

$$x^* = u - iv$$

We know that A is a separating algebra, i.e., for any s_1, s_2 in M_A there exists $x \in A$ such that $\hat{x}(s_1) \neq \hat{x}(s_2)$, and A is also closed because

$$\|x^*x\| = \|x\|^2 = \|\hat{x}\bar{\hat{x}}\|$$

which implies $\|\hat{x}\| = \|x\|$. Thus $A = C(M_A)$ by the Stone-Weierstrass theorem.

We give now an extension to an algebra without unit.

Let A be a Banach algebra without unit and involutive. We consider the set of all pairs (s, x), where s is a complex number and $x \in A$. We define on this set the operations such that we obtain a Banach algebra with unit and involutive.

These operations are defined as follows:

1. $(s_1, x) + (s_2, y) = (s_1 + s_2, x + y)$
2. $(s_1, x)(s_2, y) = (s_1 s_2, s_1 y + s_2 x + xy)$
3. $(zs, zx) = z(s, x)$

for all $x, y \in A$ and s_1, s_2, z complex numbers.

It is easy to see with these operations that the set of all pairs (s, x) is an algebra which we denote by $\tilde{A}$. It is also obvious that (1, 0) is the unit for $\tilde{A}$.

It is obvious that A is commutative if and only if $\tilde{A}$ is commutative.

The following theorem gives some information about the maximal ideals for the algebras A and $\tilde{A}$.

4.4.25 PROPOSITION If $h: x \to (0, x)$ is a mapping defined on A with values in $\tilde{A}$, then h(A) is a two-sided ideal in $\tilde{A}$.

Proof. Since h is an isomorphism, h(A) is an algebra. Let (s, x) in $\tilde{A}$ and $(0, y) \in h(A)$. In this case,

$$(s, x)(0, y) = (0, sy + xy)$$
$$(0, y)(s, x) = (0, sy + yx)$$

and thus h(A) is a two-sided ideal. The maximality is proved as follows: Let A_1 such that $h(A) \subset A_1 \subset \tilde{A}$ and let (s, x) in A_1 and not in h(A). It is clear that $s \neq 0$ and thus

$$(s, x) + (0, y) = (s, x + y)$$

and thus $(s, v) \in A_1$ for all $v \in A$. Since A_1 is an algebra

$$t(s, x) = (ts, tx) \in A_1$$

this gives that $A = A_1$ which is a contradiction and the proposition is proved.

We prove now that with an appropriate norm, $\tilde{A}$ is a Banach algebra.

4.4.26 THEOREM If we define

$$(s, x) \to \sup_{\|a\|=1} \|sa + xa\|$$

and the involution

$$*: (s, x) \to (s^*, x^*)$$

then $\tilde{A}$ is an involutive Banach algebra.

Proof. It is easy to see that

$$(s, x) \to \sup \|sa + xa\|$$

is in fact a norm. All the properties of the norm are obvious except possibly $\|(s, x)\| = 0$ if and only if $s = 0$ and $x = 0$. To prove this suppose first that $(s, x) \neq 0$.

Case 1. $s \neq 0$ and $x = 0$. The relation $\|(s, 0)\| = 0$ implies that $\|sa\| = 0$ for all $a \in A$. If a is nonzero, this clearly gives $s = 0$.

Case 2. $s = 0$ and $x \neq 0$. We obtain, in this case, $\|ax\| = 0$ for all $a \in A$, and particularly for $a = x^*$. This gives $x = 0$.

Case 3. $s \neq 0$ and $x \neq 0$. From the relation $\|(s, x)\| = 0$ we obtain that for all $a \in A$, $sa + xa = 0$; i.e., $sa + xa = 0$. Since $s \neq 0$, it follows that there exists an element $z = -x/s$ and this element satisfies the relation $yz = zy = y$ for all $y \in A$ which is a contradiction since A is not with unit. This proves that $s = 0$ and $x = 0$.

We prove now that A is an algebra with involution such that it is a B*-algebra. We suppose, without loss of generality, that $(s, x) \neq 0$.

For $\varepsilon > 0$ in $(0, 1)$ and for some $s,t \in A$, $\|s\| = \|t\| = 1$, we have

$$\varepsilon\|(v, x)\| \leq \|vs + xs\| \qquad \varepsilon\|(v^*, x^*)\| \leq \|v^*t + x^*t\|$$

and thus

$$\varepsilon^3\|(v, x)\| \cdot \|(v, x)^*\| \le \varepsilon\|vs + xs\| \cdot \|v^*t + x^*t\|$$
$$= \varepsilon\|vs + xs\| \cdot \|vt^* + xt^*\|$$
$$\le \varepsilon\|vs + xs\| \cdot \|(v, x)\|$$

which gives further

$$\varepsilon^3\|(v, x)\| \cdot \|(v, x)^*\| \le \|vs + xs\| \cdot \|vs + xs\|$$
$$= \|vs + xs\| \cdot \|(x^*v^* + x^*s^*\|$$

But

$$\|vs + xs\| \cdot \|s^*v^* + x^*s^*\| = \|vv^*ss^* + vs^*ss^* + v^*xss^* + xx + ss\|$$
$$\le \|(v, x)(v, x)^*\|$$

since

$$\|ss^*\| = \|s\| = \|s^*\| = 1$$

and from these it follows that

$$\|(v, x)(v, x)^*\| = \|(v, x)\|^2$$

and the assertion is proved.

The fact that $\tilde{A}$ is a Banach space can be easily proved and we omit this.

We are now in a position to give the representation theorem for the commutative Banach algebra without unit.

Let S be a locally compact Hausdorff space and f be a continuous complex-valued function on S. We say that f has the value zero at infinity if for each $\varepsilon > 0$ there exists a compact K_ε such that for all $s \in S \setminus K_\varepsilon$, $|f(s)| \le \varepsilon$.

It is easy to see that the set of all continuous functions which are zero at infinity is a Banach algebra without unit and the norm is defined as follows:

$$f \mapsto \|f\|_S = \sup \{|f(x)| \mid s \in S\}$$

4.4.27 THEOREM If A is an involutive B*-algebra without unit and commutative, then A is isomorphic and isometric with an algebra of all continuous complex-valued functions on a locally compact Hausdorff space, having the property that these functions are with value zero at infinity.

Notation. For any locally compact Hausdorff space the set of all continuous complex-valued functions having the value zero at infinity is denoted by $C_0(S)$.

Proof. We can construct the Banach algebra $\tilde{A}$ and apply the representation theorem for this algebra. Let $S = M_{\tilde{A}}$. Clearly $\tilde{A}$ is isomorphic and isometric with C(S). Now A can be considered as an ideal of $\tilde{A}$ which is of course maximal, say m_A. In this case $x \in A$ implies that $\hat{x}$ is zero at m_A and conversely. Thus A is isomorphic and isometric with all functions in C(S) which are zero at m_A. The induced topology on $S \setminus m_A$ gives a locally compact Hausdorff space and A is isometric with $C_0(M_{\tilde{A}} \setminus m_A)$.

4.5 REPRESENTATION THEOREMS FOR NONCOMMUTATIVE BANACH ALGEGRAS

We give now a representation theorem for B*-algebras and this representation is known as the Gelfand-Naimark theorem.

4.5.1 DEFINITION Let H be a Hilbert space and L(H) be the set of all bounded linear operators on H. A closed subalgebra of L(H), A_1, is called self-adjoint if $x^* \in A_1$ whenever $x \in A_1$.

4.5.2 EXAMPLE The closed subalgebra generated by the operators of the form {sI + C}, where s is a complex number and C a compact operator is self-adjoint.

4.5.3 THEOREM If A is a B*-algebra with unit, then it is isometric and *-isomorphic with a self-adjoint subalgebra of operators on a Hilbert space.

We recall that an isomorphism between two involutive algebras is called an **-isomorphism* if whenever x is a hermitian element of an algebra, then the image of x throughout isomorphism is also a hermitian element.

For the proof of the theorem we need some preliminary results, which also have their own interest.

4.5.4 DEFINITION If A is a B*-algebra, then

1. $x \in A$ is called *positive* if $x = x^*$ and the spectrum of x is positive, i.e., $\sigma(x) \subset R^+$.
2. An element $\phi \in A^*$ is called a *state* of A if

$$\phi(e) = \|\phi\| = 1$$

   where e is the unit of A.
3. A state of a is called a *pure state* if it is an extreme point of the set $\{x^* \mid \|x^*\| \leq 1\}_{x^* \in A^*}$.

The following theorem gives useful information about the states of A and the hermitian elements of A.

4.5.5 THEOREM Suppose that A is a B*-algebra with unit. In this case the following assertions hold:

1. For any hermitian element $x \in A$ and any $s \in \sigma(x)$ there exists a state ϕ of A such that $\phi(x) = s$.
2. A hermitian element is positive if and only if for any state ϕ, $\phi(x) \geq 0$.
3. If x and y are positive elements, then x + y is positive.
4. Any hermitian element x is of the form $x = x_1 - x_2$, where x_i are positive elements.
5. The element x is positive if and only if $x = a^*a$ for some $a \in A$.

Proof. We consider the B*-algebra A_1 generated by the element x and the unit e. Since this is a commutative B*-algebra with unit by the structure theorem, A_1 is isometric and *-isomorphic with the algebra of all continuous complex-valued functions on a compact space, which in this case is $\sigma(x)$. Let $s \in \sigma(x)$ and we define the functional on $C(\sigma(x))$ as the value of the function at the point s. Using the Hahn-Banach extension theorem, assertion 1 follows.

For assertion 2 we remark that if ϕ is a state of A, then since

$$\sigma(x) = \{\psi(x) \mid \psi \text{ state of A}\}$$

it follows that for any state ϕ of A, $\phi(x) \geq 0$ when x is positive.

Conversely, if we consider the B*-algebra generated by x and e, it follows that this is isomorphic and isometric with $C(\sigma(x))$.

If ϕ is a state of A, then the restriction of ϕ to A_1 is a linear functional F with the following properties:

1. $F(e) = 1$
2. $\|F\| = 1$

and in this case F is positive and thus $0 \leq F(x) \leq \phi(x)$.

We recall that a linear and continuous functional is called *positive* if $f(x) \geq 0$ when x is positive.

For property 3 we remark that since we have the above expression for $\sigma(x)$, the property follows immediately.

Property 4 follows from the isometric isomorphism between A_1 and $C(\sigma(x))$ as well as the decomposition of a real function as a difference of positive functions.

For property 5 we remark that from the isometry and *-isomorphism between A_1 and $C(\sigma(x))$, it follows that the condition is necessary.

We prove now the sufficiency of the condition. Let $x = a^*a$ and since

$$x^* = (a^*a)^* = a^*a$$

it follows that x is a hermitian element. It remains to prove that it is positive. By property 4 we have the decomposition

$$x = x_1 - x_2$$

where x_i are positive elements. Now since $x_1x_2 = x_2x_1 = 0$, we have

$$(xx_2)^*(xx_2) = x_2x^*xx_2 = x_2(x_1 - x_2)x_2 = x_2x_1x_2 - x_2^3$$

$$= x_1x_2^2 - x_2^3 = (x_1 - x_2)x_2^2 = -x_2^2$$

and thus since $x_2 \geq 0$, we have

$$(xx_2)^*(xx_2) \leq 0$$

In a similar way we have

$$(xx_2)(xx_2)^* \le 0$$

and if $xx_2 = y_1 + iy_2$, where y_i are hermitian elements, it follows that

$$0 \ge (xx)^*(xx) = 2(y_1^2 + y_2^2) \ge 0$$

which yields $y_1 = y_2 = 0$. From this it follows that

$$0 = (xx_2)^*(xx_2) = -x_2^3$$

and thus $x_2 = 0$.

The next theorem gives the basic construction used in the proof of the Gelfand-Naimark theorem.

4.5.6 THEOREM Let A be a B*-algebra with unit and ϕ be a state of A. In this case the following assertions hold:

1. $\phi(x^*x) \ge 0$ for all $x \in A$.
2. $\phi(x) = \overline{\phi(x^*)}$.
3. The application $[\ ,\ ]_\phi\colon A \times A \to C$ defined by the relation $[x, y]_\phi = \phi(y^*x)$ is a bilinear form and semidefinite function.
4. $[xy, z]_\phi = [y, x^*z]_\phi$ for all $x,y,z \in A$.

Proof. Assertions 1 and 2 are obvious from the above discussion. Property 3 follows from the definition of the state of A. For property 4 we remark that

$$[xy, z]_\phi = \phi(z^*xy) = \phi((x^*z)^*y) = [y, x^*z]_\phi$$

and thus the theorem is proved.

4.5.7 THEOREM If A is a B*-algebra and ϕ is a state of A, then the following assertions hold:

1. $I_\phi = \{x, [x, x]_\phi = 0\}$ is a left ideal in A.
2. $H_\phi = A/I_\phi$ is a prehilbert space with the scalar product $[\ ,\ ]_\phi$.
3. The operator $T_a^\phi\colon H_\phi \to H_\phi$, defined by the relation

$$T_a^\phi(x + I_\phi) = ax + I_\phi$$

is well defined and

a. $\|T_a^\phi\| \le \|a\|$

b. $[a, a]_\phi \le \|T_a^\phi\|^2$

and where a is arbitrary in A.

Proof. For 1 it is clear that I_ϕ is a linear subspace and even an algebra. The fact that I_ϕ is a left ideal can be seen as follows: Let $x \in I_\phi$ and $y \in A$. In this case,

$$0 \le [yx, yx]_\phi = [x, y^*yx]_\phi \le \|x\|_\phi \cdot \|y\|_\phi$$

where $[z, z]_\phi = \|z\|_\phi^2$, and the assertion follows.

For 2 we remark that it follows from Theorem 4.5.6.

We prove now 3. The fact that T_a^ϕ is a linear operator and that it is well defined follows from 1.

Now since for any $x \in A$

$$[ax, ax]_\phi = \phi(x^*z^*ax) = \|ax\|_\phi^2$$

and

$$\|a\|^2\|x\|_\phi^2 = \phi(\|a\|^2x^*x)$$

we obtain

$$\|a\|^2\|x\|_\phi^2 - \|ax\|_\phi^2 = \phi(x^*(\|a\|^2 - a^*a)x)$$

But $\|a\|^2 - a^*a \ge 0$ and thus it is of the form u^*u for some $u \in A$. In this case we have

$$\phi(x^*(\|a\|^2 - a^*a)x) = \phi(x^*u^*ux) \ge 0$$

and thus

$$\|ax\|_\phi^2 \le \|a\|^2\|x\|_\phi^2$$

which gives us that

$$\|T_a^\phi\| \le \|a\|$$

But

$$\|ae\|_\phi^2 = \phi(a^*a)$$

which implies the relation

$$\|T_a^\phi\|_\phi^2 \geq [a, a]_\phi$$

and the theorem is proved.

We remark that if A is an algebra as above and ϕ is a state of A and H_ϕ is the corresponding prehilbert space, then the properties of T_a^ϕ gives us the possibility of extending it to an operator on the completion of H_ϕ, which we denote by $\tilde{T}_a^\phi$ (and the completion of H_ϕ by $\tilde{H}_\phi$).

The properties of the operator $\tilde{T}_a^\phi$ are the following:

1. $a \to \tilde{T}_a^\phi$ is an *-homomorphism of A into a self-adjoint algebra of operators on $\tilde{H}_a^\phi$.
2. $(\tilde{T}_a^\phi)^* = \tilde{T}_{a^*}^\phi$.
3. $\|\tilde{T}_a^\phi\|_\phi \leq \|a\|$.

The proof of the Gelfand-Naimark theorem is as follows: We consider the set S of all states of A, and for each $\phi \in S$ we consider the Hilbert space $\tilde{H}_\phi$. Define the space of all sequences $x = (x_\phi)_{\phi \in S}$ which are square summable, i.e.,

$$\sum_\phi \|x_\phi\|_\phi^2 < \infty$$

and it is known and easy to verify that the space of all sequences with the above property is a Hilbert space, which is denoted by H. Clearly the scalar product on H is defined as

$$\langle x, y\rangle = \Sigma_\phi [x, y]_\phi$$

and we define now for each element $a \in A$ an operator on H by the relation

$$T_a(x) = \tilde{T}_a^\phi x_\phi)_{\phi \in S}$$

From the definition of H it is clear that

$$\|T_a x\|^2 = \Sigma \|\tilde{T}_a^{\phi} x_{\phi}\|^2 \leq \|a\|^2 \|x\|^2$$

and this gives the important fact that the application

$$a \mapsto T_a$$

has the property

$$\|T_a\| \leq \|a\|$$

From property 2 it follows that for all $a \in A$, $(T_a)^* = T_{a^*}$ and also the fact that

$$a \mapsto T_a$$

is an *-homomorphism. To prove the theorem it remains to prove that the above *-homomorphism is in fact an isometry.

First, we remark that if $x = x^*$, there exists a state of A such that

$$\phi(x^2) = \|x\|^2$$

and in this case we have

$$\|x\|^2 = \phi(x^2) \leq \|\tilde{T}_x^{\phi}\|^2 \leq \|x\|^2$$

which proves the fact that the *-homomorphism is isometric on the set of all hermitian elements.

The general fact follows easily from this. Indeed, for any $x \in A$ the element x^*x is hermitian, and thus

$$\|T_x\|^2 = \|T_x^* T_x\| = \|T_{x^*} T_x\| = \|T_{x^*x}\| = \|x^*x\| = \|x\|^2$$

and this implies that

$$\|T_x\| = \|x\|$$

Chapter 5

SPECTRAL REPRESENTATION OF OPERATORS ON HILBERT SPACES

In this chapter we give some general theorems about spectral families and also applications to spectral representation of three classes of operators on Hilbert spaces: hermitian operators, unitary operators, and normal operators. In these applications the representation theorems for commutative B*-algebras are of fundamental importance.

5.1 SEMISPECTRAL AND SPECTRAL FAMILIES OF RADON MEASURES

We suppose that M is a locally compact Hausdorff space and H is a Hilbert space. We denote by K(M) the space of all continuous complex-valued functions on M which are with compact support, i.e., for any $f \in K(M)$ the set

$$\operatorname{supp}(f) = \overline{\{m \mid f(m) \neq 0\}}$$

is a compact set in M.

5.1.1 DEFINITION A family of Radon measures defined on M,

$$\{\mu_{x,y}\}_{x,y\in H}$$

is called a *semispectral* family if the following holds:

1. $x \mapsto \mu_{x,y}$ is linear for each $y \in H$.
2. For each $x,y \in H$, $\mu_{x,y} = \overline{\{\mu_{y,x}\}}$.
3. $\operatorname{var} \mu_{x,y} \leq \|x\| \cdot \|y\|$ for each $x,y \in H$.

For each $\mu_{x,y}$ in a semispectral family we define a linear functional on the space K(M) by the relation, for $f \in K(M)$,

$$[x,y](f) = \int_M f(s)\, d\mu_{x,y}(m)$$

and from property 3 we obtain that

$$\|[x,y]\| \leq \|x\| \cdot \|y\|$$

and thus it is a continuous linear functional on the space H.

By Riesz's representation theorem of continuous linear functionals on Hilbert spaces, we can define the linear operator

$$U_f: H \to H$$

by the relation

$$\langle U_f x, y \rangle = \int_M f(m)\, d\mu_{x,y}(m)$$

and it is clear that U_f has the following properties:

1. U_f is in L(H).
2. $\|U_f\| \leq \|f\|$.

In this way we can define an application of the space K(M) in the space L(H) by the relation

$$f \mapsto U_f$$

and the properties of this application are given in the following lemma.

5.1.2 LEMMA The application

$$U: K(M) \to L(H)$$

defined by the formula

$$U(f) = U_f$$

has the following properties:

1. It is a linear mapping.
2. It is continuous mapping.

3. If $S \in L(H)$, then S commutes with U_f for all $f \in K(M)$ if and only if

$$\mu_{Sx,y} = \mu_{x,S^*y}$$

holds for all x and y in H.

Proof. Properties 1 and 2 are clear from the above remarks. To prove 3, let $S \in L(H)$ and f arbitrary in $K(M)$. Then we have

$$\langle U_f Sx, y\rangle = \int_M f(m)\, d\mu_{Sx,y}(m) = \langle SU_f x, y\rangle = \langle U_f x, S^*y\rangle$$

$$= \int_M f(m)\, d\mu_{x,S^*y}(m)$$

and this gives property 3.

We remark that for any x and y in H,

$$\langle U_{\bar{f}} x, y\rangle = \int_M \bar{f}(m)\, d\mu_{x,y}(m) = \overline{\int_M f(m)\, d\bar{\mu}_{x,y}(m)}$$

$$= \overline{\langle U_f y, x\rangle} = \langle x, U_f y\rangle = \langle U_f^* x, y\rangle$$

and the assertion is proved.

In what follows we consider the following fundamental notion: the notion of spectral family of Radon measures on a locally compact Hausdorff space.

5.1.3 DEFINITION A semispectral family of Radon measures $\{\mu_{x,y}\}_{x,y\in H}$ is called a *spectral* family of Radon measures if the following condition is satisfied: for any $g \in K(M)$ and $x,y \in H$,

$$\int_M fg(m)\, d\mu_{x,y}(m) = \int_M f(m)\, d\mu_{U_g x,y}(m) \qquad (*)$$

for all $f \in K(M)$.

5.1.4 REMARK We write in brief the equality (*) as follows

$$\mu_{U_g x, y} = g \cdot \mu_{x,y}$$

The following theorem gives a basic characterization of spectral families.

5.1.5 THEOREM A semispectral family $\{\mu_{x,y}\}_{x,y\in H}$ is a spectral family if and only if for any $f,g \in K(M)$

$$U_{fg} = U_f U_g$$

Proof. The condition is necessary. Indeed, if $\{\mu_{x,y}\}_{x,y\in H}$ is a spectral family, then for any $f,g \in K(M)$ and x,y in H, we have

$$<U_{fg}x, y> = \int_M fg(m)\, d\mu_{x,y}(m) = \int_M f(m)g(m)\, d\mu_{x,y}(m)$$

$$= \int_M f(m)\, d\mu_{U_g x,y}(m) = <U_f U_g x, y>$$

which is exactly our assertion.

Suppose now that we have the relation $U_{fg} = U_f U_g$ for all $f,g \in K(M)$, and thus for any x,y in H,

$$\int_M f(m)\, d\mu_{x,y}(m) = \int_M f(m)\, d\mu_{U_g x,y}(m)$$

and since f is arbitrary in $K(M)$ this gives, using the above relation

$$g \cdot \mu_{x,y} = \mu_{U_g x,y}$$

and the theorem is proved.

The characterization of spectral families given above can be formulated in terms of representations.

5.1.6 DEFINITION A representation of a Banach algebra A with involution "*" is any *-homomorphism

$$h: A \to L(H)$$

with $h(1) = I$, where 1 denotes the unity of A. Two representations are called *unitarily equivalent* (and we write this as $h_1 \sim h_2$, where h_i, $i = 1, 2$, are the representations) if there exists a unitary operator $U: H_1 \to H_2$ such that

$$h_2(x) = Uh_1U^{-1}(x)$$

for all $x \in A$.

5.1.7 THEOREM If $\{\mu_{x,y}\}_{x,y\in H}$ is a spectral family, then the mapping

$$h: f \to U_f$$

is a representation of the algebra $\overline{K(M)}$ into $L(H)$.

It is clear that this theorem is another form of Theorem 5.1.3.

5.1.8 THEOREM If $\{\mu_{x,y}\}_{x,y\in H}$ is a spectral family, then

1. $\mu_{x,x} \geq 0$
2. For any $x \in H$, $\|U_f x\|^2 = \int_M |f(m)|^2 \, d\mu_{x,x}(m)$.

Proof. Since the proof is simple, we omit it.

The following generalization of semispectral families and spectral families is useful in the theory of operators of Hilbert-Kaplansky modules.

5.1.9 DEFINITION Let M be a locally compact Hausdorff space and A be a commutative *-Banach algebra with unit. A family of A-valued Radon measures $\{\mu_{x,y}\}_{x,y\in H}$ is called a *semispectral family* if the following conditions are fulfilled:

1. $x \mapsto \mu_{x,y}$ is a linear mapping.
2. $\mu_{x,y} = \mu^*_{x,y}$, where $\mu^*_{x,y}(f) = [\mu_{x,y}(\bar{f})]^*$.

   "*" denotes the value of the involution of A.
3. $\mathrm{var}\ \mu_{x,y} \leq \|x\| \cdot \|y\|$.

In a similar way we can define *spectral families*.

Using the representation theorem of Riesz in certain Hilbert-Kaplansky modules, we can prove the following result.

5.1.10 THEOREM A semispectral family of A-valued Radon measures is a spectral family if and only if for any $f,g \in K(M)$, $U_{fg} = U_f U_g$.

5.2 MEASURABILITY AND INTEGRABILITY WITH RESPECT TO SPECTRAL FAMILIES

Let M be a locally compact Hausdorff space and $\{\mu_{x,y}\}_{x,y\in H}$ be a spectral family defined on M, where H is a Hilbert space.

As in the case of bilinear forms on Hilbert spaces, it is easy to see that any measure $\mu_{x,y}$ satisfies the following identity:

$$\mu_{x,y} = \frac{1}{4}[\mu_{x+y,x+y} - \mu_{x-y,x-y} + i\mu_{x+iy,x+iy} - i\mu_{x-iy,x-iy}]$$

For any $x \in H$ we consider the measure $\mu^*_{x,x}$.

5.2.1 DEFINITION If $f \geq 0$, then we say that f is negligible with respect to the spectral family if $\mu^*_{x,x}(f) = 0$, and we say that an arbitrary function g is negligible with respect to the spectral family $\{\mu_{x,y}\}_{x,y\in H}$ if and only if $|g|$ is negligible.

From the properties of the measures $\mu^*_{x,x}$, it is easy to see that the following assertion holds.

5.2.2 LEMMA The set of all negligible functions is a linear space.

A subset of M is said to be negligible if and only if the function

$$F_{M_1}(m) = \begin{cases} 1 & m \in M_1 \\ 0 & m \notin M_1 \end{cases}$$

is negligible.

The measurability with respect to a spectral measure $\{\mu_{x,y}\}_{x,y\in H}$ is defined as follows.

5.2.3 DEFINITION A function f is *measurable with respect to a spectral measure* $\{\mu_{x,y}\}_{x,y\in H}$ if f is measurable with respect to the measures $\{\mu_{x,x}\}_{x\ H}$.

We consider now a subspace of the space of all measurable functions and we define a norm so that this subspace is a Banach space.

Let f be a measurable function and let

$$\|f\|_\infty = \inf \{s \mid \{m \mid |f(m)| > s\} \text{ is negligible}\}$$

and we denote by L^∞ the set of all measurable functions such that $\|f\|_\infty$ is a finite number.

It is not difficult to see that the following assertion holds; the proof does not differ essentially from the classical case.

5.2.4 THEOREM L^∞ is a Banach space.

The following theorem gives a fundamental property of operators U_f.

5.2.5 THEOREM For any f with compact support, $\|U_f\| = \|f\|_\infty$.

Proof. Since for any f and $x,y \in H$ we have

$$\langle U_f x, y\rangle = \int_M f(m)\, d\mu_{x,y}(m)$$

we obtain

$$\|U_f\| \le \|f\|_\infty$$

To prove the assertion of the theorem, it will be sufficient to show that for any $f \in L^\infty$, $\|U_f\| \ge \|f\|_\infty$, f being with compact support.

Let $\varepsilon > 0$ and consider the set

$$V_\varepsilon = \{s \mid |f(x)| > \varepsilon + \|f\|_\infty\}$$

and we show that this set is negligible with respect to the spectral family. For this purpose we show that for any g with compact support in V_ε,

$$\int_M |g(s)|^2\, d\mu_{x,x}(s) = 0$$

Indeed, we have

$$\|U_{fg}x\| = \|U_f(U_g x)\| \le \|U_f\| \left[\int_M |g(s)|^2\, d\mu_{x,x}(s)\right]^{1/2}$$

and

$$\|U_{fg}x\| = \left[\int_M |f(s)g(s)|^2 \, d\mu_{x,x}(s)\right]^{1/2}$$

$$\geq (\varepsilon + \|U_f\|) \left[\int_M |g(s)|^2 \, d\mu_{x,x}(s)\right]^{1/2}$$

which gives us

$$\|U_f\| \left[\int_M |g(s)|^2 \, d\mu_{x,x}(s)\right]^{1/2} \geq (\varepsilon + \|U_f\|) \left[\int_M |g(s)|^2 \, d\mu_{x,x}(s)\right]^{1/2}$$

and thus

$$\int_M |g(s)|^2 \, d\mu_{x,x}(s) = 9$$

This proves the theorem since this implies that

$$\|f\|_\infty \leq \varepsilon + \|U_f\|$$

for any $\varepsilon > 0$.

5.3 A REPRESENTATION THEOREM FOR L^∞

Using the properties of U_f we can construct a representation for the Banach algebra L^∞. Indeed for any $f \in L^\infty$ we define an operator by the relation

$$\langle U_f x, y\rangle = \int_M f(s) \, d\mu_{x,y}(s)$$

and the main result in this section follows.

5.3.1 THEOREM The mapping $f \mapsto U_f$ is a representation of the Banach algebra L^∞ into the Banach algebra $L(H)$.

Proof. The fact that $f \mapsto U_f$ is a linear mapping is obvious. We show now that this is multiplicative, i.e., for any f,g in L^∞ we have

$$U_{fg} = U_f U_g$$

First we consider the case when $f \in L^\infty$ and $g \in K(M)$ and thus we have

$$\langle U_{fg}x, y\rangle = \int_M f(s)\, d\mu_{x,U_g^*y}(s) = \langle U_f x, U_g^* y\rangle = \langle U_g(U_f x), y\rangle$$

and

$$\langle U_{fg}x, y\rangle = \int_M f(s)g(s)\, d\mu_{x,y}(s) = \int_M f(s)\, d\mu_{U_g x,y}(s)$$
$$= \langle U_f(U_g x), y\rangle$$

Thus in this case

$$U_{fg} = U_f U_g = U_g U_f$$

This yields

$$\int_M g(s)\, d\mu_{U_f x,y}(s) = \int_M f(s)g(s)\, d\mu_{x,y}(s)$$

and this is true also for $g \in L^\infty$. This yields, for all f,g in L^∞,

$$U_{fg} = U_f U_g$$

and the theorem is proved.

Now for $f \in L^\infty$,

$$\langle U_f x, y\rangle = \int_M f(s)\, d\mu_{x,y}(s) = \overline{\int_M f(s)\, d\overline{\mu_{x,y}}(s)}$$
$$= \langle U_f y, x\rangle = \langle x, U_{\bar f} y\rangle$$

and thus we have the important relation

$$(U_f)^* = U_{\bar f}$$

Also, for any operator $T \in L(H)$ such that for any $f \in L^\infty$,

$$U_f T = T U_f$$

then it is easy to see that $\mu_{Tx,y} = \mu_{x,T^*y}$ for all x,y in H.

If we consider any measurable subset A in M, measurable with respect to the spectral family $\{\mu_{x,y}\}_{x,y \in H}$, then the operator U_{φ_A} has the following important property:

$$U_{\varphi_A^2} = U_{\varphi_A} U_{\varphi_A} = U_{\varphi_A}$$

and so it is a projection since $\|U_{\varphi_A}\| \leq 1$.

In what follows we use the notation

$$U_{\varphi_A} = P_A$$

The following theorem gives information about projections and measurable subsets in M.

5.3.2 THEOREM Suppose that $\{A_i\}_{i=0}^{\infty}$ is a sequence of measurable subsets in M, with respect to the spectral family $\{\mu_{x,y}\}_{x,y \in H}$. In this case

1. If $\{A_i\}_{i=0}^{\infty}$ is decreasing, then

$$P_{\bigcup_{i=0}^{\infty} A_i} = \inf_i P_{A_i}$$

2. If $A_i \cap A_j = \emptyset$ for $i \neq j$, then

$$P_{\bigcap_{i=o}^{\infty} A_i} = \Sigma P_{A_i}$$

Proof. Since for any $x \in H$,

$$\|P_{A_n} x\|^2 = \int_M \varphi_{A_n}(s)\, d\mu_{x,x}(s) \geq \int_M \varphi_{A_{n+1}}(s)\, d\mu_{x,x}(s)$$
$$= \|P_{A_{n+1}} x\|^2$$

and this gives the relation

$$P_{A_n} \geq P_{A_{n+1}}$$

Now we have

$$\left\| P_{A_n} x - P_{\bigcap_{n=0}^{\infty} A_n} x \right\|^2 = \int_M \left| \varphi_{A_n} - \varphi_{\bigcap_{n=0}^{\infty} A_n} \right|^2 d\mu_{x,x}(s)$$

and since

$$\lim \varphi_{A_n} = \varphi_{\bigcap_{n=0}^{\infty} A_n}$$

we obtain that

$$\inf P_{A_n} = P_{\bigcap_{n=0}^{\infty} A_n}$$

and the first assertion is proved.

For the second assertion, for $i \neq j$ we have

$$\varphi_{\bigcup_{n=0}^{\infty} A_n} = \sum_{n=0}^{\infty} \varphi_{A_n} = \lim_{n\to\infty} \varphi_{\bigcup_{i=0}^{n} A_i}$$

and thus for any x in H,

$$\lim \left\| P_{\bigcup_{i=0}^{n} A_i} - P_{\bigcup_{i=0}^{\infty} A_i} x \right\| = 0$$

which gives

$$P_{\bigcup_{i=0}^{\infty} A_i} = \sum_{i=0}^{\infty} P_{A_i}$$

The following theorem gives a necessary and sufficient condition for an operator $T \in L(H)$ to commute with U_f for all $f \in L^\infty$, in terms of projections associated with compact subsets of M.

5.3.3 THEOREM An operator $T \in L(H)$ commutes with U_f for all $f \in L^\infty$ if and only if T commutes with P_A for all A compact subset in M

Proof. The condition is clearly necessary. We prove now that it is also sufficient. Indeed, for any compact subset $A \subset M$ and $x, y \in H$ we have the following relation:

$$\mu_{Tx,y}(A) = \mu_{x,T^*y}(A)$$

This gives us

$$\mu_{Tx,y} = \mu_{x,T^*y}$$

and this proves the theorem.

We give now a spectral decomposition of some classes of operators using projections. Let L be the set of all finite partitions of M such that all parts are measurable. Let $\{M_i\}_{i \in I}$ be a partition and define the sum

$$\sum_{i \in I} f(t_i) P_{M_i} x \qquad t_i \in M_i$$

and the result is as follows.

5.3.4 THEOREM For any $f \in L^\infty$,

$$\lim_{L} \left\| U_f x - \sum_{i \in I} f(t_i) P_{M_i} x \right\| = 0$$

Proof. Since for any $x \in H$ we have

$$\left\| U_f x - \sum_{i \in I} f(t_i) P_{M_i} x \right\|^2 = \sum_{i \in I} \int_{M_i} |f(t) - f(t_i)|^2 \, d\mu_{x,x}(t)$$

and if the partition is sufficiently fine, the convergence follows.

We give now some examples of spectral families which also have an independent interest.

5.3.5 DEFINITION Let H be a Hilbert space and P(H) be the set of all projections on H. A spectral decomposition of the identity is any function on R with values in P(H) satisfying the following axioms:

1. $s \le t$ implies $P_s \le P_t$.
2. For any s, $P_{s-0} = P_s$.
3. For any $x \in H$, $\lim_{s \to -\infty} P_s x = 0$
4. For any $x \in H$, $\lim_{s \to \infty} P_s x = x$.

We write also $P_{-\infty} = 0$ and $P_{\infty} = I$, and thus axiom 3 is equivalent to the following assertion:

$$\lim_{s\to-\infty} P_s = 0$$

Also axiom 4 is equivalent to the assertion

$$\lim_{s\to\infty} P_s = I$$

From axiom 1 it follows that

$$P_s P_t = P_{\min\{s,t\}}$$

We mention that the spectral decomposition of the identity as defined above is sometimes called the *spectral decomposition of the identity with one dimension.*

We define now the spectral family associated with the spectral decomposition of the identity as follows: for each x,y in H let

$$\tilde{\mu}_{x,y}(s) = \langle P_s x, y\rangle$$

which is clearly a monotone function when $x = y$ and is continuous on the left.

We define $\mu_{x,y}$ as the corresponding measure associated with the function $\mu_{x,y}$, and it is clear that $\{\mu_{x,y}\}_{x,y\in H}$ is a semispectral family.

5.3.6 PROPOSITION $\{\mu_{x,y}\}_{x,y\in H}$ is a spectral family.

Proof. To prove this it will be sufficient to show that for any f and g, as in the above discussion, $U_{fg} = U_f U_g$.

Indeed we have

$$\langle U_f(U_g x), y\rangle = \int_R f(s)\ d\langle P_s(U_g(x)), y\rangle$$

and since

$$\langle P_s(U_g x), y\rangle = \langle U_g x, P_s y\rangle = \int_R g(t)\ d\langle P_t x, P_s y\rangle = \int_{-\infty}^{s} g(t)\ d\langle P_t x, y\rangle$$

which gives us

$$\langle U_f(U_g x), y\rangle = \int_R f(s)g(s)\, d\langle P_s x, y\rangle = \langle U_{fg} x, y\rangle$$

and thus $U_{fg} = U_f U_g$, and the assertion is proved.

5.3.7 *REMARK* We can consider spectral decompositions of the identity with many dimensions.

5.3.8 *DEFINITION* Consider the function

$R^n \mapsto P(H)$

with the following properties:

1. $s_1 \in S_1', s_2 \leq s_2', \ldots, s_n \leq s_n' \to P_{s_{s_1},\ldots,s_n} < P_{s_1',\ldots,s_n'}$
2. $(s_1, \ldots, s_n) \in R^n \to P_{s_1-0,\ldots,s_n-0} = P_{s_1,\ldots,s_n}$

   where

   $$P_{s_1-0,\ldots,s_n-0} = \lim_{n_j\to\infty} P_{s_1-1/n_1,\ldots,s_n-1/n_n}$$
3. $\lim P_{s_1,\ldots,s_n} x = x$ when $\lim s_i = \infty$
4. $\lim P_{s_1,\ldots,s_n} x = 0$ when $\lim s_i = -\infty$

As in the one-dimension case, we define

$\{\mu_{x,y}\}_{x,y\in H}$

as well as the associated measures $\{\mu_{x,y}\}_{x,y\in H}$, and we find that this is in fact a spectral family.

5.3.9 *NOTATION* For any $s_1, s_2 \in R$ and P_{s_1,s_2} we write P_z, where z is the complex number $s_1 + is_2$.

5.3.10 *DEFINITION* Let M be a locally compact Hausdorff space and B(M) be the Borel sets of M. Any function $P: B(M) \to L(H)$ is called a *spectral measure* if the following conditions hold:

1. $P(A \cap B) = P(A)P(B)$ for all $A,B \in B(M)$.

2. If $\{A_n\}$ is a sequence of pairwise disjoint elements of $B(M)$,

$$P(\bigcup A_n) = \Sigma\, P(A_n)$$

3. $\sup_{A \in B(M)} P(A) = I.$

The spectral measure is called *regular* if for any $A \in B(M)$ the relation holds

$$P(A) = \inf_{\substack{D \supset A \\ D \text{ open}}} P(D)$$

It is easy to prove, as in the case of standard measures, that for any regular spectral measure P,

$$P(A) = \sup_{\substack{C \subset A \\ C \text{ compact}}} P(C)$$

For any $A \in B(M)$ and x,y in H we define

$$\mu_{x,y}(A) = \langle P(A)x, y\rangle$$

and for any $f \in K(M)$,

$$\mu_{x,y}(f) = \int_M f(s)\, d\mu_{x,y}(s)$$

then $\{\mu_{x,y}\}_{x,y \in H}$ is a spectral family. Since the proof of these assertions is the same as that given above, we omit it.

5.4 SPECTRAL DECOMPOSITION OF SOME CLASSES OF OPERATORS

First we give some general remarks concerning the representation of certain Banach algebras, and we apply these to some algebras generated by operators in some special classes, such as the class of hermitian operators, the class of unitary operators, and, more generally, the class of normal operators.

Let A be a commutative B*-algebra without unit. Then by the Gelfand-Neumark theorem, A is isomorphic and isometric to the algebra of all continuous complex-valued functions which are zero at infinity, for some locally compact Hausdorff space S. For any x in A we denote by f_x the element of $C_0(S)$ which corresponds to x through the Gelfand-Neumark theorem.

Since the isomorphism between these algebras is an *-isomorphism, we have that for any $x \geq 0$, $f_x \geq 0$. If F is any continuous functional on A such that for any $x \geq 0$, $F(x) \geq 0$, then

$$F(x) = \int_S f_x(s)\, d\mu_F(s)$$

for some measure μ_F and $\|F\| = \|\mu_F\|$. We suppose now that $A \subset L(H)$ and thus we can define for any $a \in A$ and x,y in H,

$$F_{x,y}(a) = \langle a(x), y\rangle$$

and

$$|F_{x,y}(a)| \leq \|a\| \cdot \|x\| \cdot \|y\|$$

This gives that $F_{x,y}(\cdot)$ is a continuous functional on A and by the above remark we find a measure $\mu_{x,y}$ such that

$$\langle a(x), y\rangle = \int_S f_a(s)\, d\mu_{x,y}(s)$$

and thus we have a family $\{\mu_{x,y}\}_{x,y \in H}$. The basic property of this family is given in the following proposition.

5.4.1 PROPOSITION $\{\mu_{x,y}\}_{x,y \in H}$ is a spectral family.

Proof. Let g be in K(S) and a_g be an element of A such that $f_{a_g} = g$. Thus for any x and y in H,

$$\langle a \cdot a_g(x), y\rangle = \int_M f_a(s) f^{a_g}(s)\, d\mu_{x,y}(s)$$

$$= \int_M f_a(s)\, d\mu_{a_g(x),y}(s)$$

and the following relation holds:

$$f^a g\mu_{x,y} = \mu_{a_g(x),y}$$

which completes the proof.

Thus we have that

$$\langle a(x), y\rangle = \int_M f_a(s)\, d\mu_{x,y}(s)$$

where $\{\mu_{x,y}\}_{x,y\in H}$ is a spectral family.

For applications we need some remarks. If A is generated by the operator $T \in L(H)$ and the identity operator I, and we suppose that A is commutative and a B*-algebra, then A is *-isomorphic and isometric with an algebra of continuous complex-valued functions on a compact space. If N is a normal element in L(H) and A is the B*-algebra generated by N and I, then A is isomorphic and isometric with the algebra of continuous complex-valued functions on a compact space and $f_N(z) = z$.

Also we remark that

1. If N is unitary, then $M = \{z \mid |z| = 1\}$.
2. If $N = N^*$, then $M \subset R$.

The spectral decomposition is

$$\langle Nx, y\rangle = \int_M z\, d\langle P(z)x, y\rangle$$

and the spectral measure is defined as follows:

$$E(A) = P(E \cap M)$$

for any Borealian set in the complex plane. It is easy to see that

$$\langle Nx, y\rangle = \int_M z\, d\langle E(z)x, y\rangle$$

5.5 SOME REMARKS ON THE SPECTRAL MAPPING THEOREM FOR HERMITIAN AND NORMAL OPERATORS

We give now some remarks concerning the spectral mapping theorem for the case of hermitian and normal operators in the sense that first we discuss the spectral mapping theorem for hermitian operators using the Weierstrass approximation theorem. Next we use these to obtain a spectral mapping theorem for normal operators. Our approach follows the line indicated by Halmos (1963).

Let A be a hermitian operator in L(H). Some of the following results are simple consequences of the results presented in Chapter 2.

However, we give here the proof to make this section self-contained.

5.5.1 LEMMA If A is a hermitian operator in L(H), then $\sigma(T) \subset R$.

Proof. For this we show that for any point s in the complex plane with the property $\operatorname{Im} s \neq 0$, $s \in \rho(T)$, the resolvent set of A.

Indeed, for any $x \in H$,

$$\|x\|^2 \cdot \|(A - s)x\|^2 \geq |\langle (A - s)x, x\rangle| = |\langle Ax, x\rangle - s\|x\|^2$$
$$\geq |\operatorname{Im} s|^2\|x\|^2$$

and thus

$$\|(A - s)x\| \geq |\operatorname{Im} s| \cdot \|x\|$$

which proves that $s \in \rho(A)$.

5.5.2 LEMMA If $T \in L(H)$ is an arbitrary element of L(H) and $p(\cdot)$ is also an arbitrary polynomial, then

$$\sigma(p(T)) = \{p(s) \mid s \in \sigma(T)\}$$

Proof. First we remark that for any s_0 the following relation holds:

$$p(s) - p(s_0) = (s - s_0)q(s)$$

and thus

$$p(T) - p(s_0)I = (T - s_0I)q(T)$$

Suppose now that s_0 is in $\sigma(T)$ and thus $(T - s_0)q(T)$ is not invertible.

Indeed, in the contrary case, if $S = (T - s_0)q(T)$, then

$$(T - s_0)q(T)S^{-1} = I = S^{-1}(T - s_0)q(T) = q(T)S^{-1}(T - s_0)$$

and thus $(T - s_0)$ is invertible. This contradiction shows that $p(s_0) \in \sigma(p(T))$.

Suppose now that $p(s_0) \in \sigma(p(T))$ and let $s_1, \ldots, s_n$ be the roots of the equation $p(s) = p(s_0)$. In this case,

$$p(s) - p(s_0) = K(s - s_1)\cdots(s - s_n)$$

and thus

$$p(T) - p(s_0) = (T - s_0)q(T) = K(T - s_1)\cdots(T - s_n)$$

From this relation it follows that there exists an index i such that $s_i \in \sigma(T)$ and $p(s_i) = p(s_0)$. The lemma is proved.

5.5.3 LEMMA A hermitian operator $A \in L(H)$ is positive if and only if $\sigma(A) \subset R_+$.

Proof. From Lemma 5.3.1, we know that $\sigma(A) \subset R$. If $A > 0$, then for any $s > 0$, $A + s > 0$ and $A + s$ is invertible. Since s is arbitrary, it follows that $\sigma(A)$ is in $R_+ = \{s \mid s > 0\}$.

5.5.4 COROLLARY Let A be a hermitian operator and $p(\cdot)$ be a polynomial with real coefficients. Then $p(A) \geq 0$ if and only if $p(s) \geq 0$ for all $s \in \sigma(A)$.

Let A be a hermitian operator in L(H) and set

$$m = \inf \{\langle Ax, x\rangle \mid \|x\| = 1\} \qquad M = \sup \{\langle ax, x\rangle \mid \|x\| = 1\}$$

and let $f\colon [m, M] \to R$ be a continuous function. The problem is to define the operator $f(A)$ in such a way that when f is a polynomial, this operator is exactly $p(A)$. Next we study $\sigma(f(A))$ as a set connected with the set $\sigma(A)$ and f.

For this we need the following result, which has its own interest.

5.5.5 *LEMMA* If $A \in L(H)$ is hermitian, then

$$\|A\| = \sup \{|s| \mid s \in \sigma(A)\}$$

Proof. We can assume, without loss of generality, that $\|A\| = 1$. Since we have the identity

$$\|(A^2 - s^2)x\|^2 = \|A^2x\|^2 - 2s^2\|Ax\|^2 + s^4\|x\|$$

for $s = \|A\|$, we obtain, for a sequence $\{x_n\}$ with $\|x_n\| = 1$, $\lim \|Ax_n\| = 1$, $\lim \|(A^2 - s^2)x_n\| = 0$. Thus $1 \in \sigma(A^2)$ and this gives that 1 or -1 is in $\sigma(A)$. The lemma is proved.

From this lemma the following result can be proved, and this is of fundamental importance in what follows.

5.5.6 *THEOREM* There exists an *-isomorphism which is also an isometry between the algebra of all continuous real-valued functions on $I = [m, M]$ and a subalgebra of $L(H)$.

Proof. First we consider the algebra of all polynomials with real coefficients which we denote by P_I. By the Weierstrass approximation theorem it is clear that P_I is dense in C_I. For any polynomial $p(\cdot)$ we consider the operator $p(A) = \Sigma a_i A^i$, where $p(t) = \Sigma a_i t^i$.

Also it is clear that the mapping $p \mapsto p(A)$ is an *-homomorphism of the algebra P_I in $L(H)$, and clearly this is an isometry. Using this fact, we can extend this *-homomorphism to the algebra C_I in the following way: If f is any $f \in C_I$, then there exists a sequence of polynomials $\{p_n\} \subset P_I$ converging uniformly to f.

Since

$$\|p_n(A) - p_m(A)\| = \|p_n(\cdot) - p_m(\cdot)\|_I$$

we obtain that the sequence of operators $\{p_n(A)\}$ is a Cauchy sequence.

Thus we can define

$$f(A) = \lim p_n(A)$$

and it is easy to see that the limit is independent of the sequence.

Also it is clear that

$$\|f(A)\| = \|f\|_I$$

and the theorem is proved.

The following lemma gives an important property of the isometric *-isomorphism in the above theorem.

5.5.7 *LEMMA* If $f \in C_I$ and $f \geq 0$, then $F(A) \geq 0$.

Proof. To prove the assertion, it will be sufficient to show that for any $\varepsilon > 0$, $f(A) + \varepsilon > 0$. Let $\{p_n\}$ be a sequence of elements in P_I such that

$$\|(f + \varepsilon) - p_n\| \to 0 \qquad \|p_n - (f + \varepsilon)\| \leq \frac{\varepsilon}{2}$$

and in this case $p_n \geq \varepsilon/2$ on $\sigma(A)$. Since $\sigma(p_n(A)) = \{p_n(\sigma(A))\}$, we obtain that $p_n(A) \geq \varepsilon/2$. From the inequality,

$$\|p_n(A) - (f(A) + \varepsilon)\| \leq \frac{\varepsilon}{2}$$

we obtain that $f(A) + \varepsilon \geq \varepsilon/2$ and this proves the lemma.

The following result is the so-called spectral mapping theorem for hermitian operators.

5.5.8 *THEOREM* Let $A \in L(H)$ be a hermitian operator and f be any continuous real-valued function on $[m, M]$. Then, if $f(A)$ is the operator defined in Theorem 5.5.6,

$$\sigma(f(A)) = f(\sigma(A)) = \{f(s) \mid s \in \sigma(A)\}$$

Proof. First we prove that $f(\sigma(A)) \subset \sigma(f(A))$.

Indeed, if $s_0 \in \sigma(A)$, we show that $f(A) - f(s_0)$ is not invertible. Let $\{p_n\}$ be a sequence of polynomials in P_I converging uniformly to f. In this case, for n large, $p_n(A) - p_n(s_0)$ is not invertible. This gives us

$$\lim (p_n(A) - p_n(s_0)) = f(A) - f(x_0)$$

which is a noninvertible operator. Thus $f(x_0) \in \sigma(f(A))$.

Suppose now that $r \notin f(\sigma(A))$. We prove that $r \notin \sigma(f(A))$. Suppose that the function $g = f - r$ is different from zero on $\sigma(A)$. In this case $g(A)$ is an invertible operator.

Indeed, let $\varepsilon > 0$ such that $g^2 \geq \varepsilon^2$ on $\sigma(A)$. In this case we have that

$$g^2(A) \geq \varepsilon^2 I$$

and thus $g(A)$ is an invertible operator. Thus $r \notin \sigma(f(A))$.

From this theorem it follows that there exists an *-isomorphism of C_I into $L(H)$ such that

$$\|f(A)\| = \|f\|_I$$

5.5.9 REMARK Using Tietze's extension theorem, it is easy to see that we can restrict our consideration to $C_{\sigma(A)}$.

Using the results obtained for hermitian operators, we can prove similar results for the case of normal operators.

First we give the following lemma.

5.5.10 LEMMA If $N \in L(H)$ is a normal operator and if $0 \in \sigma(N)$, then for any $\varepsilon > 0$ there exists a closed subspace $V \neq \{0\}$ such that

1. V is invariant for any operator commuting with N^*N.
2. $\|N/V\| \leq \varepsilon$.

Proof. Since $0 \in \sigma(N)$, we can find a sequence $\{x_n\}$, $\|x_n\| = 1$, such that $\lim Nx_n = 0$, and thus $\lim N^*Nx_n = 0$. Thus $0 \in \sigma(N^*N)$. Let $\varepsilon > 0$ and since N^*N is hermitian, we can define the operator $f(N^*N)$, where

$$f(t) = \begin{cases} 1 & |t| \leq \frac{\varepsilon}{2} \\ 2(1 - |\frac{t}{3}|) & \frac{\varepsilon}{2} \leq |t| \leq \varepsilon \\ 0 & |t| \geq \varepsilon \end{cases}$$

is continuous on $[-\|A\|, \|A\|]$, we can define

$$V = \{x \mid f(A)x = x\} \qquad N^*N = A$$

and it is clear that this is a closed invariant subspace for all operators commuting with N^*N. It remains to show that $V \neq \{0\}$

and that $\|N/V\| \leq \varepsilon$.

Let $x \in V$, with $\|x\| = 1$. In this case, we have

$$\|Ax\| = \|Af(N^*N)x\| = \|f(N^*N)Ax\| \leq \|f(N^*N)A\| \leq \sup_{s\in\sigma(A)} |sf(s)| \leq \varepsilon$$

and since

$$\|Nx\|^2 = \langle Ax, x\rangle$$

it follows that $\|N/V\| \leq \varepsilon^{1/2}$.

We prove now that V is nonzero. Indeed,

$$\|(I - f(A))f(2A)\| = \sup |(1 - f(t))f(2t)| = 0$$

and thus $f(2A)x$ is in V for all $x \in H$; furthermore this is not zero because

$$\|f(2A)\| = \sup_{t\in\sigma(A)} |f(2t)| \geq f(0) = 1$$

and the lemma is proved.

The following lemma is an essential step in proving the spectral mapping theorem for normal operators.

5.5.11 LEMMA If $N \in L(H)$ is a normal operator and $p(t, s)$ is a polynomial, then

$$\sigma(p(N^*, N)) = \{p(z^*, z) \mid z \in \sigma(N)\}$$

Proof. Let $p(s, t) = \Sigma\, a_{n,m} s^n t^n$ and thus the operator $p(N^*, N)$ is equal to

$$\Sigma\, a_{n,m} N^{*n} N^m$$

Suppose that $s \in \sigma(N)$; thus there exists a sequence $\{x_n\}$, $\|x_n\| = 1$, such that

$$(N - s)x_n \to 0 \qquad (N^* - S^*)x_n \to 0$$

and this implies that

$$p(N^*, N)x_n - p(s^*, s)x_n \to 0$$

Thus for any s in $\sigma(N)$, $p(s^*, s)$ is in $\sigma(p(N^*, N))$. Let $r \in \sigma(p(N^*, N))$ and since the operator $B = p(N^*, N) - r$ is normal and $0 \in \sigma(B)$ by the above lemma, for $\varepsilon = 1/n$ we can find a closed invariant subspace V_n such that $\|B/V_n\| < 1/n$.

Let $s_n \in \sigma(N/V_n)$ and since N commutes with B, V_n reduces N, and we can find y_n, $\|y_n\| = 1$, such that

$$\|(s_n - N)y_n\| < \frac{1}{n}$$

We can suppose, without loss of generality, that $s_n \to s \in \sigma(N)$ and also

$$\|(N^* - s_n^*)y_n\| < \frac{1}{n}$$

Thus we have that

$$[p(N^*, N) - p(s^*, s)]y_n \to 0$$

and since $p(N^*, N)y_n - p(s^*, s)y_n \to 0$, $r = p(s^*, s)$ and the lemma is proved.

5.5.12 *LEMMA* If $N \in L(H)$ is normal, then

$$\|N\| = \sup \{|s| \mid s \in W(N)\}$$

Proof. We denote by $w(T)$ the number

$$w(T) = \sup \{|s| \mid s \in W(T)\}$$

where $W(T)$ denotes the numerical range of T, i.e., the set

$$W(T) = \{\langle Tx, x\rangle \mid \|x\| = 1\}$$

and from the inequality given in the proof of Theorem 3.6.13,

$$\|Tx\|^2 + |\langle T^2x, x\rangle| \le 2w(T)\|Tx\| \cdot \|x\|$$

we obtain that

1. $\|T\| \le 2w(T)$
2. $w(T^2) \le (w(T))^2$

Indeed, we can write

$$\|Tx\|^2 \le 2w(T)\|Tx\| \cdot \|x\|$$

which gives the first property. For the second property, we have

$$[\|Tx\| - w(T)]^2 + |\langle T^2x, x\rangle| \le (w(T))^2$$

and our assertion follows. We note that these assertions are true for arbitrary elements of L(H). We apply these relations for our case and obtain

$$w(T^{2^n}) \le (w(T))^{2^n}$$

and since N is normal,

$$\|T^{2^n}\| = \|T\|^{2^n}$$

which gives

$$\|T\| = \|T^{2^n}\|^{1/2^n} \le 2^{1/2^n} w(T)$$

and thus for $n \to \infty$ we obtain that $w(T) \ge \|T\|$. This yields

$$w(T) = \|T\|$$

when T is normal since the inequality $w(T) \le \|T\|$ is true for any element of L(H).

Now since the spectral radius of N is

$$r_N = \lim \|N^n\|^{1/2}$$

we obtain that for the case of normal operators,

$$r_N = w(N) = \|N\|$$

5.5.13 LEMMA If $N \in L(H)$ is a normal operator, then there exists an isometric *-isomorphism between the algebra of all complex-valued functions with the norm

$$\|f\| = \sup |f(s)|_{\sigma(N)}$$

and a subalgebra of L(H).

Proof. For any polynomial $p(s^*, s) = \Sigma\, a_{n,m} s^{*n} s^m$, we define the operator $p(N^*, N)$ by the relation

$$p(N^*, N) = \Sigma\ a_{n,m} N^{*n} N^m$$

and it is obvious that the mapping

$$p \mapsto p(N^*, N)$$

is an *-isomorphism between the Banach algebra $C_{\sigma(N)}$ and a subalgebra of $L(H)$. From Lemmas 5.5.11 and 5.5.12 this is in fact an isometry.

We know that the set of all polynomials $\{p(s^*, s)\}$ is dense in the set $C_{\sigma(N)}$ and this allows us to extend the above map to all continuous complex-valued functions by the relation

$$f(N) = \lim p_n(N^*, N)$$

where $\{p_n\}$ is any sequence of polynomials converging to f.

From the above results it follows that the map $f \mapsto f(N)$ satisfies the lemma.

5.5.14 THEOREM (The Spectral Mapping Theorem for Normal Operators) Let $N \in L(H)$ be a normal operator and f be a continuous complex-valued function on $\sigma(N)$. Then the following relation holds:

$$\sigma(f(N)) = f(\sigma(N))$$

Proof. Let s in $\sigma(N)$ and thus there exists a sequence $\{x_n\}$, $\|x_n\| = 1$, such that

$$\lim (N - s)x_n = 0 \qquad \lim (N^* - s^*)x_n = 0$$

and this gives us, as in the case of hermitian operators, that $f(s) \in \sigma(f(N))$.

This converse statement can be proved in the same way as for hermitian operators.

Chapter 6

THE NUMERICAL RANGE

In the last fifteen years much progress in operator theory was obtained by using numerical ranges. Our aim in this chapter is to present the basic results as well as some applications. Also, we give recent results concerning the application of numerical ranges to operator theory.

6.1 THE NUMERICAL RANGE FOR BOUNDED LINEAR OPERATORS ON HILBERT SPACES

Let H be a Hilbert space and $T \in L(H)$. The numerical range of T is introduced in the following definition.

6.1.1 DEFINITION The numerical range of T is the set

$$W(T) = \{\langle Tx, x\rangle \mid \|x\| = 1\}$$

The most important property of W(T) is given in the so-called Hausdorff-Toeplitz theorem.

6.1.2 THEOREM For any $T \in L(H)$, W(T) is a convex set.

There exist several proofs of this theorem, two of which we present here.

Proof 1. Let r and s be in W(T). Thus there exist x_1 and x_2 such that

$$r = \langle Tx_1, x_1\rangle \qquad \langle Tx_2, x_2\rangle = s$$

and we consider the element $r(z) = x_1 + zx_2$, where z is a complex number. It is obvious that $r(z) = 0$ since in the contrary case $|z| = 1$ and then $x_1 = zx_2$, we obtain

$$\langle Tx_1, x_1\rangle = -z\langle Tx_2, x_2\rangle(-z^*) = \langle Tx_2, x_2\rangle$$

and so r = s, which is a contradiction.

We consider the function

$$z \mapsto \langle Tr(z), r(z)\rangle$$

and we show that for any $t \in [0, 1]$ there exists z such that

$$\langle Tr(t), r(t)\rangle = [tr + (1 - t)s]\|r(t)\|^2$$

and this is equivalent to the equation

$$p|z|^2 + qz + mz^* + n = 0$$

where $p = t(s - r)$, $n = (1 - t)(r - s)$ and q, m are complex numbers.

The equation written above is equivalent to the following system:

$$x^2 + y^2 + ax + by - \frac{1 - t}{t} = 0 \qquad cx + dy = 0$$

Since the first equation is a circle and the second is a line, the common points (which exist, an obvious fact from the form of the equations) represent the complex number z, which proves the theorem of Hausdorff-Toeplitz.

Proof 2. Let r and s be in W(T). First we show that we can reduce the proof to the case where r = 1 and s = 0. Indeed, let m and n be the solutions of the system

$$mr + n = 1 \qquad ms + n = 0$$

and since $W(mT + nI) = mW(T) + n = \{mz + n \mid z \in W(T)\}$, the assertion follows.

Thus we can assume, without loss of generality, that r = 1 and s = 0. In this case since $T = T_1 + iT_2$, where T_i are hermitian operators, there exist x_0 and x_1 such that

1. $\|x_0\| = \|x_1\| = 1$

2. $\langle T_1x_0, x_0\rangle = 1 \qquad \langle T_1x_1, x_1\rangle = 0$

$\langle T_2x_0, x_0\rangle = \langle T_2x_1, x_1\rangle = 0$

Furthermore, we can assume that Re $\langle Tx_0, x_1\rangle = 0$. Since x_0 and x_1 are linearly independent, the element $tx_0 + (1 - t)x_1$ is not zero for all t and thus its norm is different from zero. It is easy to see that the function

$$f(t) = \langle Tx(t), x(t)\rangle \qquad x(t) = tx_0 + (1 - t)x_1$$

is a real-valued function. This implies that

$$g(t) = \frac{f(t)}{\|x(t)\|}$$

is a real-valued function with the property

$$g(0) = 0 \qquad g(1) = 1$$

Since g is obviously a continuous function, it follows that for t_0 in (0, 1) there exists at least one t_0^* in (0, 1) such that $g(t_0^*) = t_0$, and this proves the Hausdorff-Toeplitz theorem.

In the following theorem we give results about the half planes containing the numerical range of an operator $T \in L(H)$.

6.1.3 THEOREM If K is a closed half plane, then $K \supset W(T)$ if and only if

$$(T - sI)^*(T - sI) \geq [\text{dist}(s, K)]^2 I$$

where if F is any closed set in the plane,

$$\text{dist}(s, F) = \inf_{y \in F} |s - y|$$

The proof of the theorem is a consequence of the following lemmas, which have their own interest.

6.1.4 LEMMA If $T \in L(H)$ and Re $T = (1/2)(T + T^*)$, then Re $T \geq 0$ if and only if for all $s < 0$,

$$(T - sI)^*(T - sI) \geq 0$$

Proof. First we consider the following identity:

$$(T - sI)^*(T - sI) - s^2 = T^*T - s(T + T^*)$$

Suppose now that $(T - sI)^*(T - sI) \geq 0$, and thus it is clear that

$$T^*T \geq \frac{1}{s(T^*T)}$$

and for $s \to -\infty$, we obtain that $T + T^* \geq 0$. This proves that the condition is sufficient. To prove the necessity we observe that

$$T + T^* \geq \frac{1}{s(T^*T)}$$

and the above identity implies the necessity.

6.1.5 LEMMA If $K_0 = \{s \mid \text{Re } s \geq 0\}$, then if $T \in L(H)$, $W(T) \subset K_0$ if and only if

$$(T - sI)^*(T - sI) \geq (\text{Re } s)^2 I \qquad s \in K_0$$

Proof. First we remark that from

$$\text{Re } \langle Tx, x\rangle = \langle(\text{Re } T)x, x\rangle$$

it follows that $W(T) \subset K_0$ if and only if $\text{Re } T \geq 0$. Now if $s = a + ib$ with a, b real numbers and $s \notin K_0$

$$T - sI = (T - ibI) - aI$$

$$\text{Re } T = \text{Re } (T - ibI)$$

and from Lemma 6.1.7, we obtain

$$(T - sI)^*(T - sI) \geq a^2$$

if and only if $\text{Re } (T - ibI) \geq 0$, which is equivalent to $\text{Re } T \geq 0$, and the lemma is proved.

Proof of Theorem 6.1.3. First we remark that Lemma 6.1.5 is exactly Theorem 6.1.3 for $K = K_0$. If we consider the function $f(s) = as + b$, $|a| = 1$, which transforms K in K_0 (for some a and b) and $f(T) = aT + bI = aT + b$, then we have

$$f(W(T)) = \{f(s) \mid s \in W(T)\}$$

and thus $W(T) \subset K$ if and only if $f(W(T)) \subset K_0$. Now since

$$f(T) - f(s)I = a(T - s)$$

and

$$(f(T) - f(s))^*(f(T) - f(s)) = (T - sI)^*(T - sI)$$

using Lemma 6.1.5 and the fact that

$$\text{dist}\ (f(s), K_0) = \text{dist}\ (f(s), f(T)) = \text{dist}\ (s, K)$$

the assertion of the theorem follows.

In what follows we examine the relationship between the quantities associated with the numerical range and other quantities associated with an operator, such as the norm and the spectral radius. First we define the numerical radius of an operator.

6.1.6 DEFINITION The *numerical radius* of an operator $T \in L(H)$ is the number defined by the relation

$$w(T) = \sup \{|s| \mid s \in W(T)\}$$

The basic properties of the numerical radius of an operator are given in the following theorem.

6.1.7 THEOREM 1. The function defined on L(H) by

$$T \mapsto w(T)$$

is a norm and is equivalent to the standard norm of L(H),

$$T \mapsto \|T\|$$

2. For any $T \in L(H)$, $w(T) = w(T^*)$.
3. For any $T \in L(H)$, $w(T^*T) = \|T\|^2$.

Proof. The properties

$$w(T + S) \geq w(T) + w(S) \qquad w(sT) = |s|w(T)$$

are obvious and we prove the following relation:

$$\frac{1}{2}\|T\| \leq w(T) \leq \|T\|$$

which obviously implies 1. For this we use the identity

$$\|Tx\|^2 + e^{2i\vartheta}\langle T^2x, x\rangle = \frac{1}{2}\langle se^{2i\vartheta}T^2x + s^{-1}e^{i\vartheta}Tx,\ se^{i\vartheta}Tx + s^{-1}\rangle$$
$$- \frac{1}{2}\langle se^{2i\vartheta}T^2x - s^{-1}e^{i\vartheta}Tx,\ se^{i\vartheta}Tx - s^{-1}x\rangle$$

which is valid for all $s > 0$. Since for all $y \in H$

$$|\langle Ty, y\rangle| \leq w(T)\|y\|^2$$

we obtain from the above identity that

$$|\,\|Tx\|^2 + e^{2i\vartheta}\langle T^2x, x\rangle| \leq w(T)(s^2\|Tx\|^2 + s^{-2}\|x\|^2)$$

and thus for $s^2 = \|x\|/\|Tx\|$ with ϑ such that $e^{2i\vartheta}\langle T^2x, x\rangle =$ $|\langle T^2x, x\rangle|$,

$$\|Tx\| \leq 2w(T)\|x\|$$

and thus

$$\|T\| \leq 2w(T)$$

Since the relation $w(T) \leq \|T\|$ is obvious, assertion 1 is proved.

Since properties 2 and 3 follow easily from the definition, we omit the proof.

6.1.8 REMARK We now give an example to show the existence of

1. An operator $T \in L(H)$ for which $(1/2)\|T\| = w(T)$.
2. An operator for which $\|T\| = w(T)$.

We remark that we can find examples even in finite dimensional spaces. Indeed, let H_2 be the two-dimensional Hilbert space and $\{e_1, e_2\}$ be an orthonormal basis for H_2. We define $T: H_2 \to H_2$ by the relation

$$Te_1 = e_2 \qquad Te_2 = 0$$

In this case, for any $x \in H_2$, $\|x\| = 1$ and $x = ae_1 + be_2$, we have

1. $\langle Tx, x\rangle = ab^*$
2. $\|x\|^2 = 1 = |a|^2 + |b|^2$

and thus

$$w(T) \leq \frac{1}{2}$$

Indeed,

$$|\langle Tx, x\rangle| = |ab^*| \le \frac{1}{2}(|a|^2 + |b|^2) \le \frac{1}{2}$$

and for $a = b = 1/2^{1/2}$, we obtain that

$$w(T) \ge \left(\frac{1}{2^{1/2}}\right)^2 = \frac{1}{2}$$

We consider now the operator $T\colon H_2 \to H_2$ defined by the relation

$$Te_1 = e_1 \qquad Te_2 = 0$$

and it is easy to see that for this operator, $w(T) = \|T\|$.

6.1.9 REMARK There exists a large class of operators $T \in L(H)$ for which the relation

$$w(T) = \|T\|$$

holds; this class contains all normal operators.

6.2 THE NUMERICAL RANGE AND THE SPECTRUM

If T is an operator in L(H), it is important to know the relations between W(T) and $\sigma(T)$, the spectrum of T. We give now some results about the connections between W(T) and $\sigma(T)$; some of these are valid for arbitrary operators.

6.2.1 THEOREM For any operator $T \in L(H)$

$$\sigma(T) \subset \overline{W(T)} \qquad \text{(the bar denotes the closure)}$$

Proof. First we remark that the theorem is true for points in the boundary of $\sigma(T)$. Indeed, if s is such a point, then there exists a sequence $\{x_n\}$, $\|x_n\| = 1$, such that

$$Tx_n - sx_n \to 0$$

and this clearly implies that $s \in \overline{W(T)}$. Let s be an arbitrary point in $\sigma(T)$ and L be an arbitrary line through s and $L_T = L \cap \sigma(T)$. Let s_1 and s_2 be the ends of the last closed interval containing L_T.

Clearly $s_i \in \partial\sigma(T)$, and by the above remark, s_i are in $\overline{W(T)}$. Since $W(T)$ is a convex set, it follows that $\overline{W(T)}$ is also a convex set. This implies that s is in $\overline{W(T)}$ and the theorem is proved.

6.2.2 REMARK The set W(T) is not generally closed. To see this we consider the operator $T: \ell^2_+ \to \ell^2_+$ defined in

$$T(x_1, x_2, x_3, \ldots) = (x_1, \frac{1}{2} x_2, \frac{1}{3} x_3, \ldots)$$

which has the property that $W(T) = (0, 1]$.

The following theorem gives a sufficient condition for an operator to have the numerical range closed.

6.2.3 THEOREM If $T \in L(H)$ is a compact operator and $0 \in W(T)$, then W(T) is a closed set.

Proof. If s is a cluster point of W(T), then there exists a sequence $\{x_n\}$, $\|x_n\| = 1$, such that $\lim \langle Tx_n, x_n\rangle = s$. We can choose a weakly convergent subsequence and may assume, without loss of generality, that $\{x_n\}$ is weakly convergent to x. By the compactness of T, $Tx_n \to Tx$. Since we have the relations

$$\begin{aligned} |\langle Tx_n, x_n\rangle - \langle Tx, x\rangle| &< |\langle Tx_n, x_n\rangle - \langle Tx, x_n\rangle| + |\langle Tx, x_n\rangle \\ &\quad - \langle Tx, x\rangle| < \|x_n\| \cdot \|Tx_n - Tx\| \\ &\quad + |\langle x_n - x, Tx\rangle| \end{aligned}$$

we have $s = \lim \langle Tx_n, x_n\rangle = \langle Tx, x\rangle$.

Now if s = 0 the assertion of the theorem is clear; if $s \neq 0$

$$\frac{s}{\|x\|^2} = \left\langle \frac{Tx}{\|x\|}, \frac{x}{\|x\|} \right\rangle \in W(T)$$

which proves the theorem since s is on the interval $[0, s/\|x\|]$.

The following theorem gives a result concerning some special points in the boundary of the numerical range for arbitrary operators.

6.2.4 THEOREM Let $T \in L(H)$ and $W(T)$ be a closed set. Every point s in the boundary of $W(T)$ at which the boundary is not a differentiable arc, it is an eigenvalue for T.

First we give a lemma from which the theorem follows easily; also from the lemma, the convexity of $W(T)$ follows, and thus we have a third proof of the Hausdorff-Toeplitz theorem.

6.2.5 LEMMA Let T be a linear operator on the two-dimensional Hilbert space H_2. Then $W(T)$ is an ellipse and the foci of this ellipse are the eigenvalues of T; if the eigenvalues are distinct, the eccentricity of the ellipse is $\sin \vartheta$, where

$$\cos \vartheta = \frac{|\langle x, y \rangle|}{\|x\| \cdot \|y\|} \qquad 0 \le \vartheta \le \frac{\pi}{2}$$

and where x and y are the eigenvectors. Also, if there is only one eigenvalue, then the ellipse is a circle with the center at the eigenvalue s and diameter $\|T - sI\| = \|T - s\|$.

Proof. Since H_2 has the set $\{x \mid \|x\| = 1\}$ as a compact set and the function $x \mapsto \langle Tx, x \rangle$ is continuous, it follows that $W(T)$ is a compact set. First we show that the boundary of $W(T)$ is an ellipse and that $W(T)$ contain all the interior points of that ellipse.

Suppose now that T has only one eigenvalue s. In this case $T_1 = T - s$ has the property that $\sigma(T_1) = \{0\}$ and also $T_1^2 = 0$. If $T_1 = 0$, we have that $W(T_1) = \{0\}$ and thus $W(T) = \{s\}$. This clearly is a circle with center s and radius 0. If $T_1 \neq 0$, then there exists an orthonormal basis $\{e_1, e_2\}$ of H_2 such that

$$T_1 e_1 = a e_2 \qquad T_1 e_2 = 0$$

and $\|T\| = |a|$. Also $W(T_1)$ is a circle of radius $|a|/2$. This implies that $W(T)$ is a circle with radius $\|T - s\|/2$ and center s.

Now if T has two eigenvalues, say s_1 and s_2, the operator

$$T_1 = \frac{T - s_1}{s_2 - s_1}$$

has the eigenvalues 0 and 1. Also from the definition of the numerical range, the set $W(T_1)$ is obtained from $W(T)$ by a rigid motion and homothetic transformation of the plane and conversely.

Let $\{e_1, e_2\}$ be an orthonormal basis of H_2 such that

$$T_1 e_1 = 0$$

and we may choose these such that

$$T_1 u = u \qquad \|u\| = 1$$

where $u = (\cos \vartheta)e_1 + (\sin \vartheta)e_2$ and ϑ is the angle between u and e_1. Now since

$$T_1 e = (\cot \vartheta)e_1 + e_2$$

then for any $x = ae_1 + be_2$, $\|x\| = 1$, we have

$$\langle T_1 x, x\rangle = |b|^2 + a^* b \cot \vartheta = |b|^2 + |a| \cdot |b| e^{iw} \cot \vartheta$$

If w varies, then we obtain a circle in the complex plane with the center at the point $|b|^2$ and with radius $r = [t(1 - t)]^{1/2} \cot \vartheta$ and $W(T_1)$ is the union of the circles

$$(x - t)^2 + y^2 = (t - t^2) \cot^2 \vartheta$$

The envelope of this family is the ellipse

$$\frac{(x - 1/2)^2}{[\csc(\vartheta/2)]^2} + \frac{y^2}{[\cot(\vartheta/2)]^2} = 1$$

which obviously has the foci at (0, 0) and (1, 0), as well as the eccentricity $\sin \vartheta$.

To prove that any interior point s_0 of the ellipse is in $W(T)$, we remark that there exists a circle C_{t_0} containing s_0 which is tangent to the ellipse at the foot of the perpendicular from s_0 to the ellipse and it is exterior to one of the circles C_0 and C_1. Since these circles vary continuously with t, it follows that there is a circle through s_0, and this proves that $W(T)$ is an ellipse.

6.2.6 REMARK From this lemma we can prove the Hausdorff-Toeplitz theorem as follows: If a and b are two points in W(T), then there exists x and y in H such that

$$a = \langle Tx, x\rangle \qquad b = \langle Ty, y\rangle \qquad \|x\| = \|y\| = 1$$

We consider the projection $P_{x,y}$ on the space generated by x and y. In this case it is easy to see that

$$W(P_{x,y}TP_{x,y}) \subset W(T)$$

and since $W(P_{x,y}TP_{x,y})$ is an ellipse the assertion follows.

Proof of Theorem 6.2.4. It is well known that the boundary of W(T), being a convex function, is differentiable, except perhaps at a countable set of points. Let s be a point of nondifferentiability and x, $\|x\| = 1$, such that $s = \langle Tx, x\rangle$. Also, at s there exists a left and right tangent such that the angle between these tangents is smaller than π. Let y be arbitrary in H and $P_{x,y}$ be the projection on the space generated by x and y. The operator $T_1 = P_{x,y}TP_{x,y}$ has an ellipse as the numerical range, and since no circle contained in W(T) can pass through s, it follows that the ellipse $W(T_1)$ is a line segment or a point; thus s is an eigenvalue with x as an eigenvector.

For the following result we need the notion of a bare point of a set in the complex plane.

6.2.7 DEFINITION For a closed set F in the complex plane a point s is called a *bare point* if

1. s is in F.
2. s is on the boundary of a circle containing F.

6.2.8 DEFINITION A point s of a closed set F in the complex plane is called a *semibare point* if

1. s is on the boundary of a circle C.
2. The circle C contains no other points of F.

6.2.9 DEFINITION If C is a closed convex set in the complex plane, then C has a *corner* with vertex $s \in C$ if

1. $s \in C$.
2. C is contained in a half plane with vertex s and angle less than π.

6.2.10 LEMMA Let $T \in L(H)$ and

1. $1 \in W(T)$
2. $w(T) = 1$

Then $s \in \sigma_p(\mathrm{Re}\, T)$, $\mathrm{Re}\, T = (1/2)(T + T^*)$. [$\sigma_p(\cdot)$ denotes the point spectrum of $(\cdot)$.]

Proof. Since $W(\mathrm{Re}\, T) = \mathrm{Re}\, W(T)$ it follows that

$$1 = \langle \mathrm{Re}\, Tx, x\rangle$$

and this gives $\mathrm{Re}\, Tx = x$.

6.2.11 LEMMA If $T \in L(H)$ and $s \in \overline{W(T)}$ is a bare point of $\overline{W(T)}$, then $(e^{-i\vartheta}T + e^{i\vartheta}T^*)x = (e^{-i\vartheta}s + e^{i\vartheta}s^*)x$ for some $x \in H$, $\|x\| = 1$, and $0 \le \vartheta < 2\pi$.

Proof. Since s is a bare point of $\overline{W(T)}$, we can find $r > 0$ such that

$$W(T) \subset \{z \mid |z - s_0| \le r\}$$

for some s_0 and s is on the boundary of this circle. Let $s - s_0 = re^{i\vartheta}$ with $0 < \vartheta < 2\pi$ and $T_1 = r^{-1}e^{i\vartheta}(T - s_0)$. In this case, we have that the numerical range of T_1 is contained in the closed unit disk, and if $x \in H$, $\|x\| = 1$, is such that $s = \langle Tx, x\rangle$, we obtain

$$1 = \langle T_1x, x\rangle \in W(T_1)$$

and by Lemma 6.2.10,

$$\frac{1}{2}(T_1 + T_1^*)x = x$$

From the form of T_1 it follows that

$$\frac{1}{2}[r^{-1}e^{-i\vartheta}(T - s_0) + r^{-1}e^{i\vartheta}(T - s_0)^*]x = x$$

and thus further,

$$\frac{1}{2}[e^{-i\vartheta}T + e^{i\vartheta}T^*]x = \frac{1}{2}[e^{-i\vartheta}sx + e^{i\vartheta}s^*x]$$

and the lemma is proved.

6.2.12 THEOREM If $T \in L(H)$ and $s \in W(T)$ is the vertex of a corner of $\overline{W(T)}$, then s is an eigenvalue of T.

Proof. Since s is a vertex of a corner of the closure of the numerical range, then s is a bare point of this set. Thus we find r_1 and r_2, positive numbers, and two complex numbers a_0 and b_0 such that $a_0 \neq b_0$ and

$$W(T) \subset \{z \mid |z - a_0| \leq r_1\} = D_{r_1}$$

$$W(T) \subset \{z \mid |z - b_0| \leq r_2\} = D_{r_2}$$

$$s \in \overline{W(T)} \cap D_{r_1} \cap D_{r_2}$$

As in the proof of Lemma 6.2.10, we find ϑ_1 and ϑ_2 such that $\vartheta_i \in (0, 2\pi)$, $|\vartheta_1 - \vartheta_2| < \pi$ and

$$\frac{1}{2}[e^{-i\vartheta_j}T + e^{i\vartheta_j}T^*]x = \frac{1}{2}[e^{-i\vartheta_j}s + e^{i\vartheta_j}s^*]x$$

which gives

$$\frac{1}{2}[e^{-2i\vartheta_1} + e^{-2i\vartheta_2}]Tx = \frac{1}{2}[e^{-2i\vartheta} - e^{-2i\vartheta}]x$$

and since $\vartheta_1 \neq \vartheta_2$, it follows that $Tx = sx$.

6.2.13 THEOREM If $L(H)$ contains T and T is a compact operator and if $0 = s \in W(T)$ is the vertex of a corner of $W(T)$, then s is an eigenvalue of T.

Proof. Since T is compact, then $s \in W(T)$ and the result follows from Theorem 6.2.11.

6.2.14 COROLLARY Let $T \in L(H)$ and $W(T)$ be a closed polygon; in this case T is convexoid, i.e.,

$$\text{con}\,\sigma(T) = W(T)$$

Proof. If $\{s_i\}$ are the vertices of the convex polygon, then s_i are eigenvalues, and thus

$$\text{conv}\,\sigma(T) \supset W(T)$$

Since the contrary inclusion is true for all operators, the corollary follows.

In the following corollary we give a necessary and sufficient condition for a polygon to be the numerical range of a bounded linear operator on finite dimensional spaces.

6.2.15 COROLLARY If C is a closed convex polygon with m vertices, then C is the numerical range of an operator on a finite dimensional space if and only if $m \leq n$.

Proof. If C is the numerical range which is a closed polygon with vertices $s_1, \ldots, s_m$, then by Theorem 6.2.11, each s_i is an eigenvalue and this clearly implies $m \leq n$. Conversely, if $s_1, \ldots, s_m$ are the vertices of a polygon C, then the normal operator N with the matrix

$$a_{ij} = \begin{cases} s_i \delta_{ij} & 1 \leq i \leq m \\ 0 & m < i \leq n \end{cases}$$

has the property that $W(T) = C$.

We can obtain an extension of these results using the notion of a semibare point. It is easy to see that every bare point is a semibare point. Our first result in this direction is the extension of Lemma 6.2.11 to the case of semibare points.

6.2.16 LEMMA If $T \in L(H)$ and $s \in W(T)$ is a semibare point of $W(T)$, then

$$[e^{-i}T + e^{i\vartheta}T^*]x = [e^{-i\vartheta}s + e^{i\vartheta}s^*]x$$

for some $x \in H$ and $\vartheta \in [0, 2]$.

Proof. Since s is a semibare point of $W(T)$, then there exists $r > 0$ and a_0, a complex number, such that $\text{dist}\,(a_0, W(T)) = r$, and $W(T) \cap \{z \mid |z - a_0| > r\} = \{s\}$.

Let $s - a_0 = re^{i\vartheta}$ and consider the operator $T_1 = r^{-1}e^{-i\vartheta}(T - a_0)$. In this case,

$$W(T_1) = \{r^{-1}e^{i\vartheta}[\langle Tx, x\rangle - a_0] \mid \|x\| = 1\}$$

which gives that $W(T_1)$ is contained in the unit disk. Also, for $s = \langle Tx, x\rangle$, $1 \in W(T_1)$ and Lemma 6.2.10 implies that

$$\text{Re } T_1x = x$$

As in Lemma 6.2.11 we obtain our assertion.

Using this result we can generalize the result obtained in Theorem 6.2.12.

6.2.17 THEOREM Let $T \in L(H)$ and $s \in W(T)$ be a semibare point of $W(T)$. Then s is an eigenvalue of T.

Proof. Since s is a semibare point of $W(T)$, we can find positive numbers r_1 and r_2 and two complex numbers a_0 and b_0 such that

$$W(T) \cap \{z \mid |z - a_0| < r_1\} = \{s\}$$
$$W(T) \cap \{z \mid |z - b_0| < r_2\} = \{s\}$$

We can find ϑ_i in $[0, 2\pi)$ such that

$$\frac{1}{2}[e^{-i\vartheta_j}T + e^{i\vartheta_j}T^*]x = \frac{1}{2}[e^{-i\vartheta_j}s + e^{i\vartheta_j}s^*]x$$

and since $\vartheta_1 \neq \vartheta_2$, we obtain $Tx = sx$.

6.3 THE NUMERICAL RANGE AND ITS CLOSURE

As we know there exist simple examples of elements in $L(H)$ for which the numerical range is not a closed set. We give now a construction, following Berberian, to show that for any operator $T \in L(H)$ we can define a new operator, say $T^{\sim}$, such that

1. $W(T^{\sim}) = \overline{W(T)}$
2. $\sigma_p(\tilde{T}) = \sigma_{ap}(T)$

where $\sigma_p(\cdot)$ denotes the point spectrum of $(\cdot)$ and $\sigma_{ap}(\cdot)$ denotes the approximate point spectrum of (•).

The basic result due to Berberian is the following.

6.3.1 THEOREM Let $T \in L(H)$, where $L(H)$ is the set of all bounded linear operators on H. Then there exists another Hilbert space, $H^{\sim}$, and an application, $T \mapsto T^{\sim}$, such that

$$W(T^{\sim}) = \overline{W(T)} \qquad \sigma_p(T^{\sim}) = \sigma_{ap}(T)$$

Proof. The proof of this result uses the notion of generalized limit or Banach limit. First we consider the space of all sequences $x = (x_n)$ such that

$$(\|x_n\|) \in \ell^{\infty}$$

and we define a bilinear form,

$$\mathcal{B}(x, y) = \text{glim } \{\langle x_n, y_n\rangle\}$$

Since $|\langle x_n, y_n\rangle| \le \|x_n\| \cdot \|y_n\|$, the function $\mathcal{B}$ is well defined. Also it is clear that $\mathcal{B}$ is a positive symmetric functional on $L(H)$, and by the Cauchy-Schwarz inequality we see that the set

$$N(\mathcal{B}) = \{x \mid \mathcal{B}(x, x) = 0\}$$

is a linear subspace. Now we can define the space S(H)/N with an inner product defined in a standard way: if $x^{\sim} = x + N$, then

$$\langle x^{\sim}, y^{\sim}\rangle = \text{glim } \langle x_n, y_n\rangle$$

and if S(H) denotes the space of all sequences $x = (x_n)$, as above, we have an isometric linear mapping of H into a closed linear subspace of S(H)/N. Since S(H)/N is a linear subspace of a Hilbert space (its completion), we have the relation

$$H' \subset S(H)/N \subset K \qquad \text{(the completion of } S(H)/N)$$

where H' denotes the image of H into S(H).

For any $T \in L(H)$ we can define an operator on K as follows: if $x = (x_n) \in S(H)$, then we set

$$T^{\sim}(x) = (Tx_n)$$

It is easy to see that $T^{\sim}$ is bounded and N is an invariant subspace for $T^{\sim}$, and this allows us to define an operator on the space S(H)/N in a standard way. Denote this operator by $T_0^{\sim}$. In this case we have that $\|T_0^{\sim}\| \le \|T\|$, and moreover, we have in fact the equality of norms. From the density of S(H)/N in the space K it follows that $T_0^{\sim}$ extends uniquely to a bounded operator on K. This extension we denote by T^+. The following properties are easy to prove and we list these here:

$$T \mapsto T^+$$

satisfies the relations

$$T + S \mapsto T^+ + S^+$$
$$zT \mapsto zT^+$$
$$ST \mapsto S^+T^+$$
$$T^* \mapsto (T^+)^*$$
$$I \mapsto I$$
$$\|T\| = \|T^+\|$$

and

$$T \ge 0 \text{ implies } T^+ \ge 0$$

From the properties of the mapping $T \to T^+$ we obtain the assertion of the theorem in the following way: If s is not in the approximate point spectrum of T, we find an $\varepsilon > 0$ such that

$$(T - sI)^*(T - sI) > \varepsilon I$$

and this implies that the corresponding operator T^+ has the following property:

$$(T^+ - sI)^*(T^+ - sI) > \varepsilon I$$

and this gives the second assertion of the theorem. The first assertion of the theorem can be obtained by observing that the closure of the numerical range of T is the same as the closure of the numerical range of T^+. But it is easy to see that the closure of W(T) is contained in $W(T^+)$ and thus the theorem is proved.

6.4 THE ESSENTIAL NUMERICAL RANGE FOR BOUNDED LINEAR OPERATORS ON HILBERT SPACES

The notion of essential numerical range was introduced by Stampfli and Williams (1968). Their definition is not the following, but is equivalent to this.

6.4.1 DEFINITION The *essential numerical range* of an operator $T \in L(H)$ is the set

$$W_{ess}(T) = \cap \overline{W(T + k)}$$

where the intersection is taken over all compact operators $k \in K$.

The following theorem gives useful information about this set.

6.4.2 THEOREM The following assertions are equivalent:

1. $0 \in \cap W(T + k)$, where k is of finite rank.
2. $0 \in \cap W(T + k)$, k compact.
3. There exists a sequence $\{x_n\}$, $\|x_n\| = 1$, such that

$$\langle Tx_n, x_n\rangle \to 0 \qquad x_n \rightharpoonup 0$$

4. There exists an orthonormal sequence $\{e_n\}$ such that

$$\langle Te_n, e_n\rangle \to 0$$

5. There exists an infinite dimensional projection P such that PTP is compact.

Proof. 5 implies 4. Suppose that 5 is true and thus for some projection P, PTP is compact. Since the accumulation point of the spectrum of PTP is the (possible) point zero, this implies 4. The fact that 4 implies 3 is trivial. We show now that 3 implies 2.

Suppose that 3 is true and let k be a compact operator. Thus $kx_n \to 0$ which gives that $0 \in \overline{W(T + k)}$ for all k. Indeed,

$$\begin{aligned}\lim \langle (T + k)x_n, x_n\rangle &= \lim \langle Tx_n, x_n\rangle + \lim \langle kx_n, x_n\rangle \\ &= \lim \langle Tx_n, x_n\rangle\end{aligned}$$

The fact that 2 implies 1 is a consequence of the following fact, which is easy to prove: Every compact operator in a separable

Hilbert space is in the closure of the set of all operators of finite rank.

To prove the theorem, first we show that 1 implies 4. Let $\varepsilon_k \to 0$ and $\{e_k\}$ have been found such that $|\langle Te_k, e_k\rangle| < \varepsilon_k$, k = 1, ..., n. Let M be the subspace spanned by $\{e_k\}_{k=1}^{n}$ and P be the orthogonal projection on M, and to find a unit vector in $M^{\perp}$ such that $|\langle Te, e\rangle| < \infty$ we show that $0 \in \overline{W((I - P)T(I - P)}$, which clearly implies the existence of e with the above property. Let $s \in W(I - P)T)/M^{\perp}$ and define

$$F = sP - PTP - (I - P)TP - PT(I - P)$$

which is of finite rank and

$$T + F = sI/M \oplus (I - P)T/M^{\perp}$$

Since for any two operators $A, B \in L(H)$ the relation

$$W(A \oplus B) = \text{conv}\ (W(A),\ W(B))$$

holds, we obtain that

$$W(T + F) = W((I - P)T/M^{\perp})$$

and property 1 gives our assertion. Thus 1 implies 4.

To complete the proof of the theorem we show that 4 implies 5. Let $\{e_n\}$ be an orthonormal sequence with the property

$$\lim \langle Te_n, e_n\rangle = 0$$

and we can assume, without loss of generality, that

$$\Sigma\ |\langle Te_n, e_n\rangle|^2 < \infty$$

Now, if $n_1 = 1$

$$\Sigma\ |\langle Te_{n_1}, e_n\rangle|^2 < \|Te_{n_1}\|^2$$

$$\Sigma\ |\langle Te_n, e_{n_1}\rangle|^2 < \|T^*e_{n_1}\|^2$$

and by the Bessel inequality we can find an integer $n_2 > n_1$ such that

$$\sum_{n=n_2} |\langle Te_{n_1}, e_n\rangle|^2 < \frac{1}{2} \sum_{n=n_2} |\langle Te_n, e_{n_1}\rangle|^2 < \frac{1}{2}$$

Iterating, we obtain a sequence of integers $\{n_k\}$ such that

1. $n_1 < n_2 < n_3 < \cdots < n_k < n_{k+1} < \ldots$
2. $\Sigma_{n=n_{k+1}} |\langle Te_{n_k}, e_n\rangle|^2 < 2^{-k}$; $\Sigma_{n=n_k+1} |\langle Te_n, e_{n_k}\rangle|^2 < 2^{-k}$

From these we obtain that

$$\sum_{i,j} |\langle Te_{n_i}, e_{n_j}\rangle|^2 < \infty$$

and if P is the projection on the subspace generated by the sequence $\{e_{n_k}\}$, the above inequality means that PTP is a Hilbert-Schmidt operator,† i.e., PTP is a compact operator.

From this theorem we obtain other information about the points in the set $W_{ess}(T)$. Some of these are given in the following theorem.

6.4.3 THEOREM The following assertions are equivalent:

1. $s \in W_{ess}(T)$.
2. There exists a sequence $\{x_n\}$, $\|x_n\| = 1$, $x_n \to 0$, with $\lim \langle Tx_n, x_n\rangle = s$.
3. There exists a sequence $\{e_n\}$, $\langle e_n, e_m\rangle = \delta_{n,m}$, $\lim \langle Te_n, e_n\rangle = s$.
4. There exists P, $P^2 = P$, such that PTP - sP is compact.

Proof. This follows from Theorem 6.4.2.

6.5 THE MAXIMAL NUMERICAL RANGE OF A BOUNDED OPERATOR ON A HILBERT SPACE

An interesting and important notion defined for arbitrary bounded linear operators on Hilbert spaces with applications in operator theory (we mention here the problem of the norm of a derivation) was introduced by J. Stampfli and named the "maximal numerical

---

†See the Appendix, C_p classes.

range." In the case of finite-dimensional spaces this numerical range is produced by the maximal vectors.† We give the definition and some properties of the maximal numerical range.

6.5.1 DEFINITION If $T \in L(H)$, then the *maximal numerical range* is the set defined by

$$W_{max}(T) = \{s \mid s = \lim \langle Tx_n, x_n\rangle, \|x_n\| = 1, \|Tx_n\| \mapsto \|T\|\}$$

The basic property of W_{max} is given in the following theorem.

6.5.2 THEOREM $W_{max}(T)$ is a closed, convex, and nonempty set contained in $\overline{W(T)}$.

Proof. For the proof we need some results which we give as a lemma.

6.5.3 LEMMA If $T \in L(H)$ and $\|x\| = 1 = \|T\|$, $\|Tx\|^2 \geq (1 - \varepsilon)$, then

$$\|(T^*T - I)x\|^2 \leq 2\varepsilon$$

Proof. Since

$$0 \leq \|(T^*T - I)x\|^2 = \|T^*Tx\|^2 - 2\|Tx\| + \|x\|^2$$

$$\leq 2(1 - \|Tx\|^2) \leq 2\varepsilon$$

and the lemma is proved.

Proof of Theorem 6.5.2. Since $\|T\| = \lim \|Tx_n\|$, for some $\{x_n\}$, $\|x_n\| = 1$, it follows that $W_{max}(T)$ is a nonempty set. Also from the definition it is clear that it is a closed set. We prove now that this set is convex. We may assume, without loss of generality, that $\|T\| = 1$. Let a and b be two points in $W_{max}(T)$. Thus there exist sequences $\{x_n\}$ and $\{y_n\}$ such that

$$a = \lim \langle Tx_n, x_n\rangle \qquad b = \lim \langle Ty_n, y_n\rangle \qquad \|x_n\| = \|y_n\| = 1$$

We consider the space V_n generated by x_n and y_n and remark that a and b are in the numerical range of P_nTP_n, where P_n is the

---

†An element $x \in H$ is called a *maximal vector* for T if

$$\|Tx\| = \|T\| \cdot \|x\|$$

projection (orthogonal projection) on the space V_n. Then, by the convexity of $W(P_nTP_n)$, there exists a sequence $\{u_n\}$, $\|u_n\| = 1$, such that

$$\langle P_nTP_nu, u\rangle \to s$$

where s is a fixed point on the segment with ends a and b. Since $u = \alpha_n x_n + \beta_n y_n$, $\|u_n\| = 1$, it follows that $|\langle x_n, y_n\rangle| < \vartheta$, where $\vartheta < 1$, and thus we can find $M < \infty$ such that

$$|\alpha_n| < M \qquad |\beta_n| < 1$$

Hence, we can assume, without loss of generality, that

$$\alpha_n \to \alpha \qquad \beta_n \to \beta$$

and since

$$\|Tu_n\| = \langle T^*Tu_n, u_n\rangle = \|u_n\|^2 - 2M\varepsilon_n \qquad \varepsilon_n \to 0$$

we obtain $\|Tu_n\| \to 1$, and since $\langle Tu_n, u_n\rangle \to s$, the assertion of the theorem is proved.

We close this section for the following

6.5.4 PROBLEM If $s \in W_{max}(T)$ is a bare point or semibare point, then it follows that s is an approximate eigenvalue of T.

6.5.5 PROBLEM If $T \in L(H)$, we define the *maximal essential numerical range* of T as the set

$$W_{max}^{ess}(T) = \bigcap W_{max}(T)$$

where the intersection is taken over all compact operators. The problem is to describe the properties of this set.

6.6 THE EXTREME POINTS OF THE NUMERICAL RANGE FOR HYPONORMAL OPERATORS AND (WN) OPERATORS

We start with the following definition.

6.6.1 DEFINITION An operator $T \in L(H)$ is called *hyponormal* if the operator $D_T = T^*T - TT^*$ is positive.

Concerning the extreme points of the numerical range of a hyponormal operator, J. Stampfli has proved the following.

6.6.2 THEOREM If $T \in L(H)$ is a hyponormal operator and $\mathrm{Re}\, T \geq 0$, then if 0 is an extreme point of $W(T)$, $\langle Tx, x\rangle = 0$ implies $Tx = 0$, and moreover, the set

$$M = \{x \mid \langle Tx, x\rangle = 0\}$$

is a reducing subspace for T.

6.6.3 DEFINITION An operator $T \in L(H)$ is called a *(WN) operator* if

$$4T^*T - (T + T^*)^2 \geq 0$$

6.6.4 REMARK From the definition it is clear that every operator which is hyponormal is also a (WN) operator.

6.6.5 THEOREM If $T \in L(H)$ is a (WN) operator and

1. $\mathrm{Im}\, T \geq 0$
2. 0 is an extreme point of $W(T)$

then the set

$$M = \{x \mid \langle Tx, x\rangle = 0\}$$

is a reducing subspace for T and for each $x \in M$, $Tx = 0$.

Proof. From the relation $\langle Tx, x\rangle = 0$ it follows that $\langle \mathrm{Im}\, Tx, x\rangle = 0$, i.e., $\langle Jx, x\rangle = 0$, where $T = H + iJ$. Now, since T is a (WN) operator, we obtain that

$$\langle (4T^*T + 4J^2)x, x\rangle > 0$$

and this implies that

$$4i\langle (HJ - JH)x, x\rangle + 4\|Jx\|^2 = 0$$

Now if

$$N = \{x \mid Jx = 0\}$$

then this subspace is invariant under J. Indeed, from the above relation we obtain that

$$\|D_T^{1/2}x\| = 0$$

which implies that $\|JHx\| = 0$. Thus our assertion is proved. From this result it follows that N is a reducing subspace of T and since the restriction of T to this subspace is H, we know (this is obvious) that for any subspace M

$$W(T/M) \subset W(T)$$

Thus 0 is the extreme point of T/N and this implies that $Hx = 0$. Clearly $Tx = 0$, and the theorem is proved.

6.6.6 REMARK This theorem contains Stampfli's result since we know that any hyponormal operator T is also a (WN) operator, as is any operator of the form $T + s$, where T is hyponormal.

6.7 THE NUMERICAL RANGE AND SOME CLASSES OF OPERATORS

As we know for any bounded operator T on a Hilbert space, we have the inclusion,

$$\sigma(T) \subset \overline{W(T)}$$

and since W(T) is a convex set, it follows that

$$\text{conv } \sigma(T) \subset \overline{W(T)}$$

In his paper, Wintner (1929) makes an attempt to characterize those operators in L(H) for which

$$\text{conv } \sigma(T) = \overline{W(T)}$$

An operator $T \in L(H)$ is called *normaloid* if

$$\|T\| = w(T)$$

and the assertion of Wintner (1929) was that an operator has the property

$$\text{conv } \sigma(T) = \overline{W(T)}$$

if and only if it is a normaloid operator. That this assertion is false is shown by the following example due to Halmos (1967).

Let

$$A = \begin{bmatrix} 0 & 1 \\ 0 & 0 \end{bmatrix}$$

and I be the one-dimensional identity operator and we consider the operator $T = A \oplus I$. It is easy to see that

$$\overline{W(A)} = \left\{ z \mid |z| \le \frac{1}{2} \right\}$$

and $\overline{W(T)}$ is the set

$$\text{conv } \{\{\overline{W(A)}\}, 1\}$$

Thus $w(T) = 1 = \|T\|$ and $\text{conv } \sigma(T) = [0, 1]$.

For other classes of operators defined using the numerical range we refer the reader to Chapter 7.

6.8 THE NUMERICAL RANGE AND TENSOR PRODUCTS

An interesting problem concerns the connection between the tensor products of two operators and the numerical range. Suppose that T and S are elements of L(H) and we consider the set

$$\overline{W(T \otimes S)} \subset \overline{W(T)} \cdot \overline{W(S)} = \{ts \mid t \in \overline{W(T)},\ s \in \overline{W(S)}\}$$

Some results in the case of normal operators indicate the relation

$$\overline{W(T \otimes S)} = \overline{W(T)} \cdot \overline{W(S)}$$

Our first result is an example of two bounded operators T and S such that the above equality fails.

6.8.1 THEOREM There exist two bounded operators T and S such that

$$\overline{W(T \otimes S)} \ne \overline{W(T)} \cdot \overline{W(S)}$$

Proof. Let A be an operator with the matrix

$$A = \begin{bmatrix} 0 & 1 \\ 0 & 0 \end{bmatrix}$$

We know that the numerical range is

$$\overline{W(A)} = \left\{z \mid |z| \leq \frac{1}{2}\right\}$$

and we define two operators, T = A* and S = A. We obtain, in this case,

$$W(T) \cdot W(S) \subset \left\{z \mid |z| \leq \frac{1}{4}\right\}$$

and since $S \otimes T$ is of the form

$$\begin{bmatrix} 0 & 0 \\ S & 0 \end{bmatrix}$$

it follows that the numerical range of this operator contains the numbers

$$\langle (S \otimes T)x, x\rangle = \frac{1}{2}$$

for $\|x\| = 1$,

$$x = \begin{bmatrix} x_1 \\ x_2 \end{bmatrix} \qquad x_1 = \begin{bmatrix} 0 \\ 2^{1/2} \end{bmatrix} \qquad x_2 = \begin{bmatrix} 2^{1/2} \\ 0 \end{bmatrix}$$

It is clear that 1/2 is not in the set

$$\left\{z \mid |z| \leq \frac{1}{4}\right\}$$

and this proves the theorem.

We give now a positive result.

6.8.2 THEOREM If T and S are elements of L(H), then

$$\overline{W(T \otimes S)} \supset \operatorname{conv}\,(\overline{W(T)} \cdot \overline{W(S)})$$

Proof. Let $t \in W(T)$ and $s \in W(S)$. Thus we can find x and y in H such that

1. $\|x\| = \|y\| = 1$
2. $\langle Tx, x\rangle = t \qquad \langle Sy, y\rangle = s$

and we have

$$ts = \langle (T \otimes S)(x \circ y, x \circ y)\rangle \in W(T \otimes S)$$

and since $\overline{W(T \otimes S)}$ is a closed convex set, the assertion of the theorem follows.

For conditions in which the equality holds we refer the reader to Chapter 7.

We close with the following problem.

6.8.3 PROBLEM What is the connection between $W_{ess}(T)$, $W_{ess}(S)$, and $W_{ess}(T \otimes S)$?

6.9 THE NUMERICAL RANGE FOR BOUNDED LINEAR OPERATORS ON BANACH SPACES

In the last 10 or 15 years much progress has been made in extending the theory of numerical range for the case of operators on Banach spaces. During the same time the theory of numerical range has been extended to nonlinear operators. In what follows we give results on the numerical ranges of operators for bounded linear operators on Banach spaces; some of the proofs can be extended with trivial modifications to the case of elements of Banach algebras. The technique is developed using the notion of semi-inner product, a notion which was defined by Lumer (1961), and we now give some results concerning this notion.

6.9.1 DEFINITION Let L be a Banach space. A *semi-inner product* on L is a function on L × L with complex values, $[\ ,\ ]: L \times L \to C$, satisfying the following properties:

1. $[x_1 + x_2, y] = [x_1, y] + [x_2, y]$
2. $[x, x] \geq 0$ and $[x, x] = 0$ if and only if $x = 0$
3. $|[x, y]| \leq \|x\| \cdot \|y\|$, where for all $z \in L$, $[z, z] = \|z\|^2$
4. $[ax, by] = ab^*[x, y]$

for all $x, x_1, x_2, y \in L$ and $a,b \in K$.

A space with a semi-inner product is called a *semi-inner product space*.

The existence on any Banach space of a semi-inner product was proved in Chapter 2.

The following theorem gives some properties connected with the structure of normed spaces.

6.9.2 THEOREM If L is a semi-inner product space with the semi-inner product [,], then

$$x \mapsto [x, x]^{1/2}$$

is a norm on L.

Proof. All the properties of a norm are obvious except the triangle inequality

$$\|x + y\| \le \|x\| + \|y\|$$

For this we have that

$$\begin{aligned}[x + y, x + y] &= [x, x + y] + [y, x + y] \\ &\le [x, x]^{1/2}[x + y, x + y]^{1/2} \\ &\quad + [y, y]^{1/2}[x + y, x + y]^{1/2} \\ &= [x + y, x + y]^{1/2}([x, x]^{1/2} + [y, y]^{1/2})\end{aligned}$$

and this clearly implies our assertion.

The connection between semi-inner product spaces and Hilbert spaces is given in the next theorem.

6.9.3 THEOREM Any Hilbert space is a semi-inner product space, and a semi-inner product space is a Hilbert space if and only if it is complete and $x \mapsto [x, x]^{1/2}$ satisfies the parallelogram law.

Proof. The first assertion is obvious since in this case we can take

$$[x, y] = \langle x, y \rangle$$

and the second assertion follows from Theorem 3.1.8.

However, in the theory of operators the following slight modification of Lumer's definition is sometimes used.

6.9.4 DEFINITION For any $x \in L$, L a normed space, we define

$$C(x) = \{x^* \in L^* \mid \|x^*\| = \|x\|,\ x^*(x) = \|x\|^2\}$$

From the extension theorems it follows that for any $x \in L$ the set $C(x)$ is nonempty. The following property is very useful and the proof is obvious.

6.9.5 PROPOSITION For any $x \in L$, $C(x)$ is a convex and compact set in L^* (the compactness is in the weak topology).

6.9.6 REMARK There exist classes of Banach spaces for which for all x, $C(x)$ is singleton.

We give some examples of $C(x)$ for some Banach spaces.

6.9.7 EXAMPLE 1. Suppose M is a compact Hausdorff space and $L = C(M)$ is the Banach space of all complex-valued functions on M with the norm

$$f \mapsto \|f\| = \sup \{|f(s)| \mid s \in M\}$$

Let $M_f = \{m \mid |f(m)| = \|f\|\}$ and μ be a regular measure with support in M_f. Then

$$F(f) = \int \bar{f}\, d\mu$$

is in $C(f)$ and $C(f) = \{F$, with a positive Borel measure satisfying the above condition$\}$.

2. Let $L = L^p$, $1 < p < \infty$ and $f \in L^p$. In this case

$$C(f) = \{\bar{f}|f|^{p-2}\|f\|^{2-p}\} \subset L^q \ (1/p + 1/q) = 1\}$$

Another useful result is given in the following.

6.9.8 PROPOSITION If $L_1, L_2, \ldots, L_n$ are semi-inner product spaces, then $L = L_1 \times L_2 \times L_3 \times \cdots \times L_n$ is also a semi-inner product space.

We are now in a position to introduce the notion of numerical range for bounded linear operators on Banach spaces.

6.9.9 DEFINITION Let T be a bounded linear operator on a Banach space L. The *numerical range* $W(T)$ of T is the set

$$W(T) = \{x^*(Tx) \mid x^* \in C(x)\}$$

6.9.10 REMARK If T is not bounded, we can introduce the numerical range as the set

$$W(T) = \{x^*(Tx) \mid x \in C(x) \cap D_T\}$$

where D_T is the linear subspace on which T is defined.

As we know in the case of Hilbert space W(T) is a convex set for all $T \in L(H)$. In the case of Banach spaces this property fails to be true for all operators. Indeed, if we consider the Banach space $L = \ell_1^2$ of all elements of the form $x = (x_1, x_2)$, x_i complex numbers, with the norm

$$\|x\| = |x_1| + |x_2|$$

and the operator with the matrix representation

$$\begin{bmatrix} 2 & 1 \\ 1 & -2 \end{bmatrix}$$

then the set W(T) contains the points

$$[i, \quad i]\begin{bmatrix} 2 & 1 \\ 1 & -2 \end{bmatrix}\begin{bmatrix} 1 \\ 0 \end{bmatrix} = 2 + i \qquad [i, \quad i]\begin{bmatrix} 2 & 1 \\ 1 & -2 \end{bmatrix}\begin{bmatrix} 0 \\ 1 \end{bmatrix} = -2 + i$$

and $i \notin W(T)$. Since $i = (1/2)(2 + i + i - 2)$ the assertion is proved.

Also as we know in the case of Hilbert space operators, $\sigma(T) \subset \overline{W(T)}$ and the proof of this fact uses in an essential way the convexity property of W(T). We present now a proof of this fact for the Banach space operator based on another idea.

6.9.11 THEOREM If T is a bounded linear operator on a Banach space, then $\sigma(T) \subseteq \overline{W(T)}$.

The proof is very simple; first we remark that the following lemma holds.

6.9.12 LEMMA If T and S are bounded linear operators on a Banach space and $0 \notin \sigma(S) \cap W(S)$, then

$$\sigma(S^{-1}T) \subset \frac{W(T)}{W(S)}$$

Proof. See Theorem 8.1.2 for the case of Hilbert space operators, and the proof can be carried to this case word by word. Next we take $S = I$.

The following theorem gives some useful facts about the parts of W(T) and the boundary points of $\sigma(T)$.

6.9.13 THEOREM If L is a uniformly Banach space and $T \in L(L)$, then

$$\{s \mid |s| = \|T\|,\ s \in \overline{W(T)}\} \subset \sigma(T)$$

Proof. We can suppose, without loss of generality, that s = 1 and thus there exists a sequence of points $\{x_n\}$, $\|x_n\| = 1$, such that

$$x_n^*(Tx_n) - 1 \to 0$$

Since

$$x_n^* \frac{x_n + Tx_n}{2} \to 1$$

we obtain that

$$1 \leq \lim x^* \frac{x_n + Tx_n}{2} \leq \frac{\|x_n\| + \|Tx_n\|}{2} \leq 1$$

and thus $\|x_n + Tx_n\| \to 2$. Since L is uniformly convex, we obtain $\|x_n - Tx_n\| \to 0$ and the assertion is proved.

We give now an interesting application of the notion of semi-inner product to a class of operators on Banach spaces.

6.9.14 DEFINITION An operator $T \in L(L)$, L a Banach space, is called an *isometry* if

$$\|Tx\| = \|x\|$$

for all $x \in L$.

The following characterization of isometric operators in terms of semi-inner products is given by Koehler and Rosenthal (1970).

6.9.15 THEOREM A linear operator T on a Banach space is an isometry if and only if there exists a semi-inner product $\langle\ ,\ \rangle^{\sim}$ on L such that

$$\langle Tx, Ty\rangle^{\sim} = \langle x, y\rangle^{\sim}$$

for all $x, y \in L$.

Proof. It is obvious that the condition is sufficient. We prove now that this condition is also necessary. Indeed, if we define

$$\langle x, y\rangle^{\sim} = \text{glim}\ \{\langle T^n x, T^n y\rangle\}$$

where $\langle\ ,\ \rangle$ is a semi-inner product on L with the property $\|x\| = \langle x, x\rangle^{1/2}$. From the properties of glim it follows that this is indeed a semi-inner product on L having the property

$$\|x\| = \langle x, x\rangle^{\sim 1/2}$$

for all $x \in L$. From the translation invariance property of glim

$$\langle Tx, Ty\rangle^{\sim} = \text{glim}\ \{\langle T^{n+1} x, T^{n+1} y\rangle\}$$

$$= \text{glim}\ \{\langle T^n x, T^n y\rangle\} = \langle x, y\rangle^{\sim}$$

and the assertion is proved.

From this theorem we obtain easily the following result.

6.9.16 THEOREM An operator $T \in L(L)$, L a Banach space, is similar to an isometry if and only if there exists $d > 0$ and $M < \infty$ such that

$$\|T^n x\| \geq d\|x\| \qquad \|T^n\| \leq M$$

Proof. It is obvious that the conditions are necessary. We consider now a new norm on L defined by

$$\|x\|^* = \text{glim}\ \{\|T^n x\|\}$$

which is equivalent to the norm $\|\cdot\|$ and from the translation invariance of glim it follows that T is with with norm an isometry.

Suppose now that T is an isometric operator on a Hilbert space and for $s \in \sigma(T)$, $|s| = 1$, there exists an element $x \neq 0$ such that $Tx = sx$. Thus the space

$$E_s(T) = \{x \mid Tx = sx\}$$

is not equal to zero. If we have two such subspaces with t, $s \in \sigma(T)$ and $|s| = |t| = 1$, then it is easy to see that these subspaces are orthogonal. We prove now that this property extends to the case of Banach spaces with an appropriate definition of orthogonality. The definition which we use is that of Birkhoff (1935).

6.9.17 DEFINITION If L is a Banach space and x and y two elements of L, then we say that x is orthogonal to y if

$$\|x + zy\| \geq \|x\|$$

for all scalars z. We write this simply as $x \perp y$.

6.9.18 REMARK It is easy to see that this is equivalent with the notion of orthogonality in the case of Hilbert spaces.

6.9.19 THEOREM If T is an isometric operator, then

$$\{E_s(T)\}_{s \in (T)}$$

is an orthogonal family of spaces when $|s| = 1$.

Proof. Let $x \in E_s(T)$ and $y \in E_t(T)$, $|s| = |t| = 1$, and consider the function on $L \times L$ defined by

$$\langle x, y\rangle^{\sim} = \text{glim}\ \{\langle T^n x, T^n y\rangle\}$$

which gives that

$$\langle x, y\rangle^{\sim} = \langle Tx, Ty\rangle^{\sim} = st^{*}\langle x, y\rangle^{\sim}$$

and thus it is zero. This implies the relation

$$\|x + zy\| \geq \|x\|$$

and the theorem is proved.

The result can be extended to a larger class of operators on Banach spaces, the so-called unimodular contractions.

6.9.20 DEFINITION An operator $T \in L(L)$ is called a *unimodular contraction* if $\sigma(T) \subset \{z \mid |z| = 1\}$.

We can prove the following.

6.9.21 THEOREM If T is any unimodular contraction, then the family of subspaces $\{E_s(T)\}_{s \in \sigma(T)}$, $|s| = 1$, is a family of orthogonal subspaces.

Proof. We define, as above

$$\langle x, y\rangle^{\sim} = \text{glim}\ \{\langle T^n x, T^n y\rangle\}$$

which has all the above properties of a semi-inner product except $\langle x, x\rangle^{\sim} = \|x\|^2$ for all $x \in L$. However, this relation holds for elements in $E_r(T)$ and this implies the assertion.

The following theorem gives an interesting property of isometric operators in connection with nonnormal classes of operators.

6.9.22 THEOREM If T is an isometric operator, then

$$\|Tx\|^2 \le \|T^2 x\|$$

for all x, $\|x\| = 1$.

6.10 THE EXPONENTIAL FUNCTION ON THE SET OF ALL BOUNDED LINEAR OPERATORS ON A BANACH SPACE

Let L be a Banach space and L(L) be the set of all bounded linear operators on L. The following function, called the *exponential function*, is very useful in the theory of operators.

In what follows we give some results about this function as well as some applications to operators.

6.10.1 DEFINITION For any $T \in L(L)$ the operator $\exp T = e^T$ is the operator defined by the formula

$$I + \sum_{n=1}^{\infty} \frac{1}{n!} T^n$$

It is easy to see that $e^T \in L(L)$. The following theorem gives some useful properties of this function.

6.10.2 THEOREM The function $T \mapsto e^T$ has the properties:

1. e^T is an invertible operator.
2. $e^T e^{-T} = I$.
3. If $TS = ST$ [i.e., for all $x \in L$, $T(Sx) = S(Tx)$], then $e^{T+S} = e^T e^S$.

Proof. First we prove assertion 3. Let

$$a_n = I + \sum_{i=1}^{n} \frac{1}{i!} T^i$$

$$b_n = I + \sum_{i=1}^{n} \frac{1}{i!} S^i$$

$$c_n = I + \sum_{i=1}^{n} \frac{1}{i!} (T + S)^i$$

and it is obvious that

$$a_n \to e^T \qquad b_n \to e^S \qquad c_n \to e^{T+S}$$

Further, we consider

$$\tilde{a}_n = 1 + \sum_{i=1}^{n} \frac{1}{i!} \|T\|^i$$

$$\tilde{b}_n = 1 + \sum_{i=1}^{n} \frac{1}{i!} \|S\|^i$$

$$\tilde{c}_n = 1 + \sum_{i=1}^{n} \frac{1}{i!} \|(T + S)\|^i$$

and we remark that

$$a_n b_n - c_n = \sum_{i=1}^{n} a_{ik} T^i S^k \qquad a_{ik} > 0$$

which gives us

$$\|a_n b_n - c_n\| \le \sum_{i=1}^{n} a_{ik} \|T\|^i \|S\|^k = \tilde{a}_n \tilde{b}_n - \tilde{c}_n$$

and since

$$\tilde{a}_n \tilde{b}_n - \tilde{c}_n \to e^{\|T\|+\|S\|} - e^{\|T\|+\|S\|} = 0$$

and 3 is proved. Assertions 1 and 2 follow in an obvious way from this.

The following result gives, in some sense, a justification of the name "exponential function."

6.10.3 THEOREM For any $T \in L(L)$,

$$e^T = \lim_{n\to\infty} \left(I + \frac{1}{n} T\right)^n$$

Proof. Let

$$a_n = I + \sum_{i=1}^{n} \frac{1}{i!} T^i \qquad x_n = \left(I + \frac{1}{n} T\right)^n$$

$$\tilde{a}_n = 1 + \sum_{i=1}^{n} \frac{1}{i!} \|T\|^i \qquad \tilde{x}_n = \left(1 + \frac{1}{n} \|T\|\right)^n$$

and remark that the following relations hold:

$$x_n = I + T + \left(1 - \frac{1}{n}\right)\frac{1}{2!} T^2 + \cdots + \left(1 - \frac{1}{n}\right) + \cdots$$
$$+ \left(1 - \frac{n-1}{n}\right)\frac{1}{n!} T^n$$

$$a_n - x_n = \sum_{i=1}^{n} a_i T^i \qquad a_i \geq 0$$

and from these we obtain that

$$\|a_n - x_n\| \leq \sum_{i=1}^{n} a_i \|T\|^i = \tilde{a}_n - \tilde{x}_n$$

Since $\lim (\tilde{a}_n - \tilde{x}_n) = 0$ and $e^{\|T\|} = \lim (1 + (1/n)\|T\|)^n$, the assertion of the theorem follows.

The exponential function plays a fundamental role in operator theory; we use it to define several classes of operators on Banach spaces, generalizing to the Banach space setting some results in the Hilbert space theory.

6.11 THE NUMERICAL RADIUS, THE SPECTRAL RADIUS, AND THE NORM OF A BOUNDED LINEAR OPERATOR ON A BANACH SPACE

The following theorem gives an important formula for the spectral radius of an operator $T \in L(L)$, L a Banach space.

6.11.1 THEOREM For any $T \in L(L)$,

$$r_T = \max_\theta \left\{ \inf_{t>0} \frac{\log\|e^{i\theta}tT\|}{t} \right\} = \max \{\log\|e^{i\theta}T\|\}$$

This result is an easy consequence of the following result which also has its own interest.

6.11.2 LEMMA For any $T \in L(L)$,

$$\inf \frac{\log\|e^{itT}\|}{t} = \lim \frac{\|e^{-itT}\|}{t} = \sup \{\mathrm{Im}\, s \mid s \in \sigma(T)\}$$

Proof. The function

$$g(t) = \log \|e^{-itT}\|$$

is subadditive, and by a well-known result we have

$$\lim \frac{g(t)}{t} = \lim \frac{g(n)}{n} = \log \lim \|e^{-inT}\|^{1/n}$$

$$= \log \{\max\{|s| \mid s \in (\sigma e^{-iT})\}$$

$$= \log \{\max |e^{-is}| \mid s \in \sigma(T)\}$$

$$= \max \{\mathrm{Im}\, s \mid s \in \sigma(T)\}$$

and the lemma is proved.

The following theorem gives an important property of the numerical radius of an element $T \in L(L)$.

6.11.3 THEOREM The function

$$T \mapsto \|T\|_W = w(T) = \sup \{|s| \mid s \in W(T)\}$$

is a norm on $L(L)$ equivalent to the norm $\|\cdot\|$.

The number $\|T\|_W = w(T)$ is called the *numerical radius* of T.

For the proof we need some formulas which also have intrinsic interest.

6.11.4 LEMMA If $T \in L(L)$, then

$$\sup \frac{\log \|e^{-itT}\|}{t} = \lim \frac{\log \|e^{-itT}\|}{t} = \lim \frac{\|e^{-itT}\| - 1}{t}$$

$$= \lim \frac{\|I - itT\| - 1}{t} = \sup \{\mathrm{Im}\ s \mid s \in W(T)\}$$

Proof. First we remark that the first equation is a consequence of the fact that $g(t) = \log \|e^{-itT}\|$ is subadditive and bounded above. The next two equalities follow by expanding the numerator and discarding terms which vanish in the limit.

The last equation is obtained as follows: Let $k = \sup \{\mathrm{Im}\ s \mid s \in W(T)\}$ and we use the symbol $0(t)$ to denote that for any function f we have $f(t)/t \to 0$ for $t \to 0$.

Now if $x \in L$ and $x^* \in C(x)$ and also $t \in R$, we have

$$\|I \pm itT\| \geq \|(I \pm itT)x\| \geq \|x^*(I \pm itT)x\| = |1 \pm it(x^*(Tx)|$$

$$= (1 + 2t\ \mathrm{IM}(x^*(Tx) + t^2|x^*(Tx)|^{2\ 1/2}$$

$$= 1 + t\ \mathrm{Im}(x^*(Tx) + 0(t)$$

and this gives us

$$\lim \frac{\|I - itT\| - 1}{t} \geq \sup \left\{\lim \frac{1 + t\ \mathrm{Im}\ s + 0(t) - 1}{t}\right\} = k$$

and using the same inequalities in a different way, we obtain

$$\|e^{itT}x\| = \|(I + itT)x\| + 0(t) \geq 1 - tk + 0(t)$$

Thus, if t is sufficiently small, $t > 0$,

$$\|e^{-itT}\| = \sup \{\|e^{-itT}\tilde{x}\| \mid \|\tilde{x}\| = 1\} = \sup \left\{\frac{\|x\|}{\|e^{itT}x\|} \mid \|x\| = 1\right\}$$

$$= \frac{1}{(1 - tk + 0(t))} = 1 + tk + 0(t)$$

and thus,

$$\lim \frac{\|1 - itT\| - 1}{t} = \lim \frac{\|e^{-itT}\| - 1}{t} \leq k$$

which proves the equality in the last equation.

Proof of Theorem 6.11.3. We remark that from the above results

$$r_T \le \|T\|_w = w(T)$$

and since $w(T) \le \|T\|$ is trivial, it remains to prove that for a constant $M > 0$, $\|T\| \le Mw(T)$. We prove now that M can be chosen.

Indeed, we can apply the Cauchy formula for the derivative of e^{zT} and by integrating on the circle of radius $1/\|w(T)\|$ with the center in the origin, we obtain

$$\|T\| = \left\| \frac{-1}{2\pi i} \int \frac{e^{zT}}{z^2}\, dz \right\| \le e\|T\|_w = w(T)$$

and the assertion is proved.

6.11.5 REMARK It appears of interest to know whether the constant e in the above formula is "the best possible." An example of Glickfelt (1970b) shows that this is the case; also there exists an example of Crabb (1970a, 1971a, 1971b), of an operator $T \in L(L)$ such that for all n,

$$\|T^n\| = n!\left(\frac{e}{n}\right)^n w(T)$$

6.12 HERMITIAN AND NORMAL OPERATORS ON BANACH SPACES

The first successful attempt to extend the notion of a hermitian operator to Banach spaces was made by Vidav (1956). His definition is inspired by the following property of operators on Hilbert spaces: the operator $T \in L(H)$ is hermitian if and only if $\|I + itT\| = 1 + O(t)$ for $t \in R$, $t \to 0$.

Since in the above property the Hilbert space structure plays no role, this allows us to define the hermitian operators using this property as a definition. Later, Lumer observed that using the notion of semi-inner product, an equivalent definition can be given and this new definition represents the generalization of the following characterization of hermitian operators: $T \in L(H)$ is hermitian if and only if the numerical range is a subset of the reals. (See Chapter 8 for a discussion of this topic.)

In what follows we present results concerning the extensions of the notions of hermitian and normal operators to the case of Banach spaces.

6.12.1 DEFINITION An element $T \in L(L)$, L a Banach space, is called *hermitian* if for $x \in L$ and $x^* \in C(x)$, $x^*(Tx) \in R$.

The following theorem gives a characterization of hermitian operators on Banach spaces.

6.12.2 THEOREM For any $T \in L(L)$ the following assertions are equivalent:

1. T is hermitian.
2. For all $t \in R$, $\|e^{itT}\| = 1$.
3. $\|I + itT\| = 1 + 0(t)$ for $t \in R$ and $t \to 0$.

Proof. First suppose that T is hermitian; in this case the numerical range of iT is a subset of iR and thus

$$\max \operatorname{Re} W(iT) = \max \operatorname{Re} (-iT) = 0$$

and property 2 follows from Lemma 6.11.4.

Suppose now that 2 holds. In this case for any $x \in L$, we have

$$\|x\| \geq \|e^{itT}x\| = \|(I + itT)x\| + 0(t) \geq \|x\| - \|t\|\|x\|$$

and this implies that

$$\sup \|(I + itT)x\| - 1 = 0(t)$$

and 3 is proved. Suppose now that 3 holds. Thus

$$\|e^{itT}x\| = \|(I + itT)x\| + 0(t) = \|x\| + 0(t)$$

and thus for t sufficiently small and $x = 1$, we have

$$\sup \{\operatorname{Im} s \mid s \in W(T)\} = \lim \frac{\|e^{-itT}\| - 1}{t} = \lim \frac{\|I - itT\| - 1}{t} = 0$$

and thus T is hermitian.

We give now some properties of hermitian operators.

6.12.3 THEOREM Let L be a Banach space and H(L) be the set of all hermitian operators on L. The set H(L) is a closed linear real space.

Proof. The fact that H(L) is a real space follows from Theorem 6.12.1 and the closedness from Theorem 6.12.2 (also this follows from Definition 6.12.1).

6.12.4 THEOREM If $T,S \in H(L)$, then $i(TS - ST) \in H(L)$.

Proof. Since for all $t \in R$ we have

$$\|e^{itT}e^{itS}e^{-itT}e^{-itS}\| = 1$$

we obtain

$$\|I + t^2(TS - ST)\| = 1 + 0(t^3)$$

as well as

$$\|(I + t^2(TS - ST)\| = 1 + 0(t)$$

and we obtain the relation

$$\|I + it(TS - ST)\| = 1 + 0(t)$$

Thus

$$\lim \frac{\|I + it(TS - ST)\| - 1}{t} = 0$$

which proves that $i(TS - ST)$ is in H(L).

We consider now an important class of operators in H(L).

6.12.5 DEFINITION An operator $T \in H(L)$ is called *positive* if the numerical range W(T) is in R^+.

6.12.6 THEOREM The set of all positive elements in H(L) is a cone.

As is well known, in the case of Hilbert spaces every operator has a unique decomposition in the form $T_1 + iT_2$, where T_i, $i = 1, 2$, are hermitian operators. We wish to consider this set in the case of Banach spaces.

Let L be a Banach space and we denote by J(L) the set of all elements in L(L) such that there exists a decomposition of the for $T_1 + iT_2$, where T_i, i = 1, 2, are hermitian operators.

6.12.7 THEOREM If $T \in J(L)$, then it has a unique decomposition of the form $T = T_1 + iT_2$.

Proof. Clearly it is sufficient to show that if $T_1 + iT_2 = 0$, then necessarily $T_1 = T_2 = 0$. But it is clear that

$$W(T_1) = W(iT_2) = 0$$

and thus the numerical radius for these operators is zero and the assertion follows.

6.12.8 DEFINITION Every element $T \in J(L)$ is called decomposable.

In the case of Hilbert spaces, as we know, J(L) = L(L). An example of Palmer (1965) shows that this is not the case in general. Let $L = \ell_1^2$ and in this case a simple computation shows that

$$J(L) = \left\{ \begin{bmatrix} a & 0 \\ 0 & b \end{bmatrix} \;\middle|\; a,b \text{ real numbers} \right\}$$

and it is obvious that this is not L(L).

The following theorem gives some properties of the operators in J(L).

6.12.9 THEOREM If $T \in J(L)$ and $T = h + ik$, then

1. $\max\{w(h), w(k)\} \le w(T) \le [w(h)^2 + w(k)^2]^{1/2}\}$.
2. T*, the adjoint of T, is a decomposable operator.
3. J(L) is a closed subspace of L(L).

Proof. Let $\{x_n\}$, $\|x_n\| = 1$, and $x_n^* \in C(x_n)$ such that

$$\lim |x_n^*(hx_n)| = w(h)$$

and in this case we have

$$|x_n^*(hx_n)| = |\text{Re } x_n^*(Tx_n)| \le |x_n^*(Tx_n)| \le w(T)$$

and thus $w(h) \le w(T)$. Similarly, $w(k) \le w(T)$ and the first part of the inequality is proved.

Let $\{y_n\}$, $\|y_n\| = 1$, and $y_n^* \in C(y_n)$ such that

$$|y_n^*(Ty_n)| = |y_n^*(hx_n) + iy_n^*(kx_n)| \le [y_n^*(h(x_n))^2 + y_n^*(k(x_n))^2]$$

This immediately gives the relation. Thus 1 is proved.

For 2 we remark that this follows from the properties of the adjoint.

Property 3 follows from 1 since $w(\cdot)$ is a norm equivalent to the norm $\|\cdot\|$.

6.12.10 COROLLARY The set J(L) is a Banach space having the conjugation defined by the relation: If $T = h + ik$, then $T^* = h - ik$, which is an isometric operator for the norm $w(T)$.

We consider now an interesting class of operators in J(L) which represent in some sense the generalization of normal operators on Hilbert spaces.

6.12.11 DEFINITION An element $T \in J(L)$ is called *regular* if $T = h + ik$, where $hk = kh$.

The following theorem gives an important property of this class of operators.

6.12.12 THEOREM If $T \in J(L)$, then $w(T) = r_T$ if T is regular.

Proof. Let $T = h + ik$ with $hk = kh$, and let $z = re^{i\theta}$. We compute the norm of the operator e^{zT}. Since h and k are commuting hermitian operators, we have

$$\|e^{zT}\| = \|e^{sk-th}\|$$

where $z = s + it$. The assertion of the theorem follows from the following very important property of hermitian operators.

6.12.13 THEOREM If h is a hermitian element in L(L), then

$$r_h = w(h) \le \|h\| \le er_h$$

This important theorem was obtained by Vidav (1956). An interesting generalization was obtained by Browder (1971) and Sinclair.

See Bonsall and Duncan (1971).

We give further this generalization with the proof of Bonsall and Crabb (1970).

Proof of Theorem 6.12.13. Let [a,b] be the smallest interval containing the spectrum of h. Let $\varepsilon > 0$ and $x \in \ell$ with $\|x\| = 1$ and $x^* \in L^*$ with $\|x^*\| = 1$.

$$f(z) = (z + 1)^{-\varepsilon} e^{(b+\varepsilon)} \sum \frac{z^n x^*(h^n(x)}{n!}$$

is analytic in Re $z > -1$. We show now that $|f(z)| \leq 1$ for z, Re $z > 0$.

Let $z = s + it$ and thus we have

$$\left|\sum z^n \frac{x^*(h^n(x))}{n!}\right| = |x^*(e^{zh})(x)| \leq \|e^{sh}e^{ith}\| \leq \|e^{sh}\|$$

Thus $|f(z)| \leq 1$ for z, Re $z = 0$. From the definition and the formula for the spectral radius we have

$$r_e h = \lim \|e^{sh}\|^{1/s} = s^b$$

and thus for z with Re z sufficiently large $|f(z)| \leq 1$. But f is a continuous function for all z, Re $z > 0$, and thus from the maximum principle it follows that $|f(z)| \leq 1$ for all z, Re $z > 0$.

Now since $\varepsilon > 0$ is arbitrary, we obtain that

$$|x^*(e^{zh}(x))| \leq |e^{zb}|$$

for Re $z > 0$. Now we take the hermitian operator -h and we obtain

$$|x^*(e^{zh}(x))| \leq |e^{za}|$$

for z, Re $z < 0$. But

$$r_h = \max \{-a, b\}$$

and thus

$$\|e^{zh}\| \leq |e^z r_h|$$

for all $z \in C$. From the formula for the numerical radius we get

$$w(T) \le r_h$$

and clearly this implies the equality since $r_h \le w(h)$ is true for arbitrary operators. The rest of the inequality follows from Theorem 6.11.3.

The proof of Theorem 6.12.13 is as follows: As above we obtain that

$$\log \frac{\|e^{zT}\|}{r} \le r_T$$

and this is independent of r and thus it is equal to r_T. This yields the assertion of the theorem.

6.13 NORMAL OPERATORS ON BANACH SPACES

We need the following notions concerning a class of subalgebras in L(L).

6.13.1 DEFINITION Let A be a subalgebra in L(L). A is called a *V*-algebra* if the following conditions are satisfied:

1. A is a commutative algebra.
2. A is closed and every element $a \in A$ has the decomposition

   $$a = h + ik$$

   with h,k hermitian operators.

The mapping

$$a \mapsto a^* = h - ik$$

will be called a *conjugation*.

The following theorems are consequences of results about *-algebras (see Chapter 4).

6.13.2 THEOREM If A is a V*-algebra, then it is isometric and *-isomorphic with an algebra of continuous functions on a compact space.

6.13.3 THEOREM If U is a commutative algebra and every $T \in A$ has a decomposition $T = h + ik$ with h,k hermitian operators, then the closure of the set $zI + T$ in the weak topology of $L(L)$ is a V^*-algebra.

We mention that these algebras were studied first by I. Vidav.

The notion of a normal operator can be introduced as follows.

6.13.4 DEFINITION An element $T \in L(L)$ is called *normal* if T is of the form $T = h + ik$ with h and k hermitian operators in a V^*-algebra.

The following theorem was proved for the case of Hilbert space operators by Fuglede (1950).

The proof we give in what follows is that of Rosenblum (1958).

6.13.5 THEOREM If N is a normal operator on a Banach space and B is an arbitrary operator in $L(L)$ with the property $BN = NB$, then $B\tilde{N} = \tilde{N}B$, where $\tilde{N} = h - ik$ if $N = h + ik$.

Proof. From the hypothesis it follows that for all m,

$$N^m B = BN^m$$

and thus for any complex number z,

$$e^{z\tilde{N}} B e^{-z\tilde{N}} = e^{z\tilde{N}} e^{-z^*N} B e^{z^*N} e^{-z\tilde{N}}$$

and if $z = x + iy$, we get

$$e^{z\tilde{N}} B e^{-z\tilde{N}} = e^{2i(yh-xk)} B e^{-2i(yh-xk)}$$

and this clearly implies that the function

$$f(z) = \|e^{z\tilde{N}} B e^{-z\tilde{N}}\|$$

is constant and by Liouville's theorem

$$e^{z\tilde{N}} B e^{z\tilde{N}}$$

is constant. Thus we obtain the assertion by derivation.

The following extension was obtained by C. Putnam. The ingenious proof below is that of Berberian (1959).

6.13.6 THEOREM If N_1 and N_2 are two normal operators and B is arbitrary in $L(L)$ such that $BN_1 = N_2B$, then $B\tilde{N}_1 = \tilde{N}_2B$.

Proof. We consider the Banach space $L \times L$ and the operators

$$N = \begin{bmatrix} N_1 & 0 \\ 0 & N_2 \end{bmatrix} \qquad \tilde{TN} = \begin{bmatrix} 0 & 0 \\ T & 0 \end{bmatrix}$$

and it is easy to see that N is normal and TN = NT. Thus $\tilde{TN} = \tilde{NT}$ and this gives our assertion.

6.14 CLASSES OF ELEMENTS IN BANACH ALGEBRAS WITH UNIT: THE VIDAV-PALMER THEOREM

In what follows we give some results about hermitian, unitary, and normal elements in Banach algebras. Naturally, all results which can be proved, as in the case of the Banach algebra of all bounded operators on a Banach space, will only be mentioned.

We give an example of a hermitian element in a Banach algebra such that the h^2 is not hermitian. The first example was given by Lumer (1964) and we give the example of Crabb (1970a).

First we mention the definition of hermitian elements in a Banach algebra.

16.14.1 DEFINITION Let A be a Banach algebra with unit. The element h is called *hermitian* if for all $t \in R$, $\|e^{ith}\| = 1$.

The following theorem gives the characterization of hermitian elements.

6.14.2 THEOREM The element $h \in A$ is hermitian if and only if the following equivalent conditions hold:

1. $W(h) \subset R$
2. $\|(1 + ith)\| = 1 + 0(t)$ for t small

The proof of this theorem as well as for others are exactly as in the case A = L(L).

We now give Crabb's example.

6.14.3 EXAMPLE Let $L = C^3$ and let the element h be defined as $h(a, b, c) = (-a, 0, c)$. On C^3 we consider the following norm:

$$\|(a, b, c)\| = \sup_{|z|=1} |z^{-1}a + b + zc|$$

The element $h \in L(C^3)$ is hermitian since

$$e^{ith}(a, b, c) = (e^{-it}a, b, e^{it}c)$$

and thus for all $t \in R$, $\|e^{ith}\| = 1$.

Now $h^2(a, b, c) = (a, 0, c)$ and $e^{(i/2)h^2}(a, b, c) = (ia, b, ic)$ and the norm is equal to 3 for $(a, b, c) = (1, i, 1)$, which has the norm $5^{1/2}$.

This gives that h^2 is not hermitian.

An important problem in the theory of Banach algebras is to obtain characterizations of the class of B*-algebras using hermitian elements. If a is a Banach algebra, we denote by $H(A)$ the closed space of all hermitian elements of A. The following fundamental result was obtained by Vidav (1956) and Palmer (1968b, 1970).

6.14.4 THEOREM If A is a Banach algebra with unit and $A = H(A) + iH(A)$, then A is isomorphic and isometric with a C*-algebra.

The proof of this theorem will be given as a consequence of some results which are of interest in themselves.

The first result we give was obtained by Russo and Dye (1966) and extended by Palmer.

6.14.5 THEOREM If A is a B*-algebra and

$U = \{u \in A \mid u^*u = uu^* = 1\}$

$E = \{e^{ih} \mid h \in A, h^* = h\}$

$S_1 = \{x \in A \mid \|x\| < 1\}$

then the following assertions hold:

1. Russo-Dye) $S_1 = \overline{\text{conv } U}$
2. (Palmer) $S_1 = \overline{\text{conv } E}$

Proof. Let $x \in A$ and $\|x\| < 1$. As we know, we have the following relation concerning the spectral radius:

$$r_x = r_{x^*} \leq r_{x^*x}^{1/2}$$

and thus we can consider the function

$$F(z) = (1 - xx^*)^{1/2}(z + x)(1 + zx^*)(1 - x^*x)^{1/2}$$

for $|z| < \|x\|^{-1}$. This function is called the *Möbius function* of A. Interesting results and applications to Banach algebras were given by Harris (1972).

From the form of the function F it is clear that this is an analytic function, especially for $|z| < 1$.

We show that the values of F on the boundary points are unitary elements. Indeed, first we remark,

$$(1 + zx^*)^{1}(z + x) = x + z(1 + zx^*)^{-1}(1 - x^*x)$$

$$(z + x)(1 + zx^*)^{-1} = x + z(1 - xx^*)(1 + zx^*)^{-1}$$

$$(1 - xx^*)^{1/2}x = x(1 - x^*x)^{1/2}$$

and we prove that for z, $|z| = 1$, F(z) is a unitary element.

Indeed, we have

$$\begin{aligned}(F(z)^{-1})^* &= (1 - xx^*)^{1/2}(z^* + x^*)^{-1}(1 + z^*x)(1 - x^*x)^{-1/2} \\ &= (1 - xx^*)^{1/2}(1 + zx^*)^{-1}(z + x)(1 - x^*x)^{-1/2} \\ &= (1 - xx^*)^{-1/2}[x + z(1 - xx^*)(1 + zx^*)^{-1}](1 - x^*x) \\ &= F(z)\end{aligned}$$

and the assertion about the values of F on the boundary of the unit circle is proved. We prove now the result of Dye-Russo.

Since F is an analytic function, by the Cauchy formula we have

$$F(0) = \frac{1}{2\pi}\int_{-\pi}^{\pi} F(e^{is})\, ds$$

and since $F(0) = x$, the assertion follows from the result about the values of F for $|z| = 1$ and the definition of the integral.

Now we prove the result of Palmer. For this we show that any unitary element is in $\overline{\text{conv } E}$. Let us be a unitary element and $t \in (0, 1)$. In this case tu is a normal element and also the following relation holds:

$$(z + F(z))(1 + zx^*) = 2z(1 + \frac{1}{2}(z^*x + zx^*)) \qquad |z| = 1$$

and where $x = tu$. But clearly, $\|x\| = t$ and thus $z + F(z)$ is invertible. This implies that $-z \notin \sigma(F(z))$ and thus $\sigma(F(z))$ is a proper subset in $\{z, |z| = 1$, and we have the possibility to define $h = -i \log F(z)$ and this element has the property to be hermitian since $F(z)$ are unitary elements. Since $F(z) = e^{ih}$ for z, $|z| = 1$, we obtain that x is in $\overline{\text{conv } E}$. This implies that u is in $\overline{\text{conv } E}$.

The following refinement was obtained also by Palmer.

6.14.6 THEOREM For E as in Theorem 6.14.5,

$$\{x, \|x\| < 1\} \subset \text{conv } E$$

Proof. Let h be a hermitian element in A, $\|h\| < 1$. We can decompose h as follows

$$h = \frac{1}{2} (u + v)$$

with

$$u = h + i(1 - h^2)^{1/2} \qquad v = h - i(1 - h^2)^{1/2}$$

It is easy to see that u and v are unitary elements and $(u - 1)^{-1}$ and $(v - 1)^{-1}$ exist. Suppose now that $x \in A$ and $\|x\| < 1/2$. In this case, if $x = h + ik$, where h and k are hermitian elements, we have that $\|h\| < 1/2$ and $\|k\| < 1/2$. From the above remark it follows that $x \in \text{conv } E$. Let $x \in A$, $\|x\| < 1$, and put $t = \|x\|^{-1}$. In this case $tx \in \overline{\text{conv } E}$, and we find $x^{\sim}$ such that

$$\|tx - x^{\sim}\| < \frac{1}{2}(t - 1)$$

Since $tx - x^{\sim} = (t - 1)y$ for some $y \in \text{conv } E$, it follows that $x \in \text{conv } E$ since

$$x = t^{-1}x^{\sim} + (1 - t^{-1})y$$

Interesting applications of the maximum theorems to Banach algebras are given in Harris (1972); also see Chapter 14.

Suppose now that A is a Banach algebra with unit and let J(A) be the set of all elements of the form h + ik, where h and k are hermitian elements. The fundamental result obtained by Vidav asserts that if J(A) = A and $h \in H(A)$, h^2 a normal element, then there exists an isomorphism

$$T: A \to A^{\sim}$$

where $A^{\sim}$ is a B*-algebra such that Th is hermitian if h is hermitian and T is an isometry.

The fact that T is an isometric isomorphism was proved also by Berkson and Glickfeld. Later Palmer proved that it is sufficient to suppose that J(A) = A for the validity of the theorem.

6.14.7 THEOREM Let A be a Banach algebra with unit. Then the following assertions are equivalent:

1. J(A) is an algebra.
2. If h is a hermitian element, then $h^2 \in J(A)$.
3. If h is hermitic, then h^2 is hermitian.
4. If h and k are hermitian elements, then hk + kh is hermitian.

Proof. That assertion 1 implies 2 is trivial. We prove now that 2 implies 3. Let h be a hermitian element and $h^2 = p + iq$, where p and q are hermitian elements. In this case,

$$h(p + iq) = (p + iq)h$$

and thus hp - ph = i(qh - hq) are hermitian elements. This gives that hp = ph and thus $h^2 p = ph^2$. Thus we have qp = pq and, consequently, h^2 is a regular element. Since $W(h^2) = \text{conv}\ \sigma(h^2)$, the element h^2 is hermitian.

Now we show that 3 implies 4. Indeed, if h and k are hermitian elements, then h^2, k^2, and $(h + k)^2$ are hermitian elements. This obviously implies that hk + kh is hermitian.

To prove that assertion 4 implies 1, it is sufficient to show that J(A) is closed for multiplication.

Let a and b be elements in J(A); thus

$$a = h + ik \qquad b = p + iq$$

Since

$$ab + b^*a^* = (hp + ph) - (kq + qk) + i(kp + pk) + i(hq - qh)$$

we find that this is a hermitian element. Also,

$$i(ab - b^*a) = i(hp - ph) - i(kp - pk) - (kp + pk) - (hp + ph)$$

is a hermitian element. Since

$$ab = \frac{1}{2}(ab + b^*a^*) - \frac{i}{2}(ab - b^*a^*) \in J(A)$$

$$(ab)^* = \frac{1}{2}(ab + b^*a^*) + \frac{i}{2}(ab - b^*a^*) = b^*a^*$$

and thus $J(A)$ is an algebra with the involution

$$a \mapsto a^* = h - ik$$

which is also continuous.

As is well known, if $A = C(X)$, the algebra of all continuous complex-valued functions, then for any f positive there exists the square root which is also positive. Also if $A = L(H)$, the Banach algebra of all bounded operators on a Hilbert space H, then for any positive operator there exists the square root which is also a positive operator. We give now a general theorem about this problem.

For this we need a formula for the spectral radius. We prove the assertion for the case of Banach space operators; the extension to Banach algebra is translated using the mapping

$$T_a(x) = ax$$

for any $x \in A$.

6.14.8 THEOREM If $T \in L(L)$, then for any $\varepsilon > 0$ there exists an equivalent norm on L, $\|\cdot\|_\varepsilon$, such that

$$\|T\|_\varepsilon < r_T + \varepsilon$$

(Here, of course, we compute the norm of T with respect to the $\|\cdot\|_\varepsilon$ norm.)

Proof. If $\varepsilon > 0$ is given, we can find N_ε such that for all $n \geq N_\varepsilon$

$$\|T^n\|_\varepsilon < (r_T + \varepsilon)^n$$

and we define the norm $\|\cdot\|_\varepsilon$ on L by the formula,

$$\|x\|_\varepsilon = \|x\| + \frac{\|Tx\|}{r_T + \varepsilon} + \cdots + \frac{\|T^{N-1}x\|}{(r_T + \varepsilon)^{N-1}}$$

Since $\|x\| < \|x\|_\varepsilon$ is obvious, we prove that for some M, $\|x\|_\varepsilon \le M\|x\|$.

Indeed, we have, from the definition of N_ε, that

$$\|x\|_\varepsilon \le \|x\| + \Sigma\, K_i\|x\| = M\|x\|$$

Also $\|T\|_\varepsilon \le r_T + \varepsilon$, as can be seen by a simple computation.

6.14.9 COROLLARY For any $T \in L(L)$ the spectral radius is equal to the infimum of the norms of T with respect to all equivalent norms on L.

6.14.10 REMARK The above construction of the norm as well as the theorem were given by Krasnoselskii (1964).

Now we are ready to prove the "square root lemma" for a class of Banach algebras without the continuity of the involution.

6.14.11 THEOREM Let A be a Banach algebra and $a \in A$ with $r_a < 1$. Then there exists a unique element $x \in A$ such that $2x - x^2 = a$ with $r_x < 1$.

Proof. By the above results we can choose a new norm $\|\cdot\|_1$ such that $\|a\|_1 < r < 1$. Let A(a) be the least closed subalgebra of A containing a and consider the set

$$C = \{x \in A(a) \mid \|x\|_1 < r\}$$

Define the mapping on C by the formula

$$Tx = \frac{1}{2}\,(a + x^2)$$

which is a contraction mapping; since A(a) is commutative, we obtain

$$\|Tx - Ty\|_1 = \frac{1}{2}\,\|x - y\|_1\|x + y\|_1 \le r\|x - y\|_1$$

and by the contraction principle we can find a unique point x such that Tx = x. But this is equivalent with the assertion of the theorem. The uniqueness can be proved as follows. Suppose that there exists y such that $2y - y^2 = a$. Since x is a limit of polynomials in a, x commutes with a and thus y commutes with x.

Thus we have that $r_{x+y} < 2$. Also,

$$\|x - y\|' = \frac{1}{2} \|x^2 - y^2\|' < \frac{1}{2} \|x + y\|'\|x - y\|'$$

for a new norm such that $\|x + y\|' < 2$. Thus we have $\|x - y\|' = 0$, and the uniqueness is proved.

The square root lemma follows immediately from this result.

6.14.12 THEOREM Let A be a Banach algebra with involution. Suppose that $a \in A$ and $a^* = a$ and $r_a < 1$. Then there exists a unique element $x \in A$ such that $2x - x^2 = a$, $r_x < 1$, $x = x^*$.

Proof. The existence and the uniqueness are consequences of the above result. Since $r_x = r_{x^*}$, we have that $r_x < 1$. Now, $a = a^* = (2x - x^2) = 2x^* - x^{*2}$ and the uniqueness gives $x = x^*$.

The following result gives information about the elements of the form $\bar{x}x$, with $x = h + ik$ and $\bar{x} = h - ik$.

6.14.13 THEOREM If J(A) is a Banach algebra and $x = h + ik$ has the property that $\bar{x}x = 0$, then $x = 0$.

Proof. First we show that if J(A) is an algebra and $z \in J(A)$ has the property that $-\bar{z}z$ is positive, then $z = 0$.

Since for any elements $a,b \in A$, $\sigma(ab) = \sigma(ba)$ except eventually 0, and since

$$\bar{z}z = 2(h^2 + k^2) - z\bar{z}$$

it follows that the numerical range of $z\bar{z}$ is zero and thus $z\bar{z} = 0$.

This clearly implies that $\bar{z}z = 0$. Thus $h = k = 0$.

To prove the assertion of the theorem let z be arbitrary in J(A) and thus there exist p and q, hermitian elements, such that

$$\bar{z}z = p - p \qquad pq = qp = 0$$

and if we take $y = zp$, we obtain that

$$-\bar{y}y = -q\bar{z}zq = q^3$$

and since it is positive, we have in fact that $y = 0$. In this case,

$$\bar{z}zq = 0 \qquad q^2 = 0$$

and this gives $q = 0$. Thus $z = p$ which is a positive element.

We can now give a new result along the lines of Theorem 6.14.7 to obtain a complete characterization of Banach algebras which are B*-algebras.

6.14.14 THEOREM If J(A) is an algebra, then there exists an isometric *-isomorphism

$$a \mapsto Ta$$

of J(A) onto a B*-algebra.

Proof. First we recall that an element $f \in A^*$ is called a *normalized state* if $\|f\| = f(1) = 1$.

From the results about elements of the form $\bar{z}z$ it follows that $f(\bar{z}z) \geq 0$ for all $z \in A$. From this relation follows the Cauchy-Schwarz inequality:

$$|f(\bar{x}y)|^2 \leq f(\bar{x}x)f(\bar{y}y)$$

Consider now a normalized state of A and for each $x,y \in A = J(A)$ we define the bilinear form

$$\langle x, y\rangle_f = f(\bar{x}y)$$

which has all the properties of the inner products except (possibly) $\langle x, x\rangle_f = 0$ if and only if $x = 0$.

Let $I_f = \{x \mid \langle x, x\rangle_f = 0\}$, which is a left ideal (closed) of A.

We obtain a Hilbert space taking the completion of A/I_f with respect to the inner product induced by $\langle\ ,\ \rangle_f$. Denote this Hilbert space by H_f.

We can define an operator on this space by

$$T_a(x + I_f) = ax + I_f$$

A computation similar to that in Chapter 4 shows that $\|T_z\|^\sim \le \|a\|$ and the exact same estimate holds for the extension of T_a to H_f. If we take the direct sum of all spaces H_f for all f, then we can define the operator $T^\sim a$ on this direct sum in a natural way. From the Cauchy-Schwarz inequality it follows that $\|T^\sim a\|^\sim = \|a\|$ for all hermitian elements. Let $E = \{e^{ih} \mid h \text{ hermitian}\}$ and let $a \in A$ be arbitrary. We can find a sequence $\{a_n\}$ in E such that $\|a_n - a\| \to 0$ and since $\|e^{ih}\| = 1$, we obtain that $\|a_n\|^\sim < 1$. Thus $\|a\|^\sim < 1$. This norm is the norm derived from the application $a \mapsto T_a$. But $\|a\| = 1$ implies that $\|a\|^\sim \le \|a\|$. If b is an arbitrary element in A and $\|b\| > \|b\|^\sim$, then we have

$$\|\bar{b}b\| < \|\bar{b}\| \cdot \|b\| < \|\bar{b}\| \cdot \|b\| = \|\bar{b}b\| = \|\bar{b}b\|$$

and this is a contradiction. Thus $a \mapsto T_a$ is an isometry.

6.14.15 REMARK The Vidav-Palmer theorem is comprised of the results stated in Theorems 6.14.7 and 6.14.14.

6.14.16 COROLLARY (Calkin-Segal-Kaplansky) If A is a B*-algebra and I is a closed two-sided ideal, then A/I is a B*-algebra.

Proof. This follows immediately from the Vidav-Palmer theorem and the definition of the norm in A/I.

6.15 SOME PROPERTIES OF HERMITIAN AND NORMAL ELEMENTS OF A BANACH ALGEBRA

In what follows we are interested in properties of hermitian and normal elements with respect to numerical radius, spectral radius, and the norm.

First we give the following definition.

6.15.1 DEFINITION An operator $T \in L(X)$ is called *transaloid* if for all complex numbers z, $T_z = T + zI = T + z$ has the property that

$$\|T_z\| = r_{T_z}$$

The basic property of hermitian elements of a Banach algebra for which the notion of a hermitian operator is defined as in the case of L(X) is the following result of Browder-Sinclair. A special case was proved by Vidav; that is, for any hermitian element the norm is equal to the spectral radius.

6.15.2 THEOREM If T is a hermitian element of a Banach algebra, then T is transaloid.

The proof of this theorem is given using some results which are of independent interest.

6.15.3 THEOREM Let t, $|t| \leq 1$, and $\arcsin t = \Sigma_{r=1}^{\infty} c_r t^r$ and

$$F_n(z) = \sum_{r=1}^{n} c_r (\sin z)^r$$

then for any compact K in $(-\pi/2, \pi/2)$ there exists U open such that $U \subset K$ and $\lim F_n(z) = z$ uniformly on U.

Proof. Let U be an open set such that for $z \in U$, $|\sin z| < 1$. In this case the series $\Sigma\, c_r(\sin z)^r$ converges uniformly on U and the limit is an analytic function. But for $t \in K$

$$\lim F_n(t) = t$$

and thus the assertion of the theorem is proved.

6.15.4 THEOREM If T is any hermitian element, then $\|T\| = r_T$.

Proof. We can assume, without loss of generality, that $r_T < \pi/2$ and thus $\sigma(T) \subset (-\pi/2, \pi/2)$. By the functional calculus we have

$$T = \lim F_n(T) = \Sigma\, c_r(\sin (T))^r$$

But T is a hermitian element and thus,

$$\|\sin T\| \leq \frac{1}{2}\, \|e^{iT}\| + \frac{1}{2}\, \|e^{-iT}\| = 1$$

and

$$\|T\| \leq \Sigma\, c_r = \frac{\pi}{2}$$

The positive homogeneity of the norm and spectral radius gives the assertion.

6.15.5 THEOREM Suppose that F is an entire function with the property $|F(z)| \le e^{|\mathrm{Im}\ z|}$, then for any $t \in R \setminus \{n\pi\}$ the following relation holds:

$$F(0) \cos t + F'(0) \sin t = \Sigma\, a_n F(n\pi + t)$$

where $a_n = (-1)^n (\sin^2 t)/(n\pi + t)$.

Proof. For any integer n let C_n be the boundary of the square with vertices in $\{\pm a_n, \pm i a_n\}$ and since

$$\lim \int_{C_n} \frac{F(z)}{z^2 \sin(t - z)}\, dz = 0$$

the residue theorem gives the assertion.

6.15.6 THEOREM Let T be a hermitian element such that $\|T\| \le 1$ and a_n as in Theorem 6.15.5. Then for any $t \in R \setminus \{n\pi\}$

$$\cos t + (i \sin t)T = \Sigma\, a_n e^{((n+t)iT)}$$

Proof. For any $f^* \in A^*$, $\|f^*\| = 1$, the function $F(z) = f^*(e^{izT})$ is an entire function and

$$\begin{aligned} |F(z)| = |f^*(e^{izT})| &\le \|f^*\| \|e^{iRe\ zT}\| \cdot \|e^{-Im\ zT}\| \\ &\le e^{|\mathrm{Im}\ z| w(T)} = e^{|\mathrm{Im}\ z|} \end{aligned}$$

and we can apply the above result to obtain

$$\begin{aligned} f^*(\cos t + (i \sin t)T) &= \Sigma\, f^*(e^{(n\pi+t)iT}) \\ &= f^*(\Sigma\, a_n e^{(n\pi+t)iT}) \end{aligned}$$

and since f^* is arbitrary the assertion follows.

If we take some special values for t and T we obtain the following identities.

6.15.7 *THEOREM* If T is a hermitian element and $\|T\| = 1$, then

$$T = \Sigma\ 4\pi^{-2}(2n + 1)^{-2} e^{((n+1/2)(T-1)\pi i)}$$

and also

$$\Sigma\ 4\pi^{-2}(2n + 1)^{-2} = 1$$

Proof. We take $t = \pi/2$, and for the second $T = 1$.

We are now in a position to prove Theorem 6.15.2.

Suppose, without loss of generality, that $\|T\| = 1$ and thus

$$\|\cos t + (i \sin t)T\| \leq \Sigma\ |a_n| = 1$$

and since for T, $\|T\| = r_T$, we obtain that $r_{(\cos t+(i \sin t)T)} > 1$ and this gives that

$$r_{(\cos t+(i \sin t)T)} = \|\cos t + (i \sin t)T\|$$

and this obviously implies the assertion of the theorem.

Bonsall and Duncan (1973) give, as a corollary, a proof of the following result of Bollobas.

6.15.8 *THEOREM* If T is a hermitian element and $\|T\| = \|T^{-1}\| = 1$, then $T = T^{-1}$.

Since an extension of this result was obtained by Dowson for the case of operators on Banach spaces using some deep results, the following generalization is a very simple consequence of a result about nilpotent elements in Banach algebras [Hille and Phillips (1957)].

6.15.9 *THEOREM* If T is an arbitrary element of a Banach algebra with the properties

1. $W(T) \subset R$
2. $\|T^{\pm n}\| = O(n)$

then $T^2 = I$. (The identity of A is I.)

Proof. From condition 2 it follows that the spectrum of T is contained in $\{z \mid |z| \leq 1\}$ and from condition 1 it follows that $\sigma(T) \subseteq \{1, -1\}$. This gives that $\sigma(T^2 - I) = \{0\}$.

Thus the element $T^2 - I$ is quasi-nilpotent and condition 2 implies that $T^2 - I = 0$.

6.15.10 REMARK Condition 2 in the theorem can be weakened to the following condition

$$\|T^{\pm n}\| = O(n^N)$$

and in this case the conclusion is that $T^N = I$.

The following theorem gives information about the numerical range of a hermitian operator. In fact, as we prove, the assertion is valid for a more general class of elements, namely, the class of transaloid elements.

6.15.11 THEOREM If T is a transaloid element, then

$$\overline{W(T)} = \operatorname{conv} \sigma(T)$$

Proof. First we remark that for any element in L(X) the closed numerical range contains the spectrum. Indeed, let us consider the smallest commutative Banach algebra containing the element T and the identity I. Since the spectrum of T is in the spectrum of T with respect to this algebra, and since the elements of the form f(T) are in $\sigma(T)$ with f in the maximal ideal, the assertion follows.

Thus to prove the assertion of the theorem it will be sufficient to show that $W(T) \subset \operatorname{conv} \sigma(T)$. For this we use in an essential way the fact that T is a transaloid element. The assertion of the theorem follows from the equality:

$$W(T) = \cap \{z \mid |z - s| \leq \|T - s\|\}$$

This can be proved as follows: if $p \in W(T)$, then

$$|p - s| \leq |f^*(T - sI)| \leq \|T - sI\|$$

for all complex numbers s. Conversely, if $p \in W(T)$, we can assume, without loss of generality, that it is in $\operatorname{Re} z > 0$ and also that $p > 0$.

But

$$\lim \|T + t\| - t = \sup \operatorname{Re} W(T)$$

It follows that

$$\|T + t\| - t < p$$

for t sufficiently large. Thus we obtain the equality for these two sets.

6.16 THE NUMERICAL RADIUS AND THE ITERATES OF AN ELEMENT

As we know, in the case of Hilbert spaces it was proved by Bernau and Smithies that for any bounded operator on a Hilbert space the following inequality holds: $w(T^2) \leq (w(T))^2$.

The conjecture of Halmos that the inequality

$$w(T^n) \leq (w(T))^n$$

holds for all n was proved by Berger (1965). The results which we give below are in the papers of Berger and Stampfli (1967).

Let A be a B*-algebra with unit and set

$$\Delta = \{z \mid |z| \leq 1\}$$
$$P = \{z \mid \text{Re } z \geq 0\}$$

As in the case of Hilbert space operators we can see that the following assertion holds.

6.16.1 LEMMA For any $T \in A$, $w(T) \leq 1$ if and only if $\text{Re}(I - zT)^{-1} \geq 0$ for all z, $|z| \leq 1$.

Proof. Since the proof is easy, we omit it.

6.16.2 THEOREM Let A be a B*-algebra and $T \in A$ with $w(T) = 1$ and F be an analytic function in a neighborhood of T with $F(\Delta) \subset P$. Then $W(F(T)) \subset P \setminus \text{Re } F(0)$.

Proof. Since F is an analytic function, we have for z, $|z| < 1$, that

$$F(z) = -F(0) + \frac{1}{\pi}\int_0^{2\pi} (\text{Re } F(e^{it})(1 - ze^{-it})^{-1}\, dt$$

and for $r \in (0, 1)$ we get

$$F(rT) = -F(0)^* + \frac{1}{\pi}\int_0^{2\pi} \text{Re } F(e^{it})(1 - re^{-it}T)^{-1}\, dt$$

and thus

$$\text{Re } F(rT) + \text{Re } F(0)I \geq 0$$

which implies the assertion of the theorem.

The following corollaries give the inequalities for the numerical radius for elements in Banach algebras.

6.16.3 COROLLARY Let A be a B*-algebra and $T \in A$ with $w(T) \leq 1$. If F is any analytic function in a neighborhood of Δ and $F(0) = 0$, $F(\Delta) \subset \Delta$, then $w(F(T)) \leq 1$.

Proof. For $\tilde{z}$, $|\tilde{z}| < 1$, let

$$\tilde{F}(z) = \frac{1 + \tilde{z}F(z)}{1 - \tilde{z}F(z)}$$

defined on a neighborhood of Δ. Since $\tilde{F}(\Delta) \subset P$ and $\tilde{F}(0) = 0$, we can apply the above result; thus

$$W(\tilde{F}(T)) \subset P - 1 \qquad \text{Re } (\tilde{F}(T) + I) > 0$$

which gives that

$$\text{Re } ((1 - zF(T))^{-1} \geq 0$$

for all z, $|z| \leq 1$. This is equivalent with the assertion of the corollary.

6.16.4 COROLLARY For any element T of a B*-algebra the power inequality holds, i.e.,

$$w(T^n) \leq (w(T))^n$$

Proof. The function $F(z) = z^n$ satisfies the above corollary and this yields the assertion.

6.16.5 REMARK In the case of Hilbert spaces an interesting proof of the power inequality can be given using the dilation theory. Using the dilation theory some results can be obtained for functions of the form F(T), F analytic, and $F(0) = 0$.

The power inequality and the fact that for any two elements of a Banach algebra A, the inequality

$$\|ab\| \le \|a\| \cdot \|b\|$$

is by definition true suggests that perhaps an inequality of the form

$$w(TS) \le w(T)w(S)$$

holds. The following example shows that the assertion is not true even for commuting operators.

6.16.6 EXAMPLE (Halmos, 1967) Let T be the operator with the matrix

$$T = \begin{bmatrix} 0 & 0 & 0 & 0 \\ 1 & 0 & 0 & 0 \\ 0 & 1 & 0 & 0 \\ 0 & 0 & 1 & 0 \end{bmatrix}$$

and define $S = T^2$. Clearly ST = TS and a simple computation shows that

$$w(T^2) = w(T^3) = \frac{1}{2} \qquad w(T) < 1$$

Now since

$$w(TS) = w(T^3) = \frac{1}{2} \qquad \text{and} \qquad w(T)w(S) = \frac{1}{2}\, w(T) < \frac{1}{2}$$

the assertion is proved.

6.17 THE NUMERICAL RANGES OF ELEMENTS OF LOCALLY M-CONVEX ALGEBRAS

Let A be an algebra which is l.m.c. algebra [for a theory of these algebras, see Michael (1952)] and suppose that there exists a family of submultiplicative seminorms $\{\|\cdot\|_\alpha \mid \alpha \in I\}$ such that $\|I\|_\alpha = 1$ for all $\alpha \in I$ and that the topology is generated by these seminorms. If $\{A, \|\cdot\|_\alpha \mid \alpha \in I\}$ is a l.m.c. algebra, we set

$$D_\alpha(A, \|\cdot\|_\alpha, 1) = \{f \in A^* \mid f(I) = 1, |f(x)| \le \|x\|_\alpha, x \in A\}$$

and

$$D(A, \{\|\cdot\|_\alpha\}, 1) = \bigcup_\alpha \{D_\alpha(A, \|\cdot\|_\alpha, 1\}$$

6.17.1 DEFINITION The *numerical range* of an element $T \in A$ is the set defined as follows:

$$W(A, \|\cdot\|_\alpha, T) = \bigcup_\alpha \{W_\alpha(A, \|\cdot\|_\alpha, T)\}$$

where

$$W_\alpha(A, \|\cdot\|_\alpha, T) = \{f(T) \mid f \in D_\alpha(A, \|\cdot\|_\alpha, 1)\}$$

For each linear functional f on A which annihilates the null space N_α of the seminorm $\|\cdot\|_\alpha$, we can define the linear functional on $A_\alpha = A/N_\alpha$ in a natural way, and in this way, as we see immediately, the numerical range can be characterized by the numerical range in the normed algebra A_α,

$$W\ (A, \|\cdot\|_\alpha, T) = W(A_\alpha, \|\cdot\|_\alpha, T_\alpha)$$

$$W(A, \{\|\cdot\|\}, T\} = \bigcup_\alpha \{W(A_\alpha, \|\cdot\|_\alpha, T_\alpha)\}$$

We define now the numerical radius of an element in A. Let $T \in A$ and put

$$w_\alpha(A, \|\cdot\|_\alpha, T) = \sup \{|s| \mid s \in W_\alpha(A, \|\cdot\|_\alpha, T)\}$$

6.17.2 DEFINITION The *numerical radius* of an element $T \in A$ is the number

$$w(A, \{\|\cdot\|_\alpha\}, T) = \sup \{|s| \mid s \in W(A, \{\|\cdot\|_\alpha\}, T\}$$

From the definition it follows that

$$w(A, \{\|\cdot\|_\alpha\}, T) = \sup w_\alpha(A, \|\cdot\|_\alpha, T)$$

The following properties of the numerical range and the numerical radius are all obvious and thus we omit the proof.

6.17.3 THEOREM For the numerical range and the numerical radius the following relations hold:

1. $W(A, \{\|\cdot\|_\alpha\}, aT + s) = aW(A, \{\|\cdot\|_\alpha\}, T) + s$

2. $w(A, \{\|\cdot\|_\alpha\}, aT + s) \le |a| w(A, \{\|\cdot\|_\alpha\}, T) + |s|$
3. $W(A, \{\|\cdot\|_\alpha\}, T + S) \subseteq W(A, \{\|\cdot\|_\alpha\}, T) + W(A, \{\|\cdot\|_\alpha\}, S)$
4. $w(A, \{\|\cdot\|_\alpha\}, T + S) \le w(A, \{\|\cdot\|_\alpha\}, T) + w(A, \{\|\cdot\|_\alpha\}, S)$

6.17.4 DEFINITION The *spectrum* of an element in A is the set

$$\sigma(T) = \{z \mid T - zI \text{ is not invertible}\}$$

In the definition we have supposed that A is complete and with unity. The following theorem gives information about the connection between the numerical range and the spectrum of an element.

6.17.5 THEOREM If A is a complete locally m-convex algebra, then the following assertions hold:

1. $\sigma(T) \subset W(A, \{\|\cdot\|_\alpha\}, T)$
2. $\text{conv}\ \sigma(T) \subseteq \{W(A, \{\|\cdot\|_\alpha\}, T \subseteq \overline{\text{conv}\ \sigma(T)}$

Proof. For the proof, see Koehler et al. (1973).

The notion of hermitian element of A can be introduced using the numerical range.

6.17.6 DEFINITION An element $T \in A$ is called *hermitian* if

$$W(A, \{\|\cdot\|_\alpha\}, T)$$

is a subset of the real numbers.

6.17.7 DEFINITION An element T in A is called *decomposable* if there exists two hermitian elements H and J such that $T = H + iJ$. The set of all decomposable elements of A is denoted by J(A).

It is obvious that several properties known for elements in Banach algebras can be extended to this setting.

Chapter 7

NONNORMAL CLASSES OF OPERATORS

If N is a normal operator in L(H), the set of all bounded linear operators on a Hilbert space H, then it is known that it has a number of very interesting properties; among them we note the following:

1. $r_N = \|N\|$.
2. conv $\sigma(N) = \overline{W(T)}$.
3. $\|Tx\| = \|T^*x\|$ for all $x \in H$.
4. For all complex numbers z, $N + zI = N + z$ is a normal operator.
5. If H_1 is an invariant subspace of N, N/H_1 has property 1.
6. If H_1 is an invariant subspace for N and N*; then N/H_1 is normal.
7. $\sigma(T)$ and $\overline{W(T)}$ are spectral sets of T.[†]
8. $r_N = w(T)$.
9. N has the G_1 property.
10. Re $\sigma(T) = \sigma(\text{Re } T)$.

The story of nonnormal classes of operators begins with a famous paper of Wintner (1929), in which an attempt is made to obtain a characterization of operators T for which $r_T = \|T\|$. The assertion of Wintner is that for this, it is necessary and sufficient that conv $\sigma(T) = \overline{W(T)}$. An example of Halmos (1967) shows that this is not the case. The example of Halmos is the following one: Let A

[†]The notion of a spectral set is defined in Section 7.2.

on the space ℓ_2^2 with the matrix

$$\begin{bmatrix} 0 & 1 \\ 0 & 0 \end{bmatrix}$$

and it is clear that

$$\sigma(A) = \{0\} \qquad \overline{W(A)} = \{z \mid |z| \le \tfrac{1}{2}\}$$

Now let I be the operator identity on an arbitrary separable Hilbert space and consider the operator $T = A \oplus I$. It is clear that

1. $\sigma(T) = \{0, 1\}$
2. $W(T) = \text{conv}\,\{\{z \mid |z| \le \frac{1}{2}\} \cup \{1\}\}$

and thus $r_T = \|T\|$ and $\text{conv}\,\sigma(T) = [0, 1] \ne W(T)$.

Thus we have that properties 1 and 2 are not equivalent.

In the last ten to fifteen years several classes of nonnormal classes of operators were intensively investigated. The purpose of this chapter is to present some of these classes, as well as their interconnections.

7.1 CLASSES OF NONNORMAL OPERATORS

Let H be a Hilbert space and L(H) be the set of all bounded linear operators on H.

7.1.1 DEFINITION An element $T \in L(H)$ is called *quasi-normal* if

$$(T^*T)T = T(T^*T)$$

7.1.2 DEFINITION An element $T \in L(H)$ is called *subnormal* if there exists a Hilbert space $K \supset H$, H is a closed subspace of K, and a normal operator $N \in L(K)$ such that

$$Nx = Tx \qquad \text{for all } x \in H$$

7.1.3 DEFINITION An element $T \in L(H)$ is called *hyponormal* if for all $x \in H$, $\|T^*x\| \le \|Tx\|$. If T^* is hyponormal, we say T is *seminormal*.

7.1.4 DEFINITION An element $T \in L(H)$ is called of *class (N)* or *paranormal* if for all $x \in H$, $\|x\| = 1$, $\|Tx\|^2 \le \|T^2x\|$.

7.1.5 DEFINITION An element $T \in L(H)$ is called of *class (N,k)* if for all $x \in H$, $\|x\| = 1$,

$$\|Tx\|^k \leq \|T^k x\|$$

7.1.6 DEFINITION An element $T \in L(H)$ is called *normaloid* if

$$r_T = \|T\|$$

The following theorem gives the most important connections between the classes considered above.

7.1.7 THEOREM The following relations hold: normal operator $\subset$ quasi-normal operator $\subset$ subnormal operator $\subset$ hyponormal operator $\subset$ operators of class (N) $\subset$ operators of class (N,k) $\subset$ normaloid operators.

Proof. It is obvious that every normal operator is quasi-normal.

We give now an example of a quasi-normal operator which is not normal. It is clear that every isometry is such an operator.

We prove now that every quasi-normal operator is subnormal. First we remark that if T is quasi-normal, then

$$\ker T = \{x \mid Tx = 0\}$$

is a reducing subspace, i.e., invariant for T and T*. Indeed, if $x \in \ker T$, we have $T^*T = 0$, and since T is quasi-normal $T^*x = 0$. Thus we can decompose T, $T = 0 \oplus T_1$ where T_1 is also quasi-normal, but $\ker T_1 = \{0\}$. Let $T = UR$ and thus U is an isometry, and $UR = RU$ gives $U^*R = RU^*$. If $E = UU^*$, then $(I - E)U = U^*(I - E) = 0$. The normal extension of T is now constructed as follows: We consider the Hilbert space $H = H \oplus H$ and the operators

$$T_1 = \begin{bmatrix} U & I - E \\ 0 & U^* \end{bmatrix} \qquad T_2 = \begin{bmatrix} R & 0 \\ 0 & R \end{bmatrix}$$

acting on H. Clearly the operator T_2 is positive, and it is easy to see that T_1 is an isometric operator. Also, a simple computation shows that $T_1T_2 = T_2T_1$. The normal operator is defined as

$$N = \begin{bmatrix} T_1 & (I - E)R \\ 0 & U^*R \end{bmatrix}$$

and it is obviously an extension of T.

We give now an example of an operator which is subnormal and not quasi-normal. First we remark that the "unilateral shift"

$$Te_i = e_{i+1} \qquad i = 1, 2, 3, \ldots$$

on the space $\ell_+^2 = \{x \mid \Sigma_{i=0}^{\infty}|x_i|^2 < \infty\}$ is subnormal. Indeed, we consider the space $\ell^2 = \{x \mid \Sigma_{-\infty}^{\infty}|x_i|^2 < \infty\}$ and the bilateral shift

$$Te_i = e_{i+1} \qquad i = \ldots, -1, 0, 1, 2, \ldots$$

which is unitary and is an extension of T.

It is also obvious that if T is subnormal, then T + z is also subnormal for all z.

We prove now that every subnormal operator is hyponormal.

Indeed, if P is the projection P: K → H, then we have for all $x, y \in H$,

$$\begin{aligned} \langle T^*, x, y\rangle &= \langle x, Ty\rangle = \langle x, Ny\rangle = \langle N^*x, y\rangle \\ &= \langle N^*x, Py\rangle = \langle PN^*x, y\rangle \end{aligned}$$

and thus the operator PN* has H as an invariant subspace. Thus

$$T^*x = PN^*x$$

We obtain that

$$\|T^*x\| = \|PN^*x\| \leq \|N^*x\| = \|Nx\| = \|Tx\|$$

and the assertion is proved.

We give now an example of an operator which is hyponormal and not subnormal. First we give a necessity condition for subnormality of an operator: If $T \in L(H)$ is subnormal, then for any x_0, $x_1, \ldots, x_n \in H$,

$$\sum_{j,i} \langle T^i x_j, T^j x_i\rangle c_j^* c_i \geq 0$$

where $\{c_i\}$ are the complex numbers.

Indeed, for T is subnormal, we have

$$\|\Sigma N^{*j}x_j\|^2 = \langle \Sigma N^{*j}x_j, \Sigma N^{*i}x_i\rangle = \Sigma_j\Sigma_i \langle N^{*j}x_j, N^{*i}x_i\rangle$$

$$= \Sigma_j\Sigma_i \langle N^iN^{*j}x_j, x_i\rangle = \Sigma_j\Sigma_i \langle N^{*j}N^ix_j, x_i\rangle$$

$$= \Sigma_j\Sigma_i \langle N^ix_j, N^jx_i\rangle = \Sigma_j\Sigma_i \langle T^ix_j, T^jx_i\rangle$$

and if we consider $x_i = c_i\tilde{x}_i$, we obtain the assertion.

Now on any separable Hilbert space with the orthonormal basis $\{e_i\}_{i=1}^{\infty}$ we consider the operator

$$Te = \alpha_i e_{i+1}$$

where $\{\alpha_i\}$ is a bounded sequence of complex numbers. This operator is the so-called weighted shift with the weighted sequence $\{\alpha_i\}$.

If $|\alpha_i| < |\alpha_{i+1}|$, then it is easy to see that T is hyponormal. If

$$\alpha_0 = \alpha \qquad \alpha_1 = \beta \qquad \alpha_2 = \alpha_3 = \cdots = 1$$

with $\alpha < \beta$, then we have a hyponormal operator which is not subnormal since for i,j = 0, 1, 2, and $x_0 = e_0$, $x_1 = e_1$, $x_2 = e_2$, the corresponding matrix is

$$M = \begin{bmatrix} 1 & \alpha & \alpha\beta \\ \alpha & \beta & \beta \\ \alpha\beta & \beta & 1 \end{bmatrix}$$

and det M < 0.

We prove now that every hyponormal operator is of class (N). Indeed, for any $x \in H$, $\|x\| = 1$,

$$\|Tx\|^2 = \langle Tx, Ty\rangle \le \langle T^*Tx, x\rangle \le \|T^2x\|$$

To prove that the class of operators with the property that for all x, $\|x\| = 1$, $\|Tx\|^2 \le \|T^2x\|$, is larger than the class of hyponormal operators, we prove first a result about operators of this class: If T is of class (N), then T^2 is also of class (N). Indeed, for any x, $\|x\| = 1$, we have

$$\|T^4x\| = \left\|\frac{T^2T^2x}{\|T^2x\|}\right\| \|T^2x\| \geq \frac{\|T^3x\|^2}{\|T^2x\|}$$

and since

$$\|T^3x\| = \left\|\frac{T^2Tx}{Tx}\right\| \|Tx\| \geq \frac{\|T^2x\|^2}{\|Tx\|}$$

which gives

$$\|T^4x\| \geq \|T^2x\|^2\|Tx\|^2\|T^2x\| = \|T^2x\|^2 \frac{\|T^2x\|}{\|Tx\|^2} \geq \|T^2x\|^2$$

and the assertion is proved.

For an example of an operator which is of class (N) and not hyponormal, we give an example of a hyponormal operator T for which T^2 is not a hyponormal operator.

Let H be a Hilbert space and consider the Hilbert space K of all sequences $x = \{x_i\}_{-\infty}^{\infty}$, $\Sigma_{-\infty}^{\infty}\|x_i\|^2 < \infty$. If $y = \{y_i\}$, then the scalar product $\langle x, y\rangle$ is defined as

$$\langle x, y\rangle = \Sigma_{-\infty}^{\infty} \langle x_i, y_i\rangle$$

We consider a bounded sequence of operators on H, $\{P_n\}$ and define the operator P on K by the formula

$$(Px)_n = P_nx_n$$

and the operator U on K by

$$(Ux)_n = x_{n+1}$$

It is easy to see that we have the relations

$$(U^*x)_n = x_{n+1} \qquad (P^*x)_n = P_nx_n$$

and thus the operator T = UP has the property

$$(T^*Tx)_n = P_n^2x_n \qquad (TT^*x)_n = P_{n-1}^2x_n$$

From these we obtain that T is hyponormal if and only if

$$P_n^2 \geq P_{n-1}^2$$

for all n. Also the operator T^2 is hyponormal if and only if

$$P_n P_{n+1}^2 P_n - P_{n-1} P_n^2 P_n \geq 0$$

for all n. Now if $H = \ell_2^2$ and the operators P_n are defined in the following manner, we obtain the desired operator. Let

$$C = \begin{bmatrix} 1 & 0 \\ 0 & 0 \end{bmatrix} \qquad D = \begin{bmatrix} 2 & 1 \\ 1 & 1 \end{bmatrix}$$

and clearly D - C is positive. We set

$$P_n = \begin{cases} C & \text{if } n \leq 0 \\ D^{1/2} & \text{if } n > 0 \end{cases}$$

and in this case $P_n^2 - P_{n-1}^2 \geq 0$ is satisfied for all n. But

$$P_0 P_{-1}^2 P_0 = C^2 \qquad P_1 P_2^2 P_1 = D^2$$

and $D^2 - C^2$ is not positive. Thus T is hyponormal and T^2 is not.

We prove now that any operator of class (N) is of class (N,k). Indeed, we prove the assertion that if T is of class (N) and of class (N,k), then it is of class (N, k + 1). It is obvious that this implies our assertion.

We have

$$\|T^{k+1}x\| = \left\|\frac{T^k Tx}{\|Tx\|}\right\| \|Tx\| > \|T^2x\|^k \|Tx\|^{1-k} \geq \|Tx\|^{2k} \|Tx\|^{1-k} = \|Tx\|^{k+1}$$

In the case Tx = 0 the assertion is obvious.

We prove now that each operator of class (N,k) is normaloid. Indeed, we can assume, without loss of generality, that $\|T\| = 1$ and let $\{x_n\}$, $\|x_n\| = 1$, such that $\|Tx_n\| \to 1$. From the definition it is clear that

$$\lim_{n\to\infty} \|T^k x_n\| = 1$$

and from this follows that for all integers $i \in [2, k]$.

$$\lim_{n\to\infty} \|T^i x_n\| = 1$$

Now from

$$\|T^{k+1}x_n\| = \left\|\frac{T^k Tx_n}{Tx_n}\right\| \frac{1}{\|Tx_n\|} \geq \frac{\|T^2 x_n\|^k}{\|Tx_n\|^{k-1}} \to 1$$

and thus

$$\lim_{n\to\infty} \|T^{k+1}x_n\| = 1$$

An induction argument proves the fact that for all m

$$\lim_{n\to\infty} \|T^m x_n\| = 1$$

and thus $\|T^m\| = 1$. The assertion is proved.

We give now an example of an operator which is normaloid and not in the class (N,k) for all k. For this we consider a nilpotent operator T_1 on a Hilbert space H_1 and with the property that $\|T_1\| = 1$. Let T_2 be a normal operator on a Hilbert space H_2 such that $\|T_2\| = 1$. Clearly the operator $T = T_1 \oplus T_2$ on $H_1 \oplus H_2$ is normaloid and is not in the class (N,k) for all k since T_1 is nilpotent.

Theorem 7.1.7 is proved.

We give now an interesting characterization of operators of class (N).

7.1.8 THEOREM $T \in L(H)$ is of class (N) if and only if for all $\lambda \in R$

$$T^{*2}T^2 - 2\lambda T^*T + \lambda^2 \geq 0$$

Proof. Since

$$\|T^2x\|\|x\| = \inf_{\lambda>0} \frac{1}{2}\{\lambda^{-1}\|T^2x\|^2 + \lambda\|x\|^2\} = \inf_{\lambda>0} \frac{1}{2}\{<(\lambda^{-1}T^{*2}T^2 + \lambda I)x, x>\}$$

and $\|T^2x\| \geq \|Tx\|^2$, we obtain that it is equivalent with

$$<(\lambda^{-1}T^{*2}T^2 + \lambda I - 2T^*T)x, x> \geq 0$$

and the assertion is proved.

7.1.9 COROLLARY If $T \in L(H)$ is of class (N) and T^{-1} exists, then T^{-1} is of class (N).

7.1.10 COROLLARY If $T \in L(H)$ is of class (N), then for all n, T^n is of class (N).

Proof. As we know, the operator T^2 is paranormal, and thus assuming that T^k is paranormal, we prove that T^{k+1} is paranormal.

By the above characterization, for all n, we have

$$T^{*(2k+1)}T^{2k+1} - 2T^{*(k+1)}T^{k+1} + \lambda T^*T$$
$$= T^*(T^{*2k}T^{2k} - 2T^{*k}T^k + \lambda I)T \geq 0$$

and this implies that

$$\|T^{2k+1}x\|^2 - 2\|T^{k+1}x\|^2 + \|Tx\|^2 \geq 0$$

and thus

$$\|T^{k+1}x\|^4 \leq \|T^{2k+1}x\|^2\|T^2x\|$$

and since T is paranormal if and only if for all n,

$$\|T^{n+1}x\|^2 \geq \|T^nx\|^2\|T^2x\|$$

this gives

$$\|T^{k+1}x\|^2 \leq \|T^{2(k+1)}x\|$$

which is equivalent to the paranormality.

The following theorem gives a sufficient condition for an operator $T \in L(H)$ to be paranormal.

7.1.11 THEOREM If $T \in L(H)$ has the property

$$T^{*k}T^k - (T^*T)^k \geq 0$$

then T is of class (N,k) (i.e., k-paranormal).

Proof. For any x, $\|x\| = 1$, we have

$$<T^{*k}T^k x, x> - <(T^*T)^k x, x> \geq \|T^k x\| - <TITx, x>^k$$
$$= \|T^k x\|^2 - \|Tx\|^{2k} \geq 0$$

and the assertion is proved.

7.1.12 *THEOREM* If $T \in L(H)$ and for all $\lambda \in R$,

$$T^{*k}T^k - 2\lambda(T^*T)^{k/2} + \lambda^2 \geq 0$$

then T is of class (N,k).

Proof. As in the case of Theorem 7.1.11.

The following result gives another connection between hyponormal operators and operators of class (N).

7.1.13 *THEOREM* The strong closure of hyponormal operators is contained in the class of paranormal operators.

Proof. For any x, $\|x\| = 1$, we have

$$\lim \|T_n x\|^2 = \|Tx\|^2 \leq \lim \|T_n^2 x\| = \|T^2 x\|$$

The following theorem gives some new properties of hyponormal operators.

7.1.14 *THEOREM* If $T \in L(H)$ is a hyponormal operator, then

1. $T + zI = T + z$ is hyponormal.
2. $aT + bT^*$ is hyponormal or seminormal.
3. If H_1 is an invariant subspace of T, then the restriction of T on H_1, $T_1 = T/H_1$, is also a hyponormal operator.

Proof. The first assertion is obvious. For the second we note that

$$(aT + bT^*)^*(at + bT^*) - (aT + bT^*)(aT + bT^*)^*$$
$$= (|a|^2 - |b|^2)(T^*T - TT^*)$$

and the assertion follows.

For 3 we remark that for all $x \in H_1$, $P_1T_1P_1x = T_1x$, where P_1 is the orthogonal projection of H onto H_1. From this we obtain

$$\|T_1^* x\| = \|P_1 T^* P_1 x\| \le \|T_1^* P_1 x\| \le \|T_1 P_1 x\| = \|P_1 T P_1 x\| = \|T_1 x\|$$

and the assertion is proved.

The following class of operators is sometimes of interest [V. I. Istrăţescu and I. I. Istrăţescu (1967)].

7.1.15 DEFINITION An operator $T \in L(H)$ is called an *invariant normaloid* if the restriction of T to any invariant subspace is a normaloid operator.

From the definition of classes (N,k) and hyponormal operators we have the following theorem.

7.1.16 THEOREM Every hyponormal or class (N,k) operator is an invariant normaloid operator.

We mention now another generalization of hyponormal operators.

7.1.17 DEFINITION An element $T \in L(H)$ is called *hyponormal of order m* if

$$H^m J - JH^m = iC_m$$

where C_m is a positive operator and $T = H + iJ$.

7.2 SPECTRAL SETS AND DILATIONS OF OPERATORS

The notion of a spectral set for an operator on a Hilbert space was introduced by J. von Neumann (1951). There exists a strong relationship between spectral sets and the unitary dilations of operators. Our purpose here is to give some results about these as well as some applications to nonnormal operators.

7.2.1 DEFINITION Let H be a Hilbert space and K be another Hilbert space with $P: K \to H$, the orthogonal projection of K onto H.

If $T \in L(H)$ is an operator such that for some $S \in L(K)$

$$Tx = PSx \qquad x \in H$$

then we say that S is the *dilation* of T. The space K is called the *dilation space* and T is called also the *compression* of S.

A classical result of Halmos (1950) asserts that every contraction (i.e., $\|T\| \leq 1$) has a unitary dilation. This result was much improved by Nagy (1952) by showing that there exists a unitary operator U such that for all n,

$$T^n x = PU^n x$$

Suppose now that X is a compact set in the complex plane and R(X) denotes the algebra of all complex-valued rational functions whose poles are not in X. Then we say that N is the *X-dilation* of T if for all $f \in R(X)$, f(T) is the compression of f(N). Of course we have assumed that f(T) and f(N) exist.

If the operator N is normal, we say that we have *X-normal dilation*.

The notion of the spectral set considered by J. von Neumann is defined as follows.

7.2.2 DEFINITION A closed set in the complex plane is called a *spectral set* for the operator $T \in L(H)$ if

1. $\sigma(T) \subseteq X$
2. For all $f \in R(X)$,

$$\|f(T)\| \leq \|f\|_X = \sup \{|f(z)| \mid z \in X\}$$

A famous result of von Neumann is that the set

$$\{z \mid |z| \leq 1\}$$

is a spectral set for any contraction.

In what follows we give some results about dilations of operators and spectral sets.

7.2.3 THEOREM Every contraction has a unitary dilation.

Proof. Let $T: H \to H$ and $K = H \times H$. The operator U with the matrix

$$\begin{bmatrix} T & S \\ -Z & T \end{bmatrix}$$

where $S = (I - TT^*)^{1/2}$ and $Z = (I - T^*T)^{1/2}$ is unitary. Clearly $PUx = Tx$ where we have identified $x \in H$ with $(x, 0) \in K$.

We see that U is unitary as follows:

$$U^*U = \begin{bmatrix} T^*T + Z^2 & T^*S - ST^* \\ ST - TZ & S^2 + TT^* \end{bmatrix}$$

$$UU^* = \begin{bmatrix} TT^* + S^2 & -TZ + ST \\ -ZT + TS & Z^2 + T^*T \end{bmatrix}$$

and $I - T^*T = Z^2$, $I - TT^* = S^2$ shows that the elements on the main diagnonal are I. To prove the assertion of the theorem, it remains to show that $ST = TZ$ since $T^*S = ZT^*$ is obvious.

But we remark that

$$S^2T = (I - TT^*)T = T(I - T^*T) = TZ^2$$

and for all n,

$$S^{2n}T = TZ^{2n}$$

Thus for any polynomial $p(\cdot)$ we have

$$p(S^2)T = Tp(Z^2)$$

and since S and Z are limits of polynomials the assertion is proved.

7.2.4 DEFINITION If S is a dilation of T and for all $n \geq 1$,

$$T^n x = PS^n x$$

S is called the *strong dilation* of T. If K is the smallest space containing H and reducing for S, then S is called the *minimal dilation* of T.

7.2.5 REMARK In Theorem 7.2.3 the dilation is not strong since

$$PU^2x = (T^2 + ZT)x$$

The following theorem gives the existence of the minimal dilation of an operator.

7.2.6 THEOREM Every contraction on a Hilbert space H as a unique minimal strong dilation.

Proof. Let $T: H \to H$ and $\|T\| \le 1$. We consider $K = \Sigma H_i$, $H_i = H$, the space of all sequences $x = (x_i)$, $\Sigma \|x_n\|^2 < \infty$. The space H is identified with the subset of K of elements of the form

$$(\ldots, \underset{-1}{0}, \underset{0}{x}, \underset{1}{0}, \underset{2}{0}, \ldots)$$

Since every operator $S: K \to K$ is defined by a matrix (S_{nm}) where $S_{nm}: H \to H$ in such a way that if $x = (x_i)$ and $y = Sx = (y_i)$, then

$$y_n = \Sigma S_{nm} x_n$$

Thus for each $x \in H$ we have

$$PSx = P(S_{n0}x) = S_{\infty} x$$

The operator $\tilde{U}: K \to K$ is defined by the matrix (U_{nm}), where

$$U_{00} = T \qquad U_{01} = S \qquad U_{-10} = Z \qquad U_{-11} = -T^*$$

$$U_{n(n+1)} = I \qquad n \neq 0, -1$$

$U_{nm} = 0$ in the contrary case.

Here $S = (I - TT^*)^{1/2}$ and $Z = (I - T^*T)^{1/2}$. First we prove that $\tilde{U}$ is a unitary operator. Let $x = (x_n)$ and $y = Ux$. Thus $y_n = \Sigma U_{nm} x_m$ and from the definition of U_{nm} we have

$$y_{-1} = Sx_0 - Tx_1$$

$$y_0 = Tx_0 - Zx_1$$

$$y_n = x_{n+1} \qquad n \neq 0, -1$$

and thus since

$$\|y_{-1}\|^2 + \|y_0\|^2 = \|Sx_0 - T^*x_1\|^2 + \|Tx_0 - Zx_1\|^2 = \|x_0\|^2 + \|x_1\|^2$$

and we obtain that U is an isometric operator. We show now that it is surjective and this proves the unitarity of $\tilde{U}$. Indeed, let $y = (y_n) \in K$ and define $x = (x_n) \in K$ by the relations

$$x_0 = Sy_{-1} + T^*y_0$$

$$x_1 = -Ty_{-1} + Zy_0$$

$$x_n = y_{n+1} \qquad n \neq 0, 1$$

and clearly $\tilde{U}x = y$. Thus $\tilde{U}$ is unitary and we prove now that $\tilde{U}$ is a strong minimal dilation of T. Let K_0 be the set of all elements of K of the form $x = (x_n)$ with

1. $x_n \in \{Sx\}_{x \in H}$ $\qquad n \leq -1$
2. $x_0 \in H$
3. $x_n \in \{Zx\}_{x \in H}$ $\qquad n \geq 1$

Let $\tilde{U}x = y$ with $x \in K_0$. Clearly $y_{-n} \in \{Sx\}_{x \in H}$, $y_0 \in H$, and $y_n \in \{Zx\}_{x \in H}$; thus $y \in K_0$. Now if $y \in K_0$, we can find $x \in K_0$ such that $\tilde{U}x = y$ and the space K_0 reduces $\tilde{U}$. Since

$$\tilde{U}^n x = (\ldots, 0, Sx, STx, ST^2x, \ldots, ST^{n-1}x, T^n x, 0, \ldots)$$

$$\tilde{U}^* x = (\ldots, 0, T^{*n}x, ZT^{*(n-1)}x, \ldots, Zx, 0, \ldots)$$

$\tilde{U}$ is clearly the strong minimal dilation of T.

7.2.7 COROLLARY The set $\{z \mid |z| \leq 1\}$ is a spectral set for $T \in L(H)$ if and only if $\|T\| \leq 1$.

Proof. Let $T \in L(H)$ such that $\|T\| \leq 1$. We find a unitary strong dilation of T, say U. In this case, for any $f \in R(X)$, we have

$$\sup_{\|x\| \leq 1} \|f(T)x\| = \|f(T)\| \leq \sup \|f(U)x\| = \|f(U)\| = \|f\|_X$$

Now, if $x = \{z \mid |z| \leq 1\}$ is a spectral set for T, we have for $f(z) = z$, $\|T\| \leq \|z\|_X = 1$.

The following measure theoretic characterization of spectral sets is very useful in some cases.

7.2.8 THEOREM Let X be a compact set in the complex plane. X is a spectral set for the operator $T \in L(H)$ if and only if for all $x, y \in H$ there exists a Borel measure $\mu_{x,y}$ with the support in ∂X and such that

$$\int f(s)\, d\mu_{x,y}(s) = \langle f(T)x, y \rangle$$

for all $f \in R(X)$ and satisfying the following conditions:

1. $\|\mu_{x,y}\| \le \|x\|\|y\|$
2. $\mu_{x,x} \ge 0 \qquad \|\mu_{x,x}\| = \|x\|^2$

Proof. First we suppose that these measures exist and have the above properties. In this case for all $f \in R(X)$ we have

$$\|f(T)x\|^2 = \langle f(T)x,\ f(T)x\rangle = \int f(s)\ d\mu_{x,f(T)x}$$
$$\le \|f\| \cdot \|\mu_{x,f(T)x}\| \le \|f(T)\| \cdot \|x\| \cdot \|f(T)x\|$$

and this implies that

$$\|f(T)x\| \le \|f(T)\|\|x\|$$

which shows that X is a spectral set of T.

Suppose now that X is a spectral set for T. In this case for any x and y in H we define the linear functional on R(X) by the relation

$$F_{x,y}(f) = \langle f(T)x,\ y\rangle$$

and of course we can consider this on the closure of R(X).

Since

$$|F_{x,y}(f)| \le \|f\| \cdot \|x\| \cdot \|y\| = \|f\| \cdot \|x\| \cdot \|y\|$$

we have that F is a continuous linear functional on R(X).

Using the Hahn-Banach theorem and the Riesz-Kakutani representation theorem, we find a measure $\mu_{x,y}$ such that

1. $\|\mu_{x,y}\| = \|F_{x,y}\| \le \|x\| \cdot \|y\|$
2. $\int f\ d\mu_{x,y} = \langle f(T)x,\ y\rangle$

and clearly $\|x\|^2 = \int d\mu_{x,x} = \|\mu_{x,x}\|$.

In what follows we give some results about maximal dilations. For this we need the notion of operator measure and the notion of spectral measure.

7.2.9 DEFINITION Let B^2 be the Borel sets in the complex plane and $F: B^2 \to L(H)$. F is called an operator measure if

1. $0 \le F(\delta) \le I$.
2. $F(\emptyset) = 0$ and $F(R \times R) = I$.
3. F is countable additive in the weak operator topology.

The *support* of an operator measure is the intersection of all closed sets δ for which $F(\delta) = I$. The operator measure is called a *spectral measure* if $F(\delta)$ are commuting projections.

The following basic result, which we give without proof, shows a connection between these notions.

7.2.10 THEOREM [Naimark (1940, 1943)] The set function $F: B^2 \to L(H)$ is an operator measure if and only if there exists a spectral measure on a Hilbert space $K \supset H$ such that

$$F(\delta) = PE(\delta)$$

for all $\delta \in B^2$. The space K can be chosen as minimal, i.e.,

$$\overline{\{E(\delta)x \mid x \in H\}} = K$$

For the applications we need the notion of a Dirichlet algebra. If A is a function algebra, then A is called a *Dirichlet algebra* if the real space Re $A = \{\text{Re } f \mid f \in A\}$ is dense in the space $C_r(\partial X)$.

We remark that the notion of Dirichlet algebra was introduced by A. Gleason [see Browder (1969)], and that there exist two classes of function algebras which contain the class of Dirichlet algebras and have many common properties: the logmodular algebras and the algebras with unique representing measures. We conjecture that all the results which follow are also valid for these algebras.

For details and further information about these algebras, we refer to Hoffman (1962) and Browder (1969).

7.2.11 THEOREM Let X be a compact set in the complex plane for which R(X) is a Dirichlet algebra. If X is a spectral set for $T \in L(H)$, then there exists a strong minimal dilation of T such that $\partial X \supset \sigma(N)$. The dilation is normal.

Proof. Since X is a spectral set for T, by Theorem 7.2.8 there exists for each pair $x,y \in H$, a Borel measure $\mu_{x,y}$ such that

$$\int f(s)\, d\mu_{x,y} = \langle f(T)x, y\rangle$$

Since $\mu_{x,x}$ is positive, we have

$$\langle f(T)^*x, x\rangle = \langle x, f(T)x\rangle = \overline{\langle f(T)x, x\rangle} = \int \bar{f}\, d\mu_{x,x}$$

Let $f \in C(\partial X)$ and $f = f_1 + if_2$, where f_i are real-valued functions, and we define

$$f(T) = f_1(T) + if_2(T)$$

if $f_i \in R(\partial X)$. We have

$$\int f\, d\mu_{x,x} = \int f_1\, d\mu_{x,x} + i \int f_2\, d\mu_{x,x} = \langle f(T)x, x\rangle$$

For all $x,y \in H$ we define the measure

$$\mu_{x,y} = \frac{1}{4} (\mu_{x+y,x+y} \mu_{x-y,x-y} + i\mu_{x+iy,x+iy} - i\mu_{x-iy,x-iy})$$

We remark that if $f = f_1 + \bar{f}_2$ with $f_i \in \overline{R(\partial X)}$, then

1. $\int f\, d\mu_{x,y} = \langle f(T)x, y\rangle$
2. $\int f\, d_{x_1+x_2,y} = \int f\, d_{x_1,y} + \int f\, d_{x_2,y}$

and since $\overline{R(X)}$ is a Dirichlet algebra, relation 2 holds for all $f \in C(\partial X)$. From the regularity of the measure $\mu_{x,x}$ we obtain that

$$\mu_{x_1+x_2 y} = \mu_{x_1,y} + \mu_{x_2,y}$$

Similarly, we can prove that

$$\mu_{x,y_1+y_2} = \mu_{x,y_1} + \mu_{x,y_2} \qquad \mu_{ax,by} = a\bar{b}\mu_{x,y}$$

From the definition of the measure $\mu_{x,y}$ we obtain that

$$\|\mu_{x,y}\| \le \frac{1}{4} (\|x + y\|^2 + \|x - y\|^2 + \|x + iy\|^2 + \|x - iy\|^2)$$
$$= \|x\|^2 + \|y\|^2$$

Thus, if $\|x\| - 1 = \|y\|$, we have $\|\mu_{x,y}\| \le 2$, and since

$$\|\mu_{x,y}\| = \|\mu_{x/\|x\|,y/\|y\|}\|x\|\|y\|\|$$

we obtain that

$$\|\mu_{x,y}\| \le 2\|x\|\|y\|$$

From this it follows that for all Borel sets δ, there exists a bounded linear operator $F(\delta)$ such that

$$\langle F(\delta)x, y\rangle = \mu_{x,y}(\delta) \qquad (*)$$

Since for all x we have

$$0 \le \mu_{x,x}(\delta) \le \langle x, x\rangle \qquad \langle F(\delta)x, x\rangle = \mu_{x,x}(\delta)$$

it follows that $F(\delta)$ is hermitian and $0 \le F(\delta) \le I$.

From (*) it follows that $\delta \mapsto F(\delta)$ is countable additive, which gives that $\delta \mapsto F(\delta)$ is an operator measure.

For any $f \in R(X)$ we have that

$$\langle (\int f\, dF)x, y\rangle = \int f(z)\, d\langle F(z)x, y\rangle = \int f\, d\mu_{x,y} = \langle f(T)x, y\rangle$$

and thus F is an X-measure for T.

Since by the Neumark theorem every X-measure is the compression of a spectral measure, say E, then we can define the normal operator

$$N = \int z\, dE(z)$$

and for any $f \in R(X)$,

$$f(T) = \int f(z)\, dF(z) = \int f(z)\, dPE(z)P = P\int f(z)\, dE(z)P$$
$$= Pf(N)P$$

Now if E is such that $E(\delta)x = 0$ for all $x \in H$, then from the minimality in Neumark's theorem we have $E(\delta) = 0$. Now if $F(\delta) = 0$, then for any $x \in H$ we have

$$\|E(\delta)x\|^2 = \langle E(\delta)x, E(\delta)x\rangle = \langle E(\delta)x, x\rangle = \langle F(\delta)x, x\rangle = 0$$

and this tells us that the supports of the two sets of functions are the same set. Since the support of F is in ∂X the spectrum of N is also in ∂X.

The following corollaries are important in applications as well as having their own interest.

7.2.12 COROLLARY If X is the boundary of one component of the complement of X, then every operator T having X as a spectral set has a normal X-dilation.

Proof. A famous theorem of J. L. Walsh [see Browder (1969)] asserts that R(X) is a Dirichlet algebra and the assertion follows from the above results.

7.2.13 COROLLARY (Berger-Foias-Lebow) If X is a compact set which is a spectral set for the operator T, then there exists a normal dilation N such that N^n is a dilation for T^n and $\sigma(N) \subset \partial X$.

Proof. We can consider the set $X^{\sim} = X \cup C_i$, C_i are all bounded components of the complement of X. Clearly $X^{\sim}$ is a spectral set for T and the boundary of this spectral set is a subset of the boundary of X. Thus Corollary 7.2.12 applies.

7.2.14 COROLLARY [Nagy (1952)] If T is a contraction on a Hilbert space, then there exists a unitary dilation U such that for all n, U^n is a dilation of T^n.

Proof. By Corollary 7.2.7, a result of von Neumann (1951), we have that for contractions the set $\{z \mid |z| \leq 1\}$ is a spectral set. The assertion of the corollary follows from the above results.

An interesting problem appears when the spectrum of the normal dilation has some additional properties or the fact that if X is a spectral set for T then X is a set with special properties.

We quote two results in this direction. For the proof, see the original papers, where other interesting questions are also treated.

7.2.15 THEOREM [Lebow (1963)] If T and N are as in Theorem 7.2.11, and $\sigma(N)$ does not separate the plane, then T is normal.

7.2.16 THEOREM Let T be a bounded operator on a Hilbert space H satisfying the following property: For any $f \in R(X)$

$$\|f(T)\| \leq k\|f\|_X$$

and $\sigma(T) \subset X$. If R(X) is dense in C(X), then T is similar to a normal operator.

7.2.17 REMARK It is possible for the spectrum of T to be a spectral set for T with T not similar to any normal operator.

Indeed, for this we remark that the monotone shift with the weight sequence bounded by 1 has the set $\{z \mid |z| \leq 1\}$ as a spectral set and clearly it is not similar to a normal operator.

We give now some results which show a connection between spectral sets, the numerical range, and the normal dilation.

7.2.18 THEOREM If $T \in L(H)$ has the closed numerical range $\overline{W(T)}$ as a spectral set, then the following relations hold:

$$\text{conv } \sigma(T) = \overline{W(T)} = \overline{W(N)}$$

where N is a strong dilation of T.

Proof. Since W(T) is a spectral set for T, we have that

$$\|T + zI\| \leq \sup \{|s + z| \mid s \in W(T)\} \leq \|T + zI\|$$

which shows that T is transaloid. The first assertion follows from Theorem 6.15.11. Now from the above result we find a strong dilation N such that $\sigma(N) \subset \partial \text{ conv } \sigma(T)$. But

$$\text{conv } \sigma(N) \subset \text{conv } \partial\sigma(T) = \text{conv } \sigma(T)$$

and since N is normal

$$\text{conv } \sigma(T) = \overline{W(T)} \subset \overline{W(N)} = \text{conv } \sigma(N) = \text{conv } \sigma(T)$$

and the assertions of the theorem are proved.

The following result gives a characterization of spectral sets which are closed numerical ranges.

7.2.19 THEOREM If $T \in L(H)$, then W(T) is a spectral set for T if and only if there exists a strong dilation of T, say N, such that $\overline{W(T)} = \overline{W(N)}$.

Proof. From Theorem 7.2.18 it follows that the condition is necessary. We now show that it is sufficient. Let z be sufficiently large that the Neumann series converges for $(T - z)^{-1}$ and $(N - z)^{-1}$. In this case, for any $x \in H$ we have

$$(T - z)^{-1}x = P(N - z)^{-1}x$$

and thus for all $y \in H$ we obtain that

$$<(T - z)^{-1}x, y> = <(N - z)^{-1}x, y>$$

and since $W(T) = W(N)$ for any $f \in R(W(T))$,

$$<f(T)x, y> = <f(N)x, y>$$

we obtain that

$$f(T)x = Pf(N)x$$

for all $x \in H$. This obviously implies that $\overline{W(T)}$ is a spectral set for T since

$$\|f(T)\| = \sup \{\|f(T)x\| \mid \|x\| = 1\} = \sup \{|f(z)| \mid z \in \sigma(N)\}$$
$$= \sup \{|z| \mid z \in W(N)\} = \sup \{|z| \mid z \in W(T)\}$$

7.3 OPERATORS WITH G_1 PROPERTY AND SOME GENERALIZATIONS

Let H be a Hilbert space and L(H) be the set of all bounded linear operators on H. If $z \in \rho(T)$, then from the spectral mapping theorem it follows that

$$\frac{1}{d(z, \sigma(T))} \le \|(T - z)^{-1}\|$$

[Here and in what follows $d(z, \sigma(T)) = \inf \{|z - t| \mid t \in \sigma(T)\}$.]

The following interesting and important class of operators is considered with respect to the growth of the resolvent.

7.3.1 DEFINITION An operator $T \in L(H)$ is said to have G_1 property if

$$\|(T - z)^{-1}\| = \frac{1}{d(z, \sigma(T))}$$

for all $z \in \rho(T)$.

As an example of the operator satisfying the G_1 property we can quote the normal operators and the hyponormal operators.

7.3.2 THEOREM If $T \in L(H)$ has G_1 property, then $\operatorname{conv} \sigma(T) = \overline{W(T)}$.

Proof. As we know for every bounded operator $\sigma(T) \subset \overline{W(T)}$. Thus to prove the theorem, we show that $\overline{W(T)}$ is in conv $\sigma(T)$. For this it is sufficient to show that every half plane containing $\sigma(T)$ contains also $W(T)$. We can assume, without loss of generality, that our half plane is $\text{Re } z \leq 0$. Let $\text{Re } T \leq 0$ and we prove that $\text{Re } W(T) < 0$. Let x, $\|x\| = 1$, and $Tx = (a + ib)x + y$ with a and b real numbers and $\langle x, y\rangle = 0$. Since T is with G_1 property, we obtain

$$c^2 \leq \|(T - cI)x\|^2 = (a - c)^2 + b^2 + \|y\|^2$$

which implies that

$$2ac \leq a^2 + b^2 + \|y\|^2$$

and since c is an arbitrary real number, we obtain a contradiction since the right part is bounded. Thus $\text{Re } \langle Tx, x\rangle = a < 0$.

The following theorem gives the structure of operators with G_1 property in the case when $\sigma(T)$ is a finite set.

7.3.3 THEOREM If $T \in L(H)$ and has the following properties

1. T has G_1 property.
2. $\sigma(T)$ is a finite set.

then in this case T is normal.

Proof. Let $\sigma(T) = \{a_1, a_2, \ldots, a_n\}$ and for each a_j let $C_{j,\varepsilon}$ be a circle with the center in a_j with radius ε; moreover, ε is so small that for all j, the intersection of the circle with $\sigma(T)$ contains no other points. We now define the projection

$$E_j = \frac{1}{2\pi i} \int_{C_{j,\varepsilon}} (T - z)^{-1} \, dz$$

which clearly commutes with T. Since

$$\|E_j\| \leq \frac{1}{2\pi} \left\| \int_{C_{j,\varepsilon}} (T - z)^{-1} \, dz \right\| < \frac{1}{2} (2)\varepsilon \frac{1}{\varepsilon} = 1$$

it follows that E_j is a hermitian projection. Let $x \in E_jH$ and thus

$$\|(T - a_j)x\| = \left\|\frac{1}{2\pi i}\int_{C_{j,\varepsilon}} (z - a_j)(T - z)^{-1}\, dz\right\| < \varepsilon$$

and for $\varepsilon \to 0$ we obtain $Tx = a_j x$.

Since $T = \Sigma\, a_j E_j$, it follows that T is normal.

7.3.4 COROLLARY On finite dimensional Hilbert spaces every operator with G_1 property is normal.

The following theorem gives interesting information about operators with G_1 property and arbitrary operators in L(H).

7.3.5 THEOREM The set of all $T \in L(H)$ with G_1 property is strongly dense in L(H).

Proof. Let $T \in L(H)$ arbitrary and $\{x_1, x_2, \ldots, x_n\} \subset H$ and let M be the subspace generated by $\{x_1, x_2, \ldots, x_n\}$. We define the operator on H as follows:

$$Ax_i = Tx_i \qquad i = 1, 2, \ldots, n$$

and $Ax = 0$ if $x \in M^{\perp}$. We can consider a normal operator on $M^{\perp}$ such that $\sigma(N) = \{z \mid |z| \leq \|A\|\}$. The operator $A \oplus N$ satisfies the G_1 property, and since it coincides with T on the x_i, the strong density is clear.

We give now the definition of several classes of operators in $T \in L(H)$ which are, in some sense, generalizations of the class of operators with G_1 property.

7.3.6 DEFINITION Let $T \in L(H)$. T satisfies the sequential G_1 property if for every $z \in \partial\sigma(T)$ there exists a sequence $\{z_n\} \subset \rho(T)$ such that

1. $z_n \to z$
2. $\|(T - z_n)^{-1}\| = 1/d(z_n, \sigma(T))$ for all n

As a simple example of an operator with this property, we note that the Volterra operator on L^2 is with the sequential G_1 property and clearly is not with the G_1 property. Another example is given in the following.

7.3.7 THEOREM If $T \in L(H)$ is quasinilpotent and $0 \in \partial W(T)$, then T is with the sequential G_1 property.

Proof. As we know $W(T)$ is a convex set and thus there is a line support passing through 0. Thus we can assume, without loss of generality, that $W(T) \subset \text{Re } z \geq 0$. But $0 \in \partial W(T)$ and $\sigma(T) = \{0\}$. Since for any operator

$$\|(T - z)^{-1}\| \leq \frac{1}{d(z, \overline{W(T)})}$$

we can find that in our case, for any real negative number z,

$$d(z, W(T)) = |z|$$

and thus for the proof of the theorem we can take $z = -1/n$.

7.3.8 THEOREM If $T \in L(H)$ and is with the sequential G_1 property and for some m, $T^m = 0$, then $T = 0$.

Proof. Since T is with the sequential G_1 property and is nilpotent, it follows that there exists a sequence $\{z_n\}$ such that

$$\|(T - z_n)^{-1}\| = \frac{1}{|z_n|}$$

for all n. Suppose now $m > 1$ and thus

$$(z_n - T)^{-1} = \sum_{i=0}^{m-1} \frac{T^i}{z_n^{i+1}}$$

and this implies that

$$\frac{\|T^{m-1}\|}{|z_n|^m} - \sum_{i=0}^{m-2} \frac{\|T^i\|}{|z_n^{i+1}|} < \frac{1}{|z_n|}$$

for all n. This gives

$$\|T^{m-1}\| < |z_n^{m-1}| + \sum_{i=0}^{m-2} \frac{\|T^i\|}{|z_n|^{m-i-1}}$$

for all n. But

$$|z_n|^{m-1} + \sum_{i=0}^{m-2} \frac{\|T^i\|}{|z_n|^{m-i-1}} \to 0 \qquad n \to \infty$$

and this gives that $\|T^{m-1}\| = 0$. An induction argument shows that $T = 0$.

The following theorem gives the structure of a class of operators with the sequential G_1 property.

7.3.9 DEFINITION An operator $T \in L(H)$ is called *algebraic* if there exists a polynomial $p(z)$, not identically zero, such that $p(T) = 0$.

7.3.10 THEOREM Suppose $T \in L(H)$ has the following properties:

1. T is algebraic.
2. T has the sequential G_1 property.

Then T is normal.

For the proof we need the following useful lemma of Putnam and Stampfli.

7.3.11 LEMMA Suppose $T \in L(H)$ and $z_0 \in \sigma(T)$ have the following properties:

1. $Tx = z_0 x$, $\|x\| = 1$.
2. There exists a sequence $\{z_n\} \in \rho(T)$ such that $z_n \to z_0$ and $\|(T - z_n)^{-1}\|/|z_n - z_0| \to 0$.

Then $T^*x = z_0^* x$.

Proof. Without loss of generality, we can assume that $z_0 = 0$.

Suppose that $T^*x = y$. In this case $(T^* - z_n^*)x = y - z_n^* x$ and thus

$$z_n^*(T^* - z_n^*)^{-1}x = x + (T^* - z_z^*)^{-1}y$$

But x is orthogonal to $(T^* - z_n^*)y$; thus we obtain

$$\|(T^* - z_n^*)^{-1}y\|^2 + \|x\|^2 = |z_n^*|^2\|(T^* - z_n^*)^{-1}x\|^2$$

and

$$\lim \|(T^* - z_n^*)^{-1}y\| = 0$$

Since

$$\|y\| = \lim \|((T^* - z_n^*)(T^* - z_n^*)^{-1}y\| = 0$$

the assertion is proved.

Proof of Theorem 7.3.10. Since T is algebraic, from the spectral mapping theorem it follows that $\sigma(T) = \{z_1, z_2, \ldots, z_m\}$ with the multiplicities $\{n_i\}$. In this case $H = \Sigma_{i=1}^m H_i$, where

$$H_i = \{x \mid (T - z_i)^{n_i}x = 0\}$$

and clearly $\sigma(T/H_i) = \{z_i\}$. Since T has the sequential G_1 property, it follows that T/H_i also has the same property. But $T - z_i$ is a nilpotent operator on H_i. From the above result, it follows that $(T - a_i)/H_i = 0$. By the lemma of Putnam and Stampfli it follows that $(T^* - a_i^*)/H_i = 0$, which implies that for $i \neq j$, $H_i \perp H_j$.

Thus we have that $H = \Sigma H_i$ and $T = \Sigma a_iP_i$, where P_i is the orthogonal projection of H onto H_i.

The following class of operators was considered by Orland (1964) and is a natural generalization of operators with the G_1 property.

7.3.12 DEFINITION An operator $T \in L(H)$ is said to be with spectral G_1 property if for all $z \notin \text{conv}\,\sigma(T)$,

$$\|(T - z)^{-1}\| \leq \frac{1}{d(z, \text{conv}\,\sigma(T))}$$

The following theorem gives the characterization of this class of operators.

7.3.13 THEOREM An operator $T \in L(H)$ is with spectral G_1 property if and only if $\text{conv}\,\sigma(T) = \overline{W(T)}$.

For the proof we need some lemmas.

7.3.14 LEMMA Let $T \in L(H)$. Then T is with spectral G_1 property if and only if

$$d(z, \text{conv}\,\sigma(T))\|y\| \leq \|Ty - zy\|$$

for all $z \notin \text{conv}\,\sigma(T)$.

Proof. Verify by computation.

7.3.15 *LEMMA* Let $T \in L(H)$ with $\|R_z(T)\| \le 1/d(z, \text{conv}\,\sigma(T))$ for some $z \notin \text{conv}\,\sigma(T)$. Then for any $a \ne 0$ and b complex numbers the transformation $z \mapsto az + b$ maps the complement of $\text{conv}\,\sigma(T)$ onto complement of $\text{conv}\,\sigma(aT = bI)$ and

$$\|R_{az+b}(aT + bI)\| \le \frac{1}{d(az + b, \text{conv}\,\sigma(aT + bI))}$$

Proof. Since $a \ne 0$ the complex plane is mapped onto and by the spectral mapping theorem, $\sigma(T)$ is mapped onto $\sigma(aT + b)$. From this it follows that $\text{conv}\,\sigma(T)$ is mapped onto $\text{conv}\,\sigma(aT + bI)$.

Taking the complements, the first assertion follows. For the second we note that by Lemma 7.3.14, for all y we have

$$d(z, \text{conv}\,\sigma(T))\|y\| \le \|(T - z)y\|$$

and this gives us

$$\begin{aligned} |a|d(z, \text{conv}\ (T))\|y\| &= d(az, a\ \text{conv}\ (T))\|y\| \\ &= d(az, \text{conv}\ (aT))\|y\| \\ &= d(az + b, \text{conv}\ (aT + bI))\|y\| \\ &= |a|\|Ty - zy\| = \|(aT + bI)y - (a + b)y\| \end{aligned}$$

and the assertion of the lemma is proved.

7.3.16 *LEMMA* If $T \in L(H)$, then $\|R_s(T)\| \le 1/s$ for all $s > 0$ if and only if $\text{Re}\, W(T)$ is in $\{x \mid x \le 0\}$.

Proof. Let $y \in H$ and $s > 0$. Then $s\|y\| \le \|(T - s)y\|$ if and only if

$$s^2\|y\|^2 \le \|Ty\|^2 + s^2\|y\|^2 - 2s\ \text{Re}\ \langle Tx, x\rangle$$

and this gives us

$$2s\ \text{Re}\ \langle Tx, x\rangle \le \|Ty\|^2$$

for all $s > 0$ if and only if $\text{Re}\ \langle Tx, x\rangle \le 0$.

We are now in a position to prove Theorem 7.3.13.

Suppose first that T is with spectral G_1 property. As we know for any operator $T \in L(H)$, $\text{conv}\,\sigma(T) \subseteq \overline{W(T)}$. To prove the assertion we show the opposite inclusion.

Let L be a line of support of conv $\sigma(T)$ and the closed half plane which contains conv $\sigma(T)$; we show that it contains $\overline{W(T)}$. If L touches conv $\sigma(T)$ at a point, say d_0, we find the complex numbers a and b, $|a| = 1$, such that

$$z \mapsto az + b$$

sends d_0 into the origin, L into the y axis, and conv $\sigma(T)$ into a part of the left closed half plane. From Lemma 7.3.15 we obtain

$$\|R_{az+b}(aT + b)\| \leq \frac{1}{d(az + b, \text{conv } \sigma(aT + b))}$$

for all $az + b \notin \text{conv } \sigma(aT + b)$. We can choose a and b such that $az + b$ is a strictly positive number and apply Lemma 7.3.16, and thus

$$\text{Re } W(aT + b) \leq 0$$

and this implies that W(T) is on the same side of L as conv $\sigma(T)$.

Conversely, suppose that conv $\sigma(T) = \overline{W(T)}$; we prove that T is with spectral G_1 property.

Indeed, let z be any point in $C_{\overline{W(T)}=\text{conv }(T)}$. We can find a unique point $d \in \overline{W(T)}$ such that $|d - z| = d(z, \overline{W(T)})$. The line through d and perpendicular to the line joining d and z is a line of support of W(T); we denote this line as L. Hence, we can find the complex numbers a and b such that $z \mapsto az + b$ carries d into the origin, L into the y axis, and z into $|z - d|$. Since $\overline{W(aT + b)} = \text{conv } (aT + b)$, we have, by Lemma 7.3.16,

$$\|R_{|z-d|}(aT + b)\| \leq \frac{1}{|z - d|} = \frac{1}{d(|z - d|, \text{conv } \sigma(aT + b))}$$

Since the inverse of the map $z \mapsto az + b$ has an inverse mapping of the same form, Lemma 7.3.15 tells us that

$$\|R_z(T)\| \leq \frac{1}{d(z, \text{conv } \sigma(T))}$$

Another class of operators can be considered by weakening the condition in the G_1 property, in fact, by supposing that z is in an open neighborhood of $\sigma(T)$.

7.3.17 DEFINITION An operator $T \in L(H)$ is said to be of class G_1^{loc} if there exists an open neighborhood U of $\sigma(T)$ such that for all $z \in U \setminus \sigma(T)$,

$$\|R_z(T)\| \leq \frac{1}{d(z, \sigma(T))}$$

We show now that this class is strictly larger than the class of operators with the G_1 property.

Let H and K be two Hilbert spaces and $H \oplus K$ their orthogonal sum, A be an arbitrary operator in L(H), and N be a normal operator in L(K). On the space $H \oplus K$ we consider the operator $A \oplus N$.

7.3.18 THEOREM The operator $A \oplus N$ is in G_1^{loc} when $\sigma(A) \subset U \subset \sigma(N)$, U open.

Proof. It is obvious that $\sigma(A \oplus N) = \sigma(T) = \sigma(N)$. If $z \in (A \oplus N)$, then

$$\|R_z(T)\| = \max \left\{\|R_z(A)\|, \|R_z(N)\|\right\} = \max \left\{\|R_z(A)\|, \frac{1}{d(z, \sigma(N))}\right\}$$

since N is a normal operator. Now since there exists an open set U containing $\sigma(A)$ and contained in $\sigma(N)$, we can find V open such that

$$V \supset \sigma(N) = \sigma(T)$$

and thus for all $z \in U \setminus \sigma(T)$

$$\|R_z(T)\| \leq \frac{1}{d(z, \sigma(T))}$$

where $A \oplus N = T$.

For the following result we need the notion of a normal eigenvalue for an operator.

For $T \in L(H)$ and z any complex number, let

$$E_z(T) = \{x \mid Tx = zx\}$$

7.3.19 DEFINITION A complex number z is called a *normal eigenvalue* for T if

$$E_{z^*}(T^*) = E_z(T)$$

7.3.20 THEOREM If $T \in G_1^{loc}$ and z_0 is an isolated point in $\sigma(T)$, then z_0 is a normal eigenvalue.

Proof. Let $\varepsilon > 0$ and C_ε such that in the circle C_ε contains no other points of $\sigma(T)$ except z_0.

Let

$$E_{z_0} = \frac{1}{2\pi i} \int_{C_\varepsilon} (T - z)^{-1} \, dz$$

which is a projection and a computation similar to that of Theorem 7.3.3 shows that E_{z_0} is hermitian. Also the same computation as in Theorem 7.3.3 shows that for $x \in E_{z_0} H$, $Tx = z_0 x$, and since

$$\|R_z(T)\| = \|R_z(T^*)\|$$

the assertion follows.

7.3.21 COROLLARY On finite dimensional Hilbert spaces the class of normal operators with G_1 property and the class G_1^{loc} are the same.

7.3.22 EXAMPLE Let A be the operator with the matrix

$$\begin{bmatrix} 0 & 2 \\ 0 & 0 \end{bmatrix}$$

and it is easy to see that

$$\overline{W(A)} = \{z \mid |z| \le 1\}$$

and consider a normal operator N on a Hilbert space such that $\sigma(N) = \{z \mid |z| \le 1/2\}$. In this case the operator $T = A \oplus N$ is in G_1^{loc} and is not with the G_1 property.

7.3.23 REMARK The operator in the example has the property that conv $\sigma(T) \neq W(T)$.

We consider now a new class of operators which is related to the class of operators with spectral G_1 property and G_1^{loc}.

7.3.24 DEFINITION An operator $T \in L(H)$ is said to be in the *class spectral* G_1^{loc} if there exists an open neighborhood of conv $\sigma(T)$, u, such that

$$\|R_z(T)\| \le \frac{1}{d(z, \operatorname{conv} \sigma(T))}$$

for all $z \in U \setminus \operatorname{conv} \sigma(T)$.

Another class can be defined as follows.

7.3.25 *DEFINITION* An operator $T \in L(H)$ is said to be of *class R* if for all $z \notin \sigma(T)$,

$$\|R_z(T)\| = \frac{1}{d(z, \overline{W(T)})}$$

The following theorem gives the characterization of this class of operators.

7.3.26 *THEOREM* The operator $T \in L(H)$ is of class R if and only if

$$\partial W(T) \subset \sigma(T)$$

Proof. First suppose that $\partial W(T) \subset \sigma(T)$ and let $z \in W(T)$. In this case we have $d(z, W(T)) = d(z, \sigma(T))$ and

$$\frac{1}{d(z, \overline{W(T)})} = \frac{1}{d(z, \sigma(T))} \le \|R_z(T)\| \le \frac{1}{d(z, \overline{W(T)})}$$

and thus T is of class R.

Suppose now that T is of class R and let $z_0 \in \partial W(T)$. We can find $\{z_n\}$ such that $|z_n - z_0| \to 0$, $d(z_n, W(T)) > 0$ for all $n = 1, 2, \ldots$. But we have

$$\|R_{z_n}(T)\| = \frac{1}{d(z_n, \overline{W(T)})}$$

and thus $z_0 \in \sigma(T)$.

7.3.27 *THEOREM* If $T \in L(H)$ is of class R, then the following assertions hold:

1. $\operatorname{conv} \sigma(T) = \overline{W(T)}$.
2. If $\sigma(T)$ is a finite set, then T is a multiple of I.

Proof. The first assertion is a simple consequence of the above characterization of operators of class R. For the second, from the Hausdorff-Toeplitz theorem on the convexity of W(T) it follows that W(T) contains only one point, which clearly implies the rest of the theorem.

7.4 OPERATORS WITH THE PROPERTY Re $\sigma(T) = \sigma(\text{Re } T)$

Let T be a bounded linear operator on a Hilbert space; we know that for any rational function with poles in $C_{\sigma(T)}$ the following assertion holds:

$$\sigma(f(T)) = f(\sigma(T)) = \{f(s) \mid s \in \sigma(T)\}$$

Also if T is a normal operator, then for any polynomial p(z, z*) the following spectral mapping theorem holds:

$$(p(T, T^*)) = \{p(z, z^*) \mid z \in \sigma(T)\}$$

and if $p(z, z^*) = 1/2(z + z^*)$, then we obtain that

$$\text{Re } \sigma(T) = \sigma(\text{Re } \sigma T)$$

Simple examples show that this remarkable relation does not hold for arbitrary operators. It is clear that for nilpotent operators or more generally for quasinilpotent operators this fails to hold.

It appears to be a natural problem to find classes of nonnormal operators for which the relation holds. The following theorem gives some classes of operators for which the relation holds.

7.4.1 THEOREM If $T \in L(H)$ and one of the following conditions holds, then T has the property Re $\sigma(T) = \sigma(\text{Re } T)$.

1. T is hyponormal.
2. T* is hyponormal.
3. $\sigma(T)$ is a spectral set for T.
4. T has the G_1 property and $\sigma(T)$ is connected.

The proof of this theorem will be given as a consequence of some lemmas.

7.4.2 LEMMA If $T \in L(H)$ is a hyponormal operator, then Re $\sigma_{ap}(T)$ is in $\sigma(\text{Re } T)$.

Proof. Let $z = x + iy$ in $\sigma_{ap}(T)$. Thus there exists a sequence of unit elements $\{x_n\}$, $\|x_n\| = 1$, such that $(T - z)x_n \to 0$, and since $T - z$ is hyponormal

$$(T^* - z^*)x_n \to 0$$

From these relations it follows that

$$(\text{Re } T)x_n - \text{Re } z\, x_n = \frac{1}{2}[T + T^* - (z + z^*)]x_n \to 0$$

and this proves our assertion.

7.4.3 LEMMA If $T \in L(H)$ is hyponormal, then $\text{Re } \sigma(T) \subset \sigma(\text{Re } T)$.

Proof. Suppose that $z_0 \in \sigma(T)$ and the line $\text{Re } z = \text{Re } z_0$ meets the spectrum in a boundary point $\tilde{z}_0$. Since every boundary point of $\sigma(T)$ is in the approximate spectrum, it follows that $\text{Re } \tilde{z}_0$ is in $\sigma(\text{Re } T)$. But $\text{Re } z_0 = \text{Re } \tilde{z}_0$, and the assertion is proved.

7.4.4 REMARK Clearly the assertion of Lemma 7.4.3 also holds when T^* is hyponormal.

We are now ready to prove assertions 1 and 2 of Theorem 7.4.1. Clearly we can consider only case 1.

As is easy to prove, the approximate point spectrum of a hermitian operator is exactly the spectrum. Using Theorem 6.3.1, we can assume, without loss of generality, that if $T = H + iJ$, then $\sigma(H) = \sigma_p(H)$.

Let $z = a + ib$ and $a \in \sigma(H)$. Consider the subspace

$$M = N(H - aI) = N(H - a)$$

Since a is an eigenvalue $M \neq \{0\}$. We show that M is an invariant subspace for J. Let $x \in M$ and thus $Hx = ax$. Since we have

$$(H - a)J - J(H - a) = \frac{1}{2} iD$$

where $D = T^*T - TT^*$, it follows that

$$-\frac{1}{2} i \langle Dx, x\rangle = \langle (H - a)Jx, x\rangle - \langle J(H - a)x, x\rangle = 0$$

and thus $\langle Dx, x\rangle = 0$.

But $D > 0$ and thus $D^{1/2}$ exists and it is hermitian. Hence we have

$$0 = \langle Dx, x\rangle = \langle D^{1/2}x, D^{1/2}x\rangle = \|D^{1/2}x\|^2$$

which implies that $D^{1/2}x = 0$. From this, $Dx = 0$ and thus

$$0 = H(Jx) - J(Hx) = (H - a)Jx = 0$$

that is, $Jx \in M$. Clearly M is an invariant subspace for T and the restriction of T to M is of the form

$$T^{\sim} = aI + J^{\sim}$$

and since $J^{\sim}$ is hermitian, $T^{\sim}$ is a normal operator. But T is a hyponormal operator whose restriction to an invariant subspace is normal. Then M is invariant also for T^*. Clearly, $T = T/M \oplus T/M^{\perp} = T_1 \oplus T_2$. Since $\sigma(T) = \sigma(T_1) \cup \sigma(T_2)$, we obtain that the same relation holds for the real parts,

$$\operatorname{Re} \sigma(T) = \operatorname{Re} \sigma(T_1) \cup \operatorname{Re} \sigma(T_2)$$

and since $\operatorname{Re} \sigma(T_1) = \{a\}$, the assertion of the theorem follows.

In Definition 7.3.12 a class of operators was considered and was characterized as those operators T for which $\operatorname{conv} \sigma(T) = \overline{W(T)}$. For this reason, every operator with this property is called *convexoid*.

7.4.5 LEMMA If $T \in L(H)$ is a convexoid operator and $[a, b]$ is the smallest segment containing $\operatorname{Re} \sigma(T)$, then a and b are in $\sigma(\operatorname{Re} T)$.

Proof. As we know, we can assume, without loss of generality, that $W(T)$ is a closed convex set and $\sigma_{ap}(T) = \sigma_p(T)$. Let $z_0 \in \sigma(T)$ such that $\operatorname{Re} z_0 = a$ and z_0 is a point on $\partial\sigma(T)$ such that $z_0 \in W(T)$.

Since T is convexoid, $W(T)$ closed, and $\operatorname{Re} \sigma(T) \geq a$, it follows that

$$\operatorname{Re} \overline{W(T)} = \operatorname{Re} (\operatorname{conv} \sigma(T)) \geq a$$

and thus z_0 is a boundary point of $W(T)$. Hence, as is easy to see, z_0 is a normal eigenvalue. As in Lemma 7.4.2, we can prove the assertion about a, and similarly for b.

7.4.6 LEMMA If $T \in L(H)$ is a convexoid operator and $[a, b]$ is the smallest segment containing $\sigma(\text{Re } T)$, then $\sigma(\text{Re } T) \subset [a, b] \subset \text{Re } \sigma(T)$ if $\sigma(T)$ is a connected set.

Proof. From the connectedness of $\sigma(T)$ it follows that $\text{Re } \sigma(T)$ is a segment on R. To prove the assertion, it suffices to show that $a,b \in \text{Re } \sigma(T)$. Suppose, for example, that $a \notin \text{Re } \sigma(T)$. Similarly we can proceed with b. Let L be the line $\text{Re } z = a$; this is disjoint from $\sigma(T)$ since in the contrary case $a \in \text{Re } \sigma(T)$. But $\sigma(T)$ is a connected set and thus it is strictly on the one side of L. Suppose that it is on the right side. We can find $\varepsilon > 0$ such that $\text{Re } \sigma(T) > a + \varepsilon$ and thus $\text{Re } T \geq (a + \varepsilon)I$ since T being convexoid, the relation obtained implies that $\text{Re } W(T) \geq a + \varepsilon$ and this gives the relation. From the relation $\text{Re } \sigma(T) \geq a + \varepsilon$ we obtain a contradiction, and thus $a \in \text{Re } \sigma(T)$.

7.4.7 LEMMA If $T \in L(H)$ and $\sigma(T)$ is a spectral set for T, then $\sigma(\text{Re } T) \subseteq \text{Re } \sigma(T)$.

Proof. From the definition of spectral sets and the spectral mapping theorem it follows that for any rational function without poles in $\sigma(T)$, $f(T)$ is a normaloid operator. Thus $T - zI$ is a normaloid operator and this says that T is transaloid. Therefore, by Theorem 6.15.11, $\text{conv } \sigma(T) = \overline{W(T)}$. Let $a \in \sigma(\text{Re } T)$ and suppose that $a \notin \text{Re } \sigma(T)$. Let L be the line $\text{Re } z = a$ which is disjoint from $\sigma(T)$. Suppose that $\sigma(T)$ is on the left side of L and thus there exists $\varepsilon > 0$ such that $\text{Re } \sigma(T) \leq a - \varepsilon$. But T is convexoid and thus we obtain that $\text{Re } W(T) \leq a - \varepsilon$. This implies that $\text{Re } \sigma(T) \leq a - \varepsilon$. Thus $\sigma(\text{Re } T) \leq a - \varepsilon$, which is a contradiction since $a \leq a - \varepsilon$ is not possible. Of course, we can proceed similarly in case $\sigma(T)$ is on the right side.

Now if $\sigma(T) = X_1 \cup X_2$, where X_1 is on the left side and X_2 on the right side, then we can decompose $T = T_1 \oplus T_2$, $\sigma(T_i) = X_i$, $i = 1, 2$, and we can apply the above result.

Assertions 3 and 4 follow from the above results and the following.

7.4.8 LEMMA If $T \in L(H)$ has the G_1 property, then $\operatorname{Re}\sigma(T) \subset \sigma(\operatorname{Re} T)$.

Proof. Assume, without loss of generality, that $\sigma_{ap}(T) = \sigma_p(T)$. This implies that $\partial\sigma(T) \subset \sigma_{ap}(T) = \sigma_p(T)$. Let $a \in \operatorname{Re}\sigma(T)$ and L be the line $\operatorname{Re} z = a$. Let z_0 be the point where L exits the spectrum; it is clear that this is a boundary point. Obviously $\operatorname{Re} z_0 = a$.

We construct a sequence of unit vectors $\{x_n\}$ such that

1. $(T - z_0)x_n \to 0$
2. $(T^* - z_0^*)x_n \to 0$

and this implies that $a \in \sigma(\operatorname{Re} T)$.

For $n = 1, 2, 3, \ldots$, let $D_n = \{|z - z_0| \le 1/n\}$ and since $z_0 \in \sigma(T)$, the set D_n contains a point a_n in the resolvent set of T such that $|a_n - z_0| \le 1/2n$. Let b_n be such that $d(a_n, \sigma(T)) = |a_n - b_n|$. In this case b_n is in $\sigma(T)$, and since T has G_1 property, it follows that

$$N(T - b_n) = N(T^* - b_n^*)$$

Now let x_n be such that

$$(T - b_n)x_n = (T^* - b_n^*)x_n = 0$$

and from the definition of b_n we obtain that

$$(T - z_0)x_n = (b_n - z_0)x_n$$

and

$$(T^* - z_0^*)x_n = (b_n^* - z_0^*)x_n$$

Thus, since $\{x_n\}$ is a sequence satisfying the requirements, the lemma is proved.

The proof of assertions 3 and 4 is as follows: First if $\sigma(T)$ is a spectral set, then $\sigma(\operatorname{Re} T) \subset \operatorname{Re}\sigma(T)$ by Lemma 7.4.7. Since T satisfies trivially the G_1 property, the opposite inclusion follows from Lemma 7.4.8.

Suppose now that T has G_1 property and $\sigma(T)$ is connected. From Lemma 7.4.8 we have one inclusion. The opposite inclusion is obtained as follows: Let [a, b] be the smallest seqment containing

$\sigma(\text{Re } T)$ and since T is convexoid, by Lemma 7.4.5 it follows that $[a, b] \subset \text{Re } \sigma(T)$. Thus we have

$$\sigma(\text{Re } T) \subset [a, b] \subset \text{Re } \sigma(T) \subset \sigma(\text{Re } T)$$

and the assertion is proved.

The following example shows that Theorem 7.4.1, part 4, does not generalize to arbitrary convexoid operators.

Let A be an operator with the matrix

$$\begin{bmatrix} 0 & 0 \\ 1 & 0 \end{bmatrix}$$

and B with the matrix

$$\begin{bmatrix} a_1 & 0 & 0 \\ 0 & a_2 & 0 \\ 0 & 0 & a_3 \end{bmatrix}$$

where these eigenvalues lie off the imaginary axis and are the vertices of a triangle containing the disk

$$D_{1/2} = \{z \mid |z| \leq 1/2\}$$

Since $W(A) = D_{1/2}$ and $W(B) = \text{conv } (a_1, a_2, a_3) \supset D_{1/2}$ it follows that T is convexoid and since the spectrum of $A \oplus B$ is $\{0, a_1, a_2, a_3\}$, it follows that $0 \in \text{Re } \sigma(A \oplus B)$. Since $\text{Re } (A \oplus B) = \text{Re } A \oplus \text{Re } B$,

$$\sigma(\text{Re } (A \oplus B) = \sigma(\text{Re } A) \cup \sigma(\text{Re } B) = \{-1/2, 1/2\} \cup \text{Re } \sigma(B)$$

and thus $0 \notin \sigma(\text{Re } (A \oplus B))$.

7.5 THE CLASS $\bar{R}_1$

Halmos, in the course of studying reducible operators, has introduced the following class of operators: the class $\bar{R}_1$.

7.5.1 DEFINITION An operator $T \in L(H)$ is in $\bar{R}_1$ if T has an approximate reducing eigenvector; i.e., there exists z_0 such that for any $\varepsilon > 0$ there exists x_ε, $\|x_\varepsilon\| = 1$, and

$\|(T - z_0)x_\varepsilon\| < \varepsilon \qquad \|(T^* - z_0^*)x_\varepsilon\| < \varepsilon$

From the definition it is clear that $\bar{R}_1$ represents the closure of the operators with one-dimensional reducing subspace.

Halmos showed (1968a) that $\bar{R}_1$ contains the normal and isometric operators.

We need the following lemma which is also useful in the study of the structure of certain classes of operators.

7.5.2 LEMMA If $T \in L(H)$ and $z \in \sigma(T)$, $|z| = \|T\|$, then

1. If $(T - z)x = 0$, then $(T^* - z^*)x = 0$.
2. If $\lim (T - z)x_n = 0$, then $(T^* - z^*)x_n \to 0$.

Proof. We prove the first assertion only. Let x be such that $(T - z)x = 0$. We have

$$\begin{aligned}\|(T^* - z^*)x\|^2 &= \langle T^*x, T^*x\rangle - \langle T^*x, z^*x\rangle - \langle z^*x, T^*x\rangle + |z|^2\|x\|^2 \\ &= \|T^*x\|^2 - |z|^2\|x\|^2 - |z|^2\|x\|^2 - |z|^2\|x\|^2 \\ &= \|T^*x\|^2 - |z|^2\|x\|^2\end{aligned}$$

which is less than or equal to zero. This gives the assertion.

7.5.3 COROLLARY If T is a normaloid operator, then $T \in \bar{R}_1$.

We define an operator $T \in L(H)$ to be *spectraloid* if the numerical radius is equal to the spectral radius. Then the above result is a consequence of the following result.

7.5.4 THEOREM If $T \in L(H)$ is spectraloid, then $T \in \bar{R}_1$.

Proof. Let z such that $|z| = r_T = w(T)$. It is sufficient to prove that if z is with the properties

1. $Tx = zx$
2. $z \in \partial W(T)$

then $T^*x = z^*x$.

Indeed, by the transformations of the form $z \mapsto az + b$, the point spectrum and the numerical range are transformed in an obvious way, and we can suppose, without loss of generality, that $z = 0$ and $\operatorname{Re} T < 0$. But it is well known (and easy to prove) that

$$\text{Re} \langle Tx, x\rangle = \langle \text{Re } Tx, x\rangle$$

and since Re T is a hermitian operator, we obtain that Re Tx = 0 and thus Tx = 0 = Re Tx + i Im Tx which gives that Im Tx = 0. Now $T^*x = \text{Re } Tx - i \text{ Im } Tx = 0$. The case of a sequence of elements in H can be treated in a similar way.

The proof of the theorem is as follows: since $z \in \sigma(T)$, we can find a sequence of elements $\{x_n\}$ such that

$$Tx_n - zx_n \to 0$$

and from the above remark

$$T^*x_n - z^*x_n \to 0$$

which proves the theorem.

7.5.5 COROLLARY If T is an operator of class (N), then $T \in \bar{R}_1$.

Proof. Every operator of class (N) is normaloid and thus spectraloid.

7.5.6 COROLLARY If T is a hyponormal operator, then $T \in \bar{R}_1$.

Proof. Every hyponormal operator is of class (N).

For the case of operators in the above classes we can prove a more general result. For this we need the following information about a class of points in the spectrum of an operator.

7.5.7 DEFINITION A point z in $\sigma(T)$ is in $\sigma_c(T)$ if $T_z = T - z$ is one-to-one and T_zH is dense in H and $T_zH \neq H$.

From the definition it follows that $z \in \sigma_c(T)$ if and only if there exists a sequence of elements $\{x_n\}$, $\|x_n\| = 1$, such that

1. $Tx_n - zx_n \to 0$
2. $x_n \rightharpoonup 0$

7.5.8 THEOREM If T is a hyponormal operator, then for any compact operator K, $T + K \in \bar{R}_1$.

Proof. Since T is hyponormal, it is easy to see that T can be written as $T = T_1 \oplus T_2$ on $H_1 \oplus H_2$, where T_1 is normal and the point spectrum of T_2 is empty. The assertion of the theorem is clear when H_1 is infinite dimensional since we can choose an infinite sequence of eigenvectors which are mutually orthogonal. In the case dim $H_1 < \infty$, we choose a point $z \in \sigma(T_2)$ and find a sequence $\{x_n\}$, $\|x_n\| = 1$, such that

$$(T_2 - z)x_n \to 0$$

and also x_n converges weakly to zero. From the hyponormality it follows that

$$(T_2^* - z^*)x_n \to 0$$

and the weak convergence to zero of $\{x_n\}$. The fact that K is a compact operator implies that for a subsequence of x_n's (and we can suppose that this is just x_n),

$$(T_2 - z + K)x_n \to 0$$

and

$$(T_2^* - z^* + K^*)x_n \to 0$$

We prove now that Theorem 7.5.8 can be extended to other classes of operators.

7.5.9 THEOREM If $T \in L(H)$ is with the sequential G_1 property, then for any compact operator $K \in L(H)$, $T + K \in \bar{R}_1$.

For the proof we need certain facts; some of them were given above and two are given below.

7.5.10 LEMMA If $T \in L(H)$ and $\{x_n\}$ is a sequence converging weakly to x_0 and $\lim \|Tx_n\| \to 0$, then $Tx_0 = 0$.

7.5.11 LEMMA If for some sequence $\{x_n\}$, $\|x_n\| = 1$, converging weakly to zero and

$$(T - z)x_n \to 0 \qquad (T^* - z^*)x_n \to 0$$

then for any compact operator K, $T + K \in \bar{R}_1$.

Proof. Follows immediately from the definition of the class $\bar{R}_1$.

Following Wadhwa (1973a) we define for any operator $T \in L(H)$ the set

$$\gamma_T = \{z \in \sigma(T) \mid \text{there exists } Hx \text{ such that } x \neq 0 \text{ and } (T - z)x = 0,\ (T^* - z^*)x = 0\}$$

It is obvious that if γ_T is a nonempty set, then $T \in \bar{R}_1$. Also if γ_T is an infinite set, then $T \in \bar{R}_1$

7.5.12 LEMMA If $z \notin \gamma_T$ and there exists a sequence of unit elements such that

$$(T - z)x_n \to 0 \qquad (T^* - z^*)x_n \to 0$$

then $T + K \in \bar{R}_1$ for any compact operator K.

Proof. We can assume, without loss of generality, that $x_n \rightharpoonup x_0$.

By the above result $(T - z)x_0 = 0$ and $(T^* - z^*)x_0 = 0$. Since $z \notin \gamma_T$, it follows that $x_0 = 0$. Thus by Lemma 7.5.11 the assertion is proved.

7.5.13 LEMMA If $T \in L(H)$ is with the sequential G_1 property and $\gamma_T = \sigma(T)$, then for any compact operator $K \in L(H)$, $T + K \in \bar{R}_1$.

Proof. Let $\sigma(T) = \{z_1, \ldots, z_n\}$ and since $\sigma(T) = \gamma_T$ each z_i is an eigenvalue for T. Also, since T is with the sequential G_1 property, it is a normal eigenvalue. Moreover, if for some i the null space of $(T - z_i)$ is infinite dimensional, then clearly $T + K \in \bar{R}_1$. In the contrary case $H = H_1 \oplus H_2$, where $H_i = \Sigma\, N(T - z_i)$. Since H_1 reduces T and $\sigma(T)$ is a finite set, the restriction of T on H_2 is an operator with the sequential G_1 property and γ_{T_2} is empty. In this case $T_2 + K \in \bar{R}_1$.

Now we are ready to prove Theorem 7.5.9. Clearly, we can consider the case γ_T a finite set and different from $\sigma(T)$. Since T is with the sequential G_1 property for any point z_0 in $(\sigma(T) \setminus \gamma_T)$ there exists a sequence $\{a_n\} \subset \sigma(T)$ such that

$$\|(T - a_n)^{-1}\| = \frac{1}{d(a_n, \sigma(T))} \qquad a_n \to z_0$$

Since $a_n - z_0 \notin \gamma_T$ for any $a_{m_0} \in \sigma(T) \setminus \gamma_T$ such that $|a_{m_0} - z_0| <$ $\min\{|z_0 - z| \mid z \in \gamma_T\}$, we have $d(a_{m_0}, \sigma(T)) = |a_{m_0} - z_{m_0}|$ for some $z_{m_0} \in \sigma(T) \setminus \gamma_T$.

Thus we obtain that

$$\|(T - a_{m_0})^{-1}\| = \frac{1}{d(a_{m_0}, \sigma(T))} \quad \frac{1}{|a_{m_0} - z_{m_0}|}$$

and $1/(z_{m_0} - a_{m_0}) \in \sigma((T - a_{m_0} 0^{-1})$.

By Lemma 7.5.2 there exists a sequence of unit elements $\{x_n\}$ such that

$$\left[(T - a_{m_0})^{-1} - \frac{1}{(z_{m_0} - a_{m_0})^{-1}}\right] x_n \to 0$$

$$\left[(T^* - a^*_{m_0})^{-1} - \frac{1}{(z^*_{m_0} - a^*_{m_0})^{-1}}\right] x_n \to 0$$

and from this we obtain easily that a_{m_0} is not in γ_T. From Lemma 7.5.12 the assertion of the theorem follows.

7.5.14 COROLLARY If $T \in L(H)$ and has the G_1 property, then for any compact operator K, $T + K \in \bar{R}_1$.

7.5.15 COROLLARY If $T \in L(H)$ is an isometric operator, then for any compact operator K, $T + K$ is in $\bar{R}_1$.

7.6 OTHER CLASSES OF BOUNDED OPERATORS

We collect here information regarding certain classes of operators related to the classes discussed previously.

7.6.1 DEFINITION An operator $T \in L(H)$ is said to be of class (C) if for any $x \neq 0$ the sequence

$$\left(\frac{\|T^n x\|}{\|T^{n-1} x\|}\right)$$

is monotone.

As we know every operator of class (N) has this property; it is also clear that any power of T is in the same class.

It is an interesting problem if operators of class (N,k) are of class (C).

The following class of operators is related to the classes of hyponormal operators as well as to operators of class (N,k).

7.6.2 DEFINITION An operator $T \in L(H)$ is said to be of class (M,k) if

$$T^{*k}T^k - (T^*T)^k \geq 0$$

7.6.3 THEOREM Every operator of class (M,k) is of class (N,k).

Proof. See Theorem 7.1.11.

7.6.4 THEOREM The class (N,k) is larger than the class (M,k).

Proof. Let K be the direct sum of a countable number of copies of H. For positive operators A and B on H and for a positive integer $n > k$ we define the operator $T = T_{a,B,n}$ on K as follows:

$$T(x_1, x_2, \ldots) = (0, Ax_1, Ax_2, \ldots, Ax_n, Bx_{n+1}, Bx_{n+2}, \ldots)$$

and it is easy to see that

1. T is hyponormal if and only if $B^2 \geq A^2$.
2. T^2 is of class (M,k) if and only if

$$AB^{4k-2}A \geq (AB^2A)^k \qquad A^{2k-m}B^{2m}A^{2k-m} \geq A^{4k}$$
$$m = 1, 2, \ldots, 2k - 2$$

Now if we take H to be a two-dimensional Hilbert space and A and B the operators which are the square roots of

$$C = \begin{bmatrix} 2 & 1 \\ 1 & 1 \end{bmatrix} \qquad D = \begin{bmatrix} 3 & 0 \\ 0 & 2 \end{bmatrix}$$

then $A^2 \geq B^2$ and thus $T_{A,B,n}$ is hyponormal and it is known that in this case T is of class (N) and of class (N,k) for all k.

We show now that T^2 is not of class (M,k).

Suppose on the contrary that it is of class (M,k). Since A is invertible, we have

$$B^{2m} \geq A^{2m} \qquad m = 1, 2, \ldots, 2k - 2$$

But $B^4 - A^4 = D^2 - C^2$ is not positive and the contradiction proves the result.

For other results on this class of operators, we refer to Patel (1975b).

The following class of operators was introduced by Wadhwa (1974).

7.6.5 DEFINITION An operator $T \in L(H)$ is called *M-hyponormal* if for all x and z

$$\|(T^* - z^*)x\| \leq M\|(T - z)x\|$$

for a constant $M > 0$.

Clearly, for $M = 1$ we have the class of hyponormal operators. The following subclass of M hyponormal operators has also been considered.

7.6.6 DEFINITION An operator $T \in L(H)$ is said to be of power M-class if for all n and z

$$\|(T - z)x\|^n \leq M\|(T - z)^n x\|$$

for a constant $M > 0$.

It has been proved that this class has the same properties as class (N).

An interesting class of operators was considered by Stampfli and Wadhwa under the name "dominant operators."

7.6.7 DEFINITION An operator $T \in L(H)$ is called *dominant* if the range of $T - z$ is in the range of $(T - z)^*$ for all $z \in \sigma(T)$.

This is equivalent to the existence of a constant M_z such that for all $z \in \sigma(T)$ and any $x \in H$,

$$\|(T - z)^*x\| \leq M_z\|(T - z)x\|$$

For properties of this class of operators, we refer to Wadhwa and Stampfli (1977).

Chapter 8

CONDITIONS IMPLYING NORMALITY

Let H be a Hilbert space and L(H) be the set of all bounded operators on H. An important class of elements of L(H) is the class of normal operators, i.e., operators $N \in L(H)$ for which

$$N^*N = NN^*$$

as well as the subclasses of unitary and hermitian operators.

Thus it is of interest to know conditions which imply that an operator is normal, unitary, or hermitian.

In what follows we give conditions implying normality; first we give results concerning hermitian operators and next unitary operators. Several results are given on normal operators. These conditions are connected in many cases with the nature of the spectrum of the operator and in others with the behavior of the operator on the space. The precise statement of these assertions is given in what follows.

8.1 CONDITIONS IMPLYING HERMITIANITY

First we begin with the characterization of hermitian elements in L(H).

8.1.1 THEOREM A necessary and sufficient condition for an element $T \in L(H)$ to be hermitian is that its numerical range $W(T) \subset R$.

For the proof we need the following lemma, which also has independent interest.

8.1.2 LEMMA If $T, S \in L(H)$ and $0 \notin \overline{W(S)} = \text{cl } W(S)$, then

$$\sigma(S^{-1}T) \subset \frac{W(S)}{W(S)} = \left\{\frac{s}{r} \mid s \in W(T),\ r \in W(S)\right\}$$

Proof. Since S^{-1} exists for any operator $\sigma(T) \subset \overline{W(T)}$, we obtain that if $s \in \sigma(S^{-1}T)$, then

$$S^{-1}T - sI = S^{-1}(T - sS)$$

which gives that $s \in \sigma(T - sS)$. From this fact, the assertion follows since

$$0 \in W(T - sS) = W(T) - sW(S)$$

or

$$s \in \frac{W(T)}{W(S)}$$

Proof of Theorem 8.1.1. If $T \in L(H)$ is hermitian, then for all $x \in H$,

$$\langle Tx, x\rangle = \langle x, Tx\rangle = \overline{\langle Tx, x\rangle}$$

and the condition is necessary.

Suppose now that $W(T) \subset R$. Without loss of generality we can suppose that T is invertible; thus $T = UR$, where U is unitary.

We prove that $U = I$ or $U = -I$. Indeed, by the lemma we have

$$\sigma(U^{-1}) = \sigma(R^{-1}T^*) = \sigma(T^*R^{-1}) = \sigma(R^{-1}T^*) \subset \frac{W(T^*)}{W(R^{-1})}$$

which implies that

$$(U^{-1}) \subset \{1, -1\}$$

and thus $U = U_1 \oplus U_2$, where $\sigma(U_1) = \{1\}$ and $\sigma(U_2) = \{-1\}$. If $\sigma(U)$ is equal to $\{1\}$ or $\{-1\}$, then $U = I$ or $U = -I$, respectively, and the theorem is proved.

The following theorem gives conditions for the spectrum of an operator to be a subset of the real numbers.

8.1.3 THEOREM Let $T \in L(H)$ and $S \in L(H)$ with the properties:

1. $ST^pS^{-1} = T^{*p}$ $\quad$ p an integer ≥ 1

2. $0 \notin W(S)$ $\quad \sum_{i=0}^{p-1} \left(\frac{s}{r}\right)^i \neq 0 \quad s, r \in \sigma(T) \setminus \{0\}$

In this case $\sigma(T) \subset R$.

Proof. It is sufficient to show that any point $s \in \partial\sigma(T)$ is real. Suppose in the contrary case that some $s \in \partial\sigma(T)$ is with Im $s \neq 0$. We can find a sequence $\{x_n\}$, $\|x_n\| = 1$, such that

$$Tx_n - sx_n \to 0$$

and clearly for all i,

$$T^ix_n - s^ix_n \to 0$$

Using property 1 of T, we obtain

$$S^{-1}T^{*p}Sx_n - s^px_n \to 0$$

and thus,

$$T^{*p}x_n - s^pSx_n \to 0$$

Since

$$(T^{*p} - s^p)Sx_n = (T^* - s)(T^{*p-1} + sT^{*p-2} + \cdots + s^{p-1})Sx_n \to 0$$

we obtain that

$$T^*x_n - sx_n \to 0$$

From the identity

$$\begin{aligned}(s^* - s)\langle Sx_n, x_n\rangle &= \langle (T^* - s)Sx_n, x_n\rangle - \langle (T^* - s^*)Sx_n, x_n\rangle \\ &= \langle (T^* - s)Sx_n, x_n\rangle - \langle (Sx_n, (T - s)x_n\rangle\end{aligned}$$

and from the properties of the sequence $\{x_n\}$, we obtain $0 \in \overline{W(S)}$ because $s^* - s \neq 0$. This contradiction proves the theorem.

If T has the property that conv $\sigma(T) = \overline{W(T)}$ and satisfies the conditions of the theorem, then clearly T is hermitian.

8.1.4 REMARK Theorem 8.1.3 is not true if we suppose only that S is invertible. Indeed, let $H = \ell^2$ and define

$$Te_i = e_{i+1} \qquad i = 0, \pm 1, \pm 2, \ldots$$

and consider the operator S defined by

$$Se_i = e_{i-1}$$

It is easy to see that

$$S^{-1}TS = T^{-1} = T^*$$

and $\sigma(S) = \{z \mid |z| = 1\}$. T is unitary.

8.1.5 THEOREM If $T, S \in L(H)$ and

1. $T^* = S^{-1}TS$
2. $0 \notin \overline{W(S)}$

then T is similar to a hermitian operator; i.e., there exists an invertible operator S_1 and a hermitian operator A such that

$$T = S_1 A S_1^{-1}$$

Proof. Since $\overline{W(S)}$ is a convex closed set and $0 \notin W(S)$, we can find $\varepsilon > 0$ such that $\text{Re } z \geq \varepsilon$ for all $z \in W(S)$. Let $A = 1/2(S + S^*)$. Clearly, $W(A) \subset R^+$ and also A is positive and invertible. From property 1 we obtain that $TA = AT^*$. Now A being positive, $A^{1/2}$ exists and the operator $A^{1/2}TA^{1/2}$ is hermitian. Since

$$T^* = A^{-1/2}(A^{1/2}TA^{1/2})A^{1/2}$$

the assertion of the theorem is proved.

The following example shows that the invertibility of S is an essential condition.

8.1.6 EXAMPLE Let H_n be a Hilbert space, $\dim H_n = n$, and

$$S_n : H_n \to H_n$$

with the matrix

$$\begin{bmatrix} a_1 & 0 & 0 & 0 & \dots & 0 \\ 1 & a_2 & 0 & 0 & \dots & 0 \\ 0 & 1 & a_3 & 0 & \dots & 0 \\ \vdots & & & & & \\ 0 & 0 & 0 & 0 & \dots & a_n \end{bmatrix} \qquad \Sigma\, a^2 i \le 1,\ a_i \in R$$

with respect to an orthonormal basis of H_n. Clearly $\|S_n\| \le 2$.

If z is such that $|z| = 1$ and

$$x = e_1 - z^{-1}e_2 + \cdots + (-1)^{n-1}z^{-n+1}e_n$$

then

$$(zI_n + S_n)x = (z + a_1)e_1 - z^{-1}e_2 + \cdots + (-1)^{n-1}z^{-n+1}a_n e_n$$

and thus $\|(zI_n + S_n)\| \le 2$.

Since $\|x\| = n^{1/2}$, we obtain that

$$\|(zI_n + S_n)^{-1}\| \ge n^{1/2}\,\frac{1}{2}$$

We consider the space $H = \oplus H_n$ and $T: H \to H$ such that T/H_n is exactly S_n. Since (S_n) is in R, we find $T_n > 0$ such that $T_nS_n = S_nT_n$. Of course, we can assume, without loss of generality, that $\|T_n\| \le 1$ and the existence of T follows. Since $T \ge 0$, TH is dense in H, and $TS = S^*T$. But $\sigma(S) \supset \{z \mid |z| = 1\}$.

The following example shows that if $S^{-1}TS = T^*$ and $0 \notin W(S)$, then T is not necessarily hermitian.

8.1.7 EXAMPLE Let $S > 0$ be arbitrary and S_1 hermitian such that $SS_1 \ne S_1S$. In this case, $T = SS_1$ has the property

$$T^* = S_1S = S_1(SS_1)S_1^{-1}$$

and T is not hermitian.

The following theorem gives information about operators which realize this similarity.

8.1.8 THEOREM If $T = S\tilde{T}S^{-1}$ where $\tilde{T}$ is hermitian, then T is similar to T*, $T = \tilde{S}T^*\tilde{S}^{-1}$, $0 \notin \overline{W(\tilde{S})}$.

Proof. Let T be as in the theorem. In this case,

$$T^* = (S\tilde{T})^{-1}TS(S\tilde{T})$$

and $0 \notin W(S\tilde{T})$ since $S\tilde{T}$ is positive and invertible.

We mention now another characterization of hermitian operators. See also the discussion of hermitian elements in Chapter 6.

8.1.9 THEOREM An operator $T \in L(H)$ is hermitian if and only if one of the following equivalent conditions are fulfilled:

1. For all $t \in R$, $\|e^{itT}\| = 1$.
2. For $t \in R$, $t \to 0$, $\|(I + itT\| = 1 + 0(t)$

We now give a characterization of hermitian operators using the resolvent.

First we recall the definition of G_1 property.

8.1.10 DEFINITION An operator $T \in L(H)$ is said to have the G_1 property if for all $z \in \rho(T)$,

$$\|R_z(T)\| = \frac{1}{d(z, \sigma(T))}$$

In connection with the G_1 property we remark that for all $z \in \rho(T)$

$$\frac{1}{d(z, \sigma(T))} \le \|R_z(T)\|$$

a relation which follows from the spectral mapping theorem.

Another interesting fact is that for any operator $T \in L(H)$ we have the following inequality:

$$\frac{1}{d(z, W(T))} \le \|R_z(T)\| \qquad z \notin \text{cl } W(T) \tag{*}$$

Indeed, for any x, $\|x\| = 1$,

$$\begin{aligned}\|(T - z)x\| &\ge |\langle (T - z)x, x\rangle| \\ &= |\langle Tx, x \rangle - z| \ge \min\{|z - w| \mid w \in W(T)\} \\ &= d(z, W(T))\end{aligned}$$

which is equivalent with the relation (*).

It is worth noting that a sort of inequality, namely,

$$\|R_z(T)\| \le \frac{1}{d(z, W(T))}$$

is connected with an important class of operators, namely the class of "convexoid operator," i.e., operators $T \in L(H)$ for which $\operatorname{conv} \sigma(T) = \overline{W(T)}$. (See the details in Chapter 7.)

8.1.11 THEOREM A necessary and sufficient condition for an operator $T \in L(H)$ to be hermitian is the following:

1. $W(T) \subset R$
2. $\|R_z(T)\| \le 1/|z| \qquad z \in \rho(T),\ \operatorname{Im} z \ne 0,\ \operatorname{Re} z = 0$

Proof. It is obvious that the condition is necessary. We prove now the sufficiency. For this we show that $W(T) \subset R$. Let $z = is$ with s a real number. In this case,

$$\|R_{is}\| \le \frac{1}{|s|}$$

which gives

$$|s|^2\|x\|^2 \le \|(T - is)x\|^2$$

or

$$|s|^2\|x\|^2 \le \|Tx\|^2 + s^2\|x\|^2 - 2s \operatorname{Im} \langle Tx, x\rangle$$

But s is an arbitrary real number and we obtain that

$$2s \operatorname{Im} \langle Tx, x\rangle \le \|Tx\|^2$$

which clearly implies $\operatorname{Im} \langle Tx, x\rangle = 0$.

8.1.12 THEOREM An operator $T \in L(H)$ is hermitian if and only if

$$(T + T^*)^2 \ge 4T^*T$$

Proof. First we remark that the condition is necessary; in fact if T is hermitian, the above inequality is an equality.

Suppose now that T satisfies the above inequality. In this case for all $x \in H$,

$$2\|Tx\| \le \|Tx + T^*x\| \le \|Tx\| + \|T^*x\|$$

which gives the relation

$$\|Tx\| \le \|T^*x\|$$

i.e., T^* is a hyponormal operator. To prove the theorem, it will be sufficient to show that the $\sigma(T)$ or $\sigma(T^*)$ is real, since for hyponormal operators conv $\sigma(T) = \overline{W(T)}$.

Since the boundary of the spectrum $\sigma(T)$ is a subset of the approximate point spectrum, to prove the assertion it is sufficient to show that any point in the approximate point spectrum is real. Suppose that this is not the case, and thus there exists a point $s = a + ib$, $b \ne 0$, which is the approximate point spectrum.

Let $\{x_n\}$, $\|x_n\| = 1$, such that $Tx_n - sx_n \to 0$ and since $T - s$ is hyponormal, we obtain that

$$\|(T^* - s^*)x_n\| \to 0$$

and this gives the relation

$$4|\text{Re } s|^2 \ge 4|s|^2$$

which implies Im $s = 0$. The theorem is proved.

In a similar way we can prove the following theorem.

8.1.13 THEOREM An operator $T \in L(H)$ is hermitian if and only if

$$4T^*T \le (T + T^*)^2 + \frac{(T - T^*)}{i}$$

8.1.14 REMARK In the above results (Theorems 8.1.12 and 8.1.13) we can change the role of T and T^*.

8.2 CONDITIONS IMPLYING UNITARITY

We give now some characterizations of the class of unitary operators. We recall that $T \in L(H)$ is unitary if and only if U is a surjective operator ($TH = H$) and for all $x,y \in H$, $\langle Tx, Ty\rangle = \langle x, y\rangle$.

8.2.1 THEOREM The following classes of operators are the same:

1. The class of unitary operators

2. The class C of operators $T \in L(H)$ satisfying the following conditions
   a. $T \in C$ implies T^{-1} exists as an element of $L(H)$ and $\|Tx\| \geq \|x\|$ for all $x \in H$.
   b. $w(T) \leq 1$.
3. The class C_1 of operators $T \in L(H)$ satisfying the following properties:
   a. $0 \in \sigma(T)$.
   b. $\|R_0(T)\| \leq 1$.
   c. If $\{t_n\}$ is an unbounded sequence, $t_n > 1$, then $\|R_{z_n}(T)\| \leq 1/(t_n - 1)$, $|z_n| = t_n$.

Proof. We prove the following inplications: $1 \to 3 \to 2 \to 1$. Indeed, let U be a unitary operator. Clearly, $0 \in \rho(U)$ and $\|R_0(U)\| \leq 1$ and the spectral mapping theorem gives, for any z, $|z| = t_n$,

$$\|R_z(U)\| = \frac{1}{d(z, (U))} \leq \frac{1}{t_n - 1}$$

Thus we have 1 implies 3. Suppose now that 3 holds and we prove 2. Let z, $|z| = t_n$, and for any $x \in H$ we have

$$\|x\| \leq (t_n - 1)^{-1}\|(T - z)x\|$$

or

$$\|x\|^2(t_n - 1)^2 \leq \|Tx\|^2 + t^2\|x\|^2 - 2(\operatorname{Re} z)\langle Tx, x\rangle$$

For x, $\|x\| = 1$, and $\langle Tx, x\rangle = \rho e^{i\rho}$, $z = t_n e^{i\theta}$ we obtain

$$2t_n[\rho \cos(\theta - \phi) - 1] \leq \|Tx\|^2 - 1$$

and thus $\rho \leq 1$. Since 3(a), (b) implies 2(a), we have 3 implies 2.

We suppose now that 2 is true and we prove that T is unitary. Since T is invertible, $U = T(T^*T)^{-1/2}$ is unitary and if $R = (T^*T)^{1/2}$, then $T = UR$. Clearly R is hermitian and $\sigma(R) \subset [1, \|R\|]$.

Thus $R = I + R_1$, where R_1 is positive. Let x, $\|x\| = 1$, and $\langle Rx, x\rangle = 1 - \varepsilon$, $\varepsilon > 0$. In this case we have

$$\langle Tx, x\rangle = \langle (U(I + R_1)x, x\rangle = \langle Ux, x\rangle + \langle UR_1x, x\rangle$$
$$= \langle Ux, x\rangle + \langle R_1x, U^{-1}x\rangle$$

Let y, $\|y\| = 1$, such that $\langle x, y\rangle = 0$ and $U^{-1}x = ax + by$. We have

$$a = \overline{\langle Ux, x\rangle} \qquad |a|^2 + |b|^2 = 1$$

and since $w(T) \leq 1$, we obtain

$$|\langle Tx, x\rangle| = |\langle Ux, x\rangle + \langle R_1x, ax + by\rangle| = |\langle Ux, x\rangle + a^*\langle R_1x, x\rangle$$
$$+ b^*\langle R_1x, y\rangle|$$
$$= |\langle Ux, x\rangle|[1 + \langle R_1x, x\rangle] + b^*|\langle R_1x, y\rangle| \leq 1$$

But R is a positive operator and by the generalized Schwarz inequality we have

$$|\langle R_1x, y\rangle|^2 < \langle R_1x, x\rangle\langle R_1y, y\rangle$$

and since $|b| < (2\varepsilon)^{1/2}$ we obtain the estimate

$$(1 - \varepsilon)(1 + \langle R_1x, x\rangle) \leq 1 + 2\|R_1\|\varepsilon\langle R_1x, x\rangle^{1/2}$$

Thus $\langle R_1x, x\rangle/\varepsilon$ is bounded when $\varepsilon \to 0$ and there exists a constant $k > 0$ such that

$$\langle R_1x, x\rangle < k(1 - |\langle Ux, x\rangle|)$$

Let $U = \int_0^1 e^{2i\theta}\, dE(\theta)$ and for any n we consider

$$P_k = E_{k/n} - E_{(k-1)/n} \qquad k = 1, 2, \ldots, n$$

Let $x \in P_kH$ and $\|x\| = 1$. In this case $\langle Ux, x\rangle \in$ conv $\{z \mid \arg z < 2\pi/n\}$ and thus

$$1 - |\langle Ux, x\rangle| \leq \frac{\pi}{n^2}$$

which gives the following relation

$$\|R_1 x\|^2 = \langle R_1^2 x, x\rangle \le \|R_1\| \frac{k}{n}$$

and thus

$$\|R_1 x\| \le c \frac{\|x\|}{n}$$

Let z be arbitrary and $y = R_1 z$. In this case we have

$$|\langle y, x\rangle| = |\langle R_1 z, x\rangle| = |\langle z, R_1 x\rangle| \le \|z\|\|R_1 x\| \le \|z\| C \frac{\|x\|}{n}$$

and if $x = P_k y$ this relation gives

$$\|P_k y\|^2 = \langle P_k y, y\rangle \le \|z\|\|P_k y\| \frac{C}{n}$$

and thus

$$\|y\|^2 = \sum \|P_k y\|^2 \le n\|z\| \frac{C^2}{n^2} = \|z\|^2 \frac{C^2}{n}$$

Now n is arbitrary and thus y = 0. But in this case, z being arbitrary, we have $R_1 = 0$ and the theorem is proved.

8.2.2 EXAMPLE This is an example of a bounded operator satisfying the G_1 property and not normal.

Let T_1 be an operator on ℓ_2^2 with the matrix

$$\begin{bmatrix} 0 & 1 \\ 0 & 0 \end{bmatrix}$$

which is nilpotent ($T_1^2 = 0$). Let T_2 be the bounded operator defined on an arbitrary Hilbert space H with a countable basis $\{e_i\}_{i=1}^{\infty}$ by

$$T_2 e_i = a_i e_i$$

and where a_i are such that

$$\min_i |z - a_i| < |z|^2$$

for all z, $|z| < 1$. The operator $T = T_1 \oplus T_2$ has the G_1 property and is clearly nonunitary (also nonnormal).

8.2.3 REMARK The preceding example shows that the restriction of an operator with G_1 property fails to have the G_1 property on the invariant subspace.

8.3 CONDITIONS IMPLYING NORMALITY

A normal operator, as we know, satisfies a number of conditions weaker than normality. Such conditions are, for example hyponormality, convexoidity, and transaloidity. Theorems in the converse direction can be obtained by supplementing or strengthening such conditions. Very often, the supplementary conditions refer to the spectrum, finite dimensionality, compactness, etc.

In this section we give some results in this direction. First we consider the case of finite dimensional spaces and next the case of infinite dimensional spaces. In the case of finite dimensional space proofs can be presented in such a way that the extension to infinite dimensional case is trivial.

First we introduce the form of normal operators on finite dimensional spaces.

Suppose that H is a finite dimensional Hilbert space and T is a linear operator on H. It is easy to see that T is continuous.

Since the spectrum is a finite set, let $\sigma(T) = \{z_1, \ldots, z_n\}$ and $C_{\sigma(T)}$ be the algebra of all continuous functions on $\sigma(T)$. For any polynomial $p(z, z^*)$ we define the operator

$$p(T, T^*) = \Sigma\, a_{ij} T^i T^{*j}$$

if $p(z, z^*) = \Sigma\, a_{ij} z^i z^{*j}$.

Suppose now that T is normal, i.e., $T^*T = TT^*$. In this case

$$p \mapsto p(T, T^*)$$

is an homomorphism of the algebra $C_{\sigma(T)}$ in the algebra of all linear operators on H with the following property:

$$\|p(T, T^*)\| = \|p(z, z^*)\| = \sup \{|p(z, z^*)| \mid z \in \sigma(T)\}$$

Now we use the following property of finite sets in the plane: there exists a sequence of polynomials $\{p_n(z, z^*)\}$ converging to z^* on the set. Thus the function

$$z \mapsto z^*$$

is in $C_{\sigma(T)}$ and from the Weierstrass-Stone theorem the closure of polynomials $p(z, z^*)$ is equal to $C_{\sigma(T)}$.

In this case it is clear that for any function f on $\sigma(T)$ we can define the operator f(T) using the above homomorphism.

Let

$$f_i(z) = \begin{cases} 1 & z = z_i \\ 0 & z \neq z_i \end{cases}$$

and since

$$f_i f_j = 0 \qquad i \neq j$$

and

$$\Sigma f_i(z) = 1$$

we obtain that the corresponding operators $f_i(T) = P_i$ have the following properties:

1. $P_i^2 = P_i$
2. $P_i P_j = 0$, $i \neq j$
3. $\Sigma P_i = I$
4. $T = \Sigma z_i P_i$

From the above representation the next lemma follows immediately.

8.3.1 LEMMA If $T \in L(H)$, H is a finite dimensional Hilbert space, and T a normal operator on H, then for any polynomial $p(z, z^*)$ the operator $p(T, T^*)$ has the property

$$\|p(T, T^*)\| = \sup \{|p(s, z^*)| \mid z \in \sigma(T)\}$$

8.3.2 COROLLARY For any z, T_z is an operator with the G_1 property and is also transaloid.

8.3.3 COROLLARY For T normal, conv $\sigma(T) = \overline{W(T)}$.

Proof. The assertion follows from Corollary 8.3.2 and Theorem 6.15.11.

Also from Lemma 8.3.1 it follows that for any polynomial $p(z, z^*)$ the relation

$$\|p(T, T^*)\|^2 = \|p(T, T^*)^2\|$$

holds.

For the following result we need the important notion related to the fine structure of operators, the notion of invariant subspace.

8.3.4 DEFINITION Let H be an arbitrary Hilbert space and T be a bounded linear operator on H. A closed subspace H_1 of H is called an *invariant subspace* for T if for any $x \in H_1$, $Tx \in H_1$.

The invariant subspace reduces T if it is invariant for T*.

The following result is a consequence of the structure theorem for normal operators.

8.3.5 LEMMA Every invariant subspace of a normal operator on a finite dimensional Hilbert space reduces T.

We give now some conditions implying normality for operators on finite dimensional Hilbert spaces.

8.3.6 THEOREM (J. von Neumann) If T is an operator on a finite dimensional Hilbert space and for any polynomial p(z) we have

$$\|p(T)\| = \sup \{|p(z)| \mid z \in F\}$$

where F is a finite set in the complex plane, then T is normal and $\sigma(T) \subset F$.

Proof. Let $F = \{z_1, \ldots, z_n\} = \sigma(T)$, which does not restrain the generality, and p_k be the polynomial of the degree $n - 1$ with the value 1 in z_k and 0 in any z_i with $i \neq k$. In this case we have for any $z \in F$,

$$z = \Sigma\, z_k p_k(z)$$

and if we consider the operators

$$p_k(T) = P_k$$

then the following relations hold:

1. $p_k(T) = P_k^2 = P_k$
2. $P_k P_j = 0, \; i \neq j$
3. $\|P_k\| = 1$

From these it is clear that

$$T = \Sigma \; z_k P_k$$

and thus T is normal and the spectrum of T is exactly $\{z_1, \ldots, z_n\}$.

8.3.7 REMARK A careful examination of the proof shows that in fact the theorem is valid for arbitrary Hilbert spaces.

For the next theorem we need the following definition.

8.3.8 DEFINITION Let T be a bounded linear operator on a Hilbert space H and P a property. We say that T has the *rest-P* if the restriction of T to every invariant subspace has the property P. We say that T has the *red-P* if the restriction of T to every reducing invariant subspace has the property P.

8.3.9 REMARK The above definition for the case of the property P to be normaloid were considered for the first time in V. I. Istrăţescu and I. I. Istrăţescu (1967).

8.3.10 THEOREM Let T be an operator on a finite dimensional Hilbert space. Then the following conditions are equivalent:

1. T is normal.
2. T has rest-transaloid.
3. T has rest-normaloid.
4. T has rest-convexoid.
5. T has rest-spectraloid.
6. T has rest-of class (N).
7. T has rest-of class (N,k).

For the proof we use the property proved in the proof of Theorem 7.5.4 which is as follows: if z is a complex number such that

1. $Tx = zx$
2. $r_T = |z|$

and T is spectraloid, then $T^*x = z^*x$. In other words, z is a normal eigenvalue.

From this the following implications hold: $1 \to 2 \to 3 \to 5$ and $2 \to 4 \to 5$.

Suppose now that 5 holds and thus for any z, $r_T = |z|$ is a normal eigenvalue if $z \in \sigma(T)$. In this case the subspace

$$E_z(T) = \{x \mid Tx = zx\}$$

is reducing and thus on the orthogonal complement, the operator T_1 which is the restriction of T to this subspace is also a spectraloid operator, and thus we can repeat the argument. We obtain that T has the form

$$Tx = \Sigma \quad z_i a_i x_i$$

and

$$T^*x = \Sigma \quad z_i^* a_i x_i$$

which gives that T is normal. Now if 6 or 7 holds, then it is clear that T is rest-normaloid and the assertion follows from 3.

8.3.11 REMARK Simple examples show that even if an operator has rest-P or red-P, T* does not necessarily have the same property.

Indeed, let T be the shift operator. In this case T is with rest-normaloid property and clearly T* does not have this property.

As is easy to see, for any normal operator T on a Hilbert space the operators T*T and TT* commute with T. In what follows we give some theorems in the opposite direction.

For this we need the following lemma, which is also of independent interest. We mention that the result was conjectured by Kaplansky and a partial result was obtained by Vidav (1955) and next by C. R. Putnam. The conjecture was proved independently by Kleinecke (1957) and Shirocov (1956).

8.3.12 LEMMA If a and b are elements of a Banach algebra with unit and $q = ab - ba$ (the commutator of a and b) commutes with a (or b), then q is a quasi-nilpotent element.

For the proof we need a result about a special type of derivation on a Banach algebra.

Define the operator $d: A \to A$ by the formula

$$d(x) = ax - xa$$

where a is the above element. A simple computation shows that d is a derivation, i.e.,

$$d(xy) = d(x)y + xd(y)$$

and an induction argument gives the "Leibniz rule" for the derivation d. Since the element b has the property that $aq = qa$, it follows that

$$d(b^2) = 0$$

From these we obtain an estimate for $d^n(b^n)$. First we remark that $d(b^n)$, by the Leibniz rule, is equal to the sum of n possible products each of which has one factor equal to d(b) and others are b. Now if we apply d to this sum $d(d(b)) = 0$ and this gives that $d(d(b^n))$ is equal to a sum of $n - 1$ products. This can be repeated and we obtain that $d^n(b^n)$ is equal to a sum of n! terms, and since each of these terms is equal to d(b), we have the formula

$$d^n(b^n) = n!(d(b))^n$$

Now we have that

$$\|(d(b))^n\| = \frac{1}{n!}\|d^n(b^n)\| \le \frac{1}{n!}\|d\|^n\|b\|^n$$

and this gives the result since

$$\left(\frac{1}{n!}\right)^{1/n} \to 0$$

From the result of Kleinecke and Shirokov we can easily derive the following theorem about normality.

8.3.13 THEOREM If $T \in L(H)$, H is a Hilbert space and is an operator satisfying the relations

1. $T(T^*T) = (T^*T)T$
2. $T(TT^*) = (TT^*)T$ or $T(T^*T - TT^*) = (T^*T - TT^*)T$

then T is a normal operator.

Proof. We consider the commutator of T and T*, $Q = T^*T - TT^*$ and the conditions of the theorem imply that Q is a quasi-nilpotent operator. Since Q is hermitian, we have that $Q = 0$.

8.3.14 COROLLARY If $T \in L(H)$ and $T^*T - TT^*$ is a compact operator, then $TT^* - T^*T$ is also a compact operator.

We give now a characterization of normality in terms of the properties of an operator such as normaloidity, convexoidity.

8.3.15 THEOREM The following conditions are equivalent:

1. T is normal.
2. Any quadratic polynomial in T and T* is normaloid.
3. Any quadratic polynomial in T and T* is convexoid.
4. Any quadratic polynomial in T and T* is with G_1 property.

Proof. It is obvious that 1 implies 2. Suppose that 2 holds. Since every quadratic polynomial in T and T* is of the form

$$p(T, T^*) = aT^2 + bTT^* + cT^*T + dT^{*2} + eT + fT^* + gI$$

where a, b, ..., g are complex numbers. This obviously implies that 3 holds. We show now that 3 implies 1. Let $T = A + iB$, where A and B are hermitian operators. We can assume, without loss of generality, that $B \geq I$, and the assertion is proved if we show that $AB = BA$.

Since $B \geq I$, the operator $B^{1/2}$ exists and since

$$\sigma(AB) = \sigma(B^{1/2}AB^{1/2})$$

except possibly zero, the operator $B^{1/2}AB^{1/2}$ is hermitian; the spectrum of AB is thus in R. But the operator AB is convexoid and this implies that the numerical range W(AB) is real. Thus AB is a hermitian operator. This gives that 3 implies 1.

Now it is clear that 1 implies 4. But 4 implies 3 and thus 4 implies 1.

We give now some conditions implying normality connected with the characteristics of the spectrum.

8.3.16 THEOREM Let $T \in L(H)$ with red-G_1 property and $\sigma(T)$ a countable set. Then T is normal.

For the proof we need the following lemma which is also of independent interest.

8.3.17 LEMMA Let $T \in L(H)$ and T with the rest-convexoid property and z_0 be an isolated point in $\sigma(T)$. Then z_0 is an eigenvalue of T.

Proof. Let C_ε be a circle with the center in z_0 and radius ε so that

$$\{z \mid |z - z_0| \leq \varepsilon\} \cap \sigma(T) = \{z_0\}$$

We define now the projection P_{z_0} by the formula

$$P_{z_0} = \frac{1}{2\pi i} \int_{C_\varepsilon} (z - T)^{-1} \, dz$$

and clearly $P_{z_0}H$ is an invariant subspace of T. Since the spectrum of the restriction of T to this subspace is $\{z_0\}$ and the restriction is a convexoid operator, it follows that it is of the form $z_0 I/P_{z_0}$. This gives the assertion.

Proof of Theorem 8.3.16. Let M be the set of all eigenvalues of T with the property that for $E_z(T) = \{Tx = zx\}$ we have

$$E_{z^*}(T^*) \subset E_z(T)$$

If H_1 is the closed linear span of all subspaces $E_z(T)$ with $z \in M$, then clearly this is a reducing subspace of T. Also it is obvious that $T/H_1 = T_1$ is a normal operator. If the orthogonal complement of H_1 is $\{0\}$, then the assertion of the theorem is proved. If this is not the case, then $T = T_1 \oplus T_2$ and the point spectrum of

T_1 is exactly M and also $\sigma(T) = \sigma(T_1) \cup \sigma(T_2)$ and the point spectrum of T_2 is $\sigma_p(T) - M$. From these it is clear that $\sigma(T_2)$ is also a countable set. In this case $\sigma(T_2)$ has an isolated point (in the contrary case it is a perfect set and thus uncountable). Since T_2 satisfies the G_1 property, it follows that the isolated point is an eigenvalue; moreover, it is a normal eigenvalue. But this contradicts the definition of M. Thus $H_1^{\perp} = \{0\}$.

8.3.18 DEFINITION A normal operator $N \in L(H)$ is called *diagonal* if the set of eigenvectors of N is total in H.

With this definition, the above result can be formulated as follows.

8.3.19 THEOREM If $T \in L(H)$ is with the rest-G_1 property and $\sigma(T)$ is a countable set, then T is a diagonal operator.

The above theorem can be generalized.

8.3.20 THEOREM If $T \in L(H)$ has the rest-convexoid property and all but a finite number of elements of $\sigma(T)$ are real, then T is normal.

Proof. We can decompose $T = T_1 \oplus T_2$, where T_1 is as above. We show now that T_2 is a hermitian operator. This follows from the hypothesis since $\sigma(T_2)$ is real and T_2 is a convexoid operator.

8.3.21 THEOREM If $T \in L(H)$ is with the rest-convexoid property and $\sigma(T)$ is a finite set, then T is a normal operator.

Proof. Let $\sigma(T) = \{z_1, z_2, \ldots, z_n\}$ be the spectrum of T. Since T is convexoid, we have that

$$\text{conv } \sigma(T) = \text{cl } W(T)$$

and thus W(T) is a closed set. Let z_1 be such that this is also an extreme point of W(T). Of course, the theorem is true if n = 1.

Suppose now that the assertion is true for all $i \leq n - 1$. Since $\sigma(T) = \sigma(T/E_{z_1}(T)) \cup \sigma(T/E_{z_1}(T)^{\perp})$, it follows that T is a normal operator, i.e., z_1 is a normal eigenvalue and on $E_{z_1}(T)^-$ is a normal operator.

The following theorem is related to Theorem 8.3.16.

8.3.22 THEOREM If $T \in L(H)$ is an operator with the rest-convexoid property and $\sigma(T)$ has only one limit point, then T is normal.

Proof. We may assume, without loss of generality, that the limit point is zero. It is then clear that we can arrange the points of $\sigma(T) - \{0\}$ in a sequence $\{z_n\}$ such that

$$|z_1| = \cdots = |z_r| > |z_{r+1}| = \cdots = |z_l| > |z_{l+1}| = \cdots$$

and $\lim z_n = 0$.

From Lemma 8.3.17 it follows that each z_n is a normal eigenvalue. Let N be the closed linear span of all $N(T - z_i)$. From this it is clear that T/N is a normal operator. From the rest-convexoid property we can show very easily that $T/N^{\perp} = 0$.

This result can be extended to the case of a finite set of limit points in $\sigma(T)$.

8.3.23 THEOREM If $T \in L(H)$ is with the rest-convexoid property and $\sigma(T)$ has only a finite set of limit points, then T is normal.

For the proof we need the following lemma.

8.3.24 LEMMA If $T \in L(H)$ is with the rest-convexoid property and s and z are two distinct eigenvalues, then

$$E_s(T) \perp E_z(T)$$

Proof. Let N be the space generated by $E_s(T)$ and $E_z(T)$. Clearly T/N is a normal operator, and the assertion follows.

Proof of Theorem 8.3.23. We use an induction argument. If the number of limit points is one, this is exactly Theorem 8.3.22. Let z_0 be a limit point of $\sigma(T)$ and C_ε be a circle such that no other limit point of $\sigma(T)$ is in this circle. We can define the projection

$$P_{z_0} = \frac{1}{2\pi i} \int_{C_\varepsilon} (z - T)^{-1} \, dz$$

and let N_{z_0} be the null space of P_{z_0} and M_{z_0} be the range of P_{z_0}.

Clearly these are invariant subspaces of T. Furthermore, T/N satisfies Theorem 8.3.22 and thus is a normal operator. If we prove that N_{z_0} and M_{z_0} are reducing, then the assertion of the theorem follows immediately from an induction argument concerning the number of limit points in $\sigma(T)$.

The property of reducing these subspaces follows from Lemma 8.3.24 and the remark that these spaces are generated by eigenvectors corresponding to elements in $\sigma(T/N_{z_0})$ and $\sigma(T/M_{z_0})$.

The following result is related to the behavior of T in the Calkin algebra.

For any operator $T \in L(H)$, $\hat{T}$ denotes the image in the Calkin algebra of H.

8.3.25 THEOREM Let $T \in L(H)$ with the following properties:

1. $\sigma(\hat{T}) = \{0\}$.
2. T is reduced by each of its finite dimensional eigenspaces.
3. T is with the red-spectraloid property.

Then T is normal (and compact).

Proof. Since for any operator

$$\partial\omega(T) \subset \sigma(\hat{T})$$

it follows that $\omega(T) = \{0\}$. Let M be the closed linear span of the finite dimensional eigenspaces of T. Let $T_1 = T/M$ and $T_2 = T/M^-$; it is clear that T_1 is a normal operator and T_2 has no eigenvalues of finite multiplicity. If T_2 is not zero, we have that

$$\omega(T) = \omega(T_2)$$

and since $\sigma(T_2) = \omega(T_2)$, we have that T = 0 since it is spectraloid.

8.3.26 DEFINITION An operator $T \in L(H)$ is called *isoloid* if every isolated point in the spectrum is an eigenvalue.

8.3.27 THEOREM Suppose $T \in L(H)$ has the following properties:

1. $\sigma(T)$ is countable.
2. T is reduced by each of its eigenspaces.
3. T is with the red-isoloid property.

Then T is normal.

Proof. From the relation

$$\partial\omega(T) \subset \sigma(\hat{T})$$

it follows that $\omega(T)$ is a countable set. Let M be the closed linear span of all eigenspaces of T. Define $T_1 = T/M$ and $T_2 = T/M^{\perp}$. As above, it is clear that T_1 is normal and T_2 has no eigenvalues.

If we show that $M^{\perp} = \{0\}$, the assertion of the theorem follows.

Suppose that this is not the case, and as above, $\sigma(T_2) = \omega(T_2) \subseteq \omega(T)$. This gives that $\sigma(T_2)$ is a countable set. Thus it has an isolated point and this is an eigenvalue. This contradiction proves the assertion.

8.3.28 REMARK In connection with some of the above results it appears of some interest to know if an operator $T \in L(H)$ with the following properties:

1. $\sigma(T)$ has at most a countable set of nonreal points
2. T is with the red-spectraloid property

is a normal operator.

An example of an operator with $\sigma(T)$ as in 1 is any $T \in L(H)$ with a compact imaginary part; also any polynomial of T is an operator with a compact imaginary part.

In what follows we give theorems about normality connected with the compactness.

8.3.29 THEOREM If $T \in L(H)$ is an operator of class (N) and for some p and q the operator $T^{*p}T^q$ is compact, then T is normal and compact.

Proof. Since T is of class (N), there exists $z \in \sigma(T)$ such that $|z| = \|T\|$. In this case there exists a sequence of unit vectors $\{x_n\}$ such that

$$Tx_n - zx_n \to 0$$

and since $|z| = \|T\|$, this implies that

$$T^*x_n - z^*x_n \to 0$$

From these we obtain easily that

$$T^{*p}T^q x_n - z^{*p}z^q x_n \to 0$$

and from the compactness of $T^{*p}T^q$ it follows that the sequence $\{x_n\}$ is compact. Let $x_{n_k} \to x_0$. In this case we have

$$Tx_0 = zx_0 \qquad T^*x_0 = zx_0$$

Consider now the subspace

$$E_{z_0}(T) = \{x \mid Tx = z_0\}$$

which does not reduce to $\{0\}$ and is clearly a reducing subspace. On the orthogonal complement of $E_{z_0}(T)$, $T/E_{z_0}(T)$ is also an operator of class (N) satisfying the same properties as T. Thus if we consider the space M with the closed linear space generated by the points in $\sigma_p(T)$ constructed as above, we have that M is a reducing subspace and T/M is normal. We show that on $M^\perp$, $T/M^\perp = 0$.

Indeed, in the contrary case, $T/M^\perp$ is an operator of class (N) satisfying the same properties as T and thus we can find a point in $\sigma_p(T)$ and an eigenspace which is reducing. But this contradicts the definition of M. Thus $T/M^\perp = 0$ and the assertion of the theorem is proved.

In the same way we can prove the following more general result.

8.3.30 THEOREM If $T \in L(H)$ is an operator of class (N) and if for some system of integers $\{p_1, q_1, \ldots, p_n, q_n\}$ the operator

$$T^{*p_1}T^{q_1}\cdots T^{*p_n}T^{q_n}$$

is compact, then T is normal and compact.

Also from the proof it is clear that the following extension of the above theorem holds.

8.3.31 THEOREM Let $T \in L(H)$ and with the properties:

1. For a system of integers $\{p_1, q_1, \ldots, p_n, q_n\}$ the operator

$$T^{*p_1}T^{q_1}\cdots T^{*p_n}T^{q_n}$$

is compact.

2. If T is with the rest-normaloid property, then T is normal and compact.

The following theorem uses the Calkin algebra of H. Also it is worth noting that the compactness of T in Theorems 8.3.29 and 8.3.30 follows easily using the Calkin algebra.

8.3.32 THEOREM If T is an operator of class (N) and is of Riesz type, then T is normal.

Proof. We consider for this image of T, $\hat{T}$, in the Calkin algebra of H. We have, for all $s \in R$,

$$\hat{T}^{*2}\hat{T}^2 - 2s\hat{T}^*\hat{T} + s^2\hat{I} \geq 0$$

and since every Riesz operator is of the form C + Q, where C is compact and Q quasi-nilpotent, it follows in our case that the image of T is quasi-nilpotent. Since it is also of class (N), it follows that $\hat{T} = 0$.

8.3.33 REMARK An extension of this result is possible if we suppose that an operator of the type considered in Theorems 8.3.30 and 8.3.31 is of Riesz type.

For the following results we need information about the spectrum of T having a special form.

8.3.34 THEOREM Let $T \in L(H)$ and suppose that there exists a hermitian operator A and an invertible operator S such that

$$T^* = S^{-1}TS + A \qquad A \text{ compact, } 0 \notin \text{cl } W(S)$$

In this case,

1. If $\sigma_p(T) = \emptyset$, then $\sigma(T)$ is real.
2. The set of points with an imaginary part different from zero in $\sigma(T)$ is finite or countable and has the limits point on R.

Proof. We consider first case 1. Since $\sigma(T)$ is a subset of the approximate point spectrum, in this case it will be sufficient to prove the following assertions: If $\{x_n\}$ is a sequence of unit vectors satisfying the following properties:

1. $x_n \rightharpoonup 0$ ($\rightharpoonup$ means weakly)
2. $Tx_n - zx_n \to 0$

then z is a real number.

First we remark the identity

$$(z^* - z)\langle S^{-1}x_n, x_n\rangle = \langle (T^* - z)S^{-1}x_n, x_n\rangle - \langle (T^* - z^*)S^{-1}x_n, x_n\rangle$$

and since T^* has the above form, we obtain

$$\begin{aligned}(z^* - z)\langle S^{-1}x_n, x_n\rangle &= \langle [S^{-1}(T - z) + A]S^{-1}x_n, x_n\rangle \\ &\quad - \langle S^{-1}x_n, (T - z)x_n\rangle \\ &= \langle S^{-1}(T - z)x_n, x_n\rangle - \langle S^{-1}x_n, (T - z)x_n\rangle \\ &\quad + \langle AS^{-1}x_n, x_n\rangle\end{aligned}$$

Now $Ax_n \to 0$ (or a subsequence) and this gives that if Im $z \neq 0$, then $0 \in \text{cl } W(S)$, and this is a contradiction.

Now we consider case 2. We define the operator

$$A^{\sim} = \frac{1}{2}(S + S^*)$$

and we remark that the hypothesis about S implies that $A^{\sim}$ is invertible (and, of course, it is hermitian).

We consider also the operator

$$T^{\sim} = A^{\sim -1/2}TA^{\sim 1/2}$$

and we prove that

$$T^{\sim *} = T^{\sim} + A_1$$

where A_1 is a compact operator.

For the proof we remark that

$A^{\sim}T^* = 1/2(S + S^*)T^* = 1/2(ST^* + S^*T^*)$

and since

$$ST^* = TS + SA$$

we have that

$$S^*T = TS^* - AS^*$$

and this implies that

$$ST^* + S^*T^* = ST^* + TS^* - AS^*$$

and thus

$$AT^* = \frac{1}{2} T(S + S^*) + \frac{1}{2} (SA - AS) = T\tilde{A} + A_1$$

It is clear that A_1 is a compact operator and since $\sigma(T) = \sigma(\tilde{T})$, the result is a consequence of the behavior of the spectrum of an operator with compact imaginary part, since

$$\tilde{T}^* = \tilde{T} + A_2$$

where A_2 is also a compact operator.

8.3.35 DEFINITION A unitary operator U is said to be with *cramped spectrum* if there exists ϑ_0 such that

$$\sigma(U) \subset \{e^{i\vartheta} \mid \vartheta_0 < \vartheta < \vartheta_0 + \pi\}$$

8.3.36 THEOREM If $T \in L(H)$ is with the property that

$$T^* = U^{-1}TU + A \qquad T \text{ normal}$$

where U is a unitary operator with cramped spectra and A is a compact hermitian operator, then T is a hermitian operator.

Proof. Since U has a cramped spectrum, it follows that 0 is not in cl W(U) and thus the spectrum of T is real.

8.3.37 THEOREM If $T \in L(H)$ is hyponormal and there exists the operators S and C such that

1. $0 \notin \text{cl } W(S)$
2. $T^p = ST^{*p}S^{-1} + C \qquad C \text{ compact}$

then T is a normal operator.

Proof. For the proof we need the following very important result of Putnam (1970).

8.3.38 THEOREM If $T \in L(H)$ is a hyponormal operator and the area of $\sigma(T)$ is zero, then T is normal.

Proof of Theorem 8.3.37. Consider the image of T in the Calkin algebra; thus we have that

$$\hat{T}^p = \hat{S}\hat{T}^{*p}\hat{S}^{-1}$$

and this implies that $\sigma(\hat{T})$ is on p lines through origin. Since $\partial\omega(T) \subset \sigma(\hat{T})$, it follows that $\omega(T)$ is also on these p lines, and since for T Weyl's theorem holds (see Chapter 11), it follows that $\sigma(T) - \omega(T)$ is countable. Thus the area of $\sigma(T)$ is zero and the assertion follows from the above theorem.

8.3.39 REMARK This theorem can be proved without Putnam's result [see Berberian (1970)].

The following theorem is an extension of Theorem 8.3.38; its proof is not essentially different, and thus we omit it.

8.3.40 THEOREM If $T \in L(H)$ is a hyponormal operator satisfying the relation

$$T^p = ST^{*p}S^{-1} + A$$

with $p \geq 1$ and $0 \notin \text{cl } W(S)$, A a compact and hermitian operator, then T is normal.

The following example shows that in some sense for normality the normaloidity for the restriction to invariant subspaces is a necessary condition.

8.3.41 EXAMPLE There exists a normaloid operator T such that for any fixed n, T^n is compact and T is not normal.

We define the operator T on a Hilbert space with the basis $\{e_i\}_1$ by the relations

$$Te_i = \begin{cases} e_i & i = 1 \\ e_{i+1} & i = 2m,\ m = 1, 2, 3, \ldots \\ 0 & i = 2m + 1,\ m = 1, 2, 3, \ldots \end{cases}$$

and clearly $\|T^i\| = 1$, T is nonnormal, and T^2 is a compact operator.

Obviously this construction can be modified to give the compactness of T^p with p given.

The following theorem gives a very special class of hermitian operators.

8.3.42 THEOREM If $T \in L(H)$ is idempotent and $w(T) \leq 1$, then T is a projection.

In what follows we give two proofs of this result.

Proof. 1. Let $y = Tx$, where $x \in R(T)^-$, and thus for any $t \in R^+$ we have

$$T(x + ty) = 1 + ty$$

and

$$t(1 + t)\|y\|^2 = \langle (1 + t)y, x + ty \rangle = \langle T(x + ty), x + ty \rangle$$
$$\leq \|x + ty\|^2 = \|x\|^2 + t^2\|y\|^2$$

which implies that $t\|y\|^2 \leq \|x\|^2$.

Since t is arbitrary it follows that $y = 0$. Thus $Tx = 0$ for all $x \in R(T)$ and $Tx = x$ for $x \in R(T)$. Thus T is a projection.

For the second proof we need the following result by T. Kato.

8.3.43 LEMMA For an operator $T \in L(H)$ the following two assertions are equivalent:

1. $w(T) \leq 1$
2. $\|e^{zT}\| \leq e^{|z|}$ for all complex numbers z

Proof. Assertion 2 is obviously equivalent to

$$\|e^{t(e^{i\vartheta}T-I)}\| \leq 1$$

and this implies that the operator T is *dissipative*, i.e.,

$$\langle Tx, x\rangle + \langle x, Tx\rangle \le 0$$

for all $x \in H$. But this is obviously equivalent to $w(T) \le 1$ and thus the lemma is proved.

Proof 2 of Theorem 8.3.42. Since T is idempotent

$$\|e^{zT}\| = \|I + (e^z - 1)T\| \le e^{|z|}$$

For $t \in R^+$ and $t \to \infty$

$$\|e^{-t}T + (1 - e^{-t})I\| \le 1$$

and this gives that $\|T\| \le 1$.

From this we obtain that

$$\begin{aligned}\|Tx - T^*Tx\|^2 &= \|Tx\|^2 - \langle Tx, T^*Tx\rangle - \langle T^*Tx, Tx\rangle + \|T^*Tx\|^2 \\ &\le \|Tx\|^2 - \|Tx\|^2 + \|T^*Tx\|^2 + \|T^*Tx\|^2 \\ &= \|T^*Tx\|^2 - \|Tx\|^2 \le 0\end{aligned}$$

and this gives the assertion.

An interesting characterization of normality was obtained by Stampfli using the class of hyponormal operators.

8.3.44 THEOREM If $T \in L(H)$ is a hyponormal operator and for some integer p, T^p is normal, then T is normal.

This result was extended to operators of Class (N) by Ando (1972).

8.3.45 THEOREM A necessary and sufficient condition for an operator $T \in L(H)$ to be normal is that for some $p > 1$ the operator T^p is of class (N).

Proof. It is clear that it is sufficient to prove that if T^p is normal, then T is normal.

The operator $(T^{*p}T^p)$ is hermitian and let $\{E_t\}$ be the spectral measure associated with this operator. From the Fuglede-Putnam theorem it is clear that E_t commutes with T and T^*. Since $TE_0 = 0$ [because TE_0 is of class (N) and $(TE_0)^p = T^pE_0 = 0$], it remains to prove that $T(I - E_t)$ is a normal operator for any $t > 0$.

Let $\varepsilon > 0$ and

$$t = t_0 < t_1 < \cdots < t_n = \|T^p\| \qquad 1 - \varepsilon < t_j^{-1} t_{j-1}$$

Define $F_j = E_{t_j} - E_{t_{j-1}}$ and we obtain

$$t_{j-1}\|F_j x\| \le \|T^p F_j x\| = \|T^{*p} F_j x\| \le t_j \|F_j x\|$$

But TF_j is an operator of class (N) and this gives

$$\|TF_j\| = \|T^* F_j\|$$

We show now that

$$(1 - \varepsilon)\|TF_j x\| \le \|T^* F_j x\| \le (1 - \varepsilon)^{-1}\|TF_j x\|$$

Indeed, if $y = F_j y$, $\|y\| = 1$, and $\|Ty\| \le (1 - \varepsilon)\|T^* y\|$, we have

$$t_{j-1}\|T^p y\| \le \|TF_j\|^{p-1}\|Ty\| \le (1 - \varepsilon)t_j$$

which contradicts the definition of t_j.

Let $x = (I - E_t)x$ and thus

$$(1 - \varepsilon)^2\|Tx\|^2 = (1 - \varepsilon)^2 \sum \|TF_j x\|^2 \le \|T^* x\|^2$$

$$\le (1 - \varepsilon)^{-2} \sum \|TF_j x\|^2 \le (1 - \varepsilon)^2\|Tx\|^2$$

and for $\varepsilon \to 0$ we obtain that

$$\|Tx\| = \|T^* x\|$$

This is equivalent with the assertion of the theorem.

Using this result we can improve Stampfli's result as follows.

8.3.46 THEOREM If $T \in L(H)$ is a hyponormal operator and for some p, T^{*p} is hyponormal, then T is normal.

Proof. As we know the operator T^p is of class (N). Also since T^{*p} is hyponormal, it is also of class (N). Now we prove that the null spaces of T^p and T^{*p} are the same.

Indeed, if $x \in N(T^p)$, then it follows that $x \in N(T)$ since T is of class (N). But T is hyponormal and this implies that $x \in N(T^*)$.

From this we obtain that

$$N(T^p) \subset N(T^{*p})$$

and using hyponormality of T^{*p}, we obtain the equality of these subspaces.

Now the result follows from the following characterization of normality.

8.3.47 THEOREM An operator $T \in L(H)$ is normal if and only if

1. T and T* are of class (N).
2. $R(T) = R(T^*)$.

For the proof of this theorem we use the following lemma.

8.3.48 LEMMA If A and B are bounded positive operators such that

$$2sA^2(A^2 + s^2)^{-1} \leq B \leq (2s)^{-1}(A^2 + s^2)$$

for all $s > 0$, then $A = B$.

Proof. It will be sufficient to show that B commutes with the spectral measure associated with A since

$$\sup_{s>0} 2sr^2(r^2 + s^2)^{-1} \leq r \leq \inf_{s>0} (2s)^{-1}(s^2 + r^2)$$

and thus it will be sufficient to show that for $t > s > 0$,

$$(I - E_{t+s})B(E_t - E_s) = 0$$

where $\{E_t\}$ is the spectral measure associated with A.

Let

$$x = (E_t - E_s)x \qquad y = (E_t - E_s)y$$

and

$$x_j = (E_{t_j} - E_{t_{j-1}})x \qquad A_j = 2t_j(A^2 + t_j^2)^{-1}$$

In this case we have

$$0 < B - A^2A_j \le A_j^{-1} - A^2A_j = (2t_j)^{-1}(A - t_j)^2(A + t_j^2)(A^2 + t_j^2)^{-1}$$
$$\le s^{-1}(A - t_j)^2$$

and since

$$\langle A^2A_jx_j, y\rangle = 0 \qquad \|(A - t_j)x_j\| < \frac{1}{n}t\|x_j\|$$

from the Schwarz inequality we obtain

$$|\langle Bx_j, y\rangle| = |\langle(B - A^2A_j)x_j, y\rangle|^2 \langle(B - A^2A_j)x_j, x_j\rangle\langle(B - A^2A_j)y, y\rangle$$
$$\le s^{-1}\|B\|\|(A - t_j)x_j\|^2\|y\|^2$$
$$\le (n^2s^{-1})^{-1}t^2\|B\| \cdot \|x_j\| \cdot \|y\|^2$$

This implies that

$$|\langle Bx, y\rangle|^2 = \left|\sum_{j=1}^{n} \langle Bx_j, y\rangle\right|^2 \le n \sum_{j=1}^{n} |\langle Bx_j, y\rangle|^2$$
$$\le (ns)^{-1}t^2\|B\| \cdot \|x_j\| \cdot \|y\|$$

and for $n \to \infty$ we have $\langle Bx, y\rangle = 0$, and the assertion of the lemma is proved.

Proof of Theorem 8.3.46. First we remark that the conditions are obviously necessary. Now suppose that T has properties 1 and 2, and consider the operators

$$A = (T^*T)^{1/2} \qquad B = (TT^*)^{1/2}$$

and the assertion of the theorem is proved if $A^2 = B^2$. For this we remark that we can assume, without loss of generality, that A and B are invertible since $R(T) = R(T^*)$. Now the operators T and T^* arc of class (N) and thus for all $s \in R$,

$$T^{*2}T^2 - 2sT^*T + s^2 \ge 0 \qquad \text{and} \qquad T^2T^{*2} - 2sTT^* + s^2 \ge 0$$

Let $S = (BA^2B)^{1/2}$ and since

$$D_S - 1 = D_{BS} - 1 \qquad \|S^{-1}x\| = \|(BS^{-1}x\|$$

we obtain that

$$\|(S^2 + s^2)S^{-2}x\|^2 = \|x\|^2 + s^2\|S^{-1}x\|^2$$
$$= \|x\|^2 + s^2\|A^{-1}B^{-1}x\|^2 \geq 2s\|B^{-1}x\|^2$$

which implies that

$$B^2 \leq (2s)^{-1}(S^2 + s^2)$$

since

$$2s^2S^2(S^2 + B^2)^{-1} \leq B^2$$

From these we have that $B^2 = S$ and thus $A = B$.

We close with some problems which we think to be of interest, and these are connected with some results given above.

8.3.49 PROBLEM If T is of class (N) and for some p, T^{*p} is of class (N), does it follow that T is normal?

8.3.50 PROBLEM If T is hyponormal [or of class (N)] and for some integers p and q the operator $T^{*p}T^q$ is hyponormal, then is T normal?

8.3.51 PROBLEM If $T \in L(H)$ has the property that for all complex numbers z the operator $T_z = T + z$ is also of class (N), then is T hyponormal?

Chapter 9

SYMMETRIZABLE OPERATORS: GENERALIZATIONS AND APPLICATIONS

As we know, one of the most important classes of operators defined on a Hilbert space is the class of symmetric (hermitian) operators. Thus the problem appears to be whether there exists a class of operators having many properties in common with this class.

There exist different classes of operators having this property and it is our purpose to give an account of the results obtained in this direction. The first class considered with this aim was that of J. Marty in 1910 [see A. Zaanen (1953)] for the case of integral transformations. Important contributions were made by Zaanen (1953), Reid (1951) and Lax (1954).

We give in Section 9.1 the theory in the case of Hilbert spaces, and in the rest of the chapter we discuss the general theory as it was proposed in I. I. Istrăţescu and V. I. Istrăţescu (1969a).

Also we give for the first time some results of the theory in the case of C*-algebras.

9.1 SYMMETRIZABLE OPERATORS ON HILBERT SPACES

Let X be a Hilbert space and L(X) be the space of all continuous linear operators on H.

9.1.1 DEFINITION If H is a hermitian operator on T and K is an element of L(X), then we say that K is *symmetrizable* (relative to H) (*left symmetrizable*) if the operator HK is hermitian.

9.1.2 EXAMPLE If A is an element of $L(H)$ such that

$$HAH = HA^*H$$

then the operator $K = AH$ is symmetrizable relative to H.

9.1.3 EXAMPLE If K is symmetrizable relative to H, then for any integer n, the operator K^n is symmetrizable (relative to H).

Indeed, for $n = 1$ this is true and suppose that for all $i \leq n$ K^i is symmetrizable. In this case for $i = n + 1$ we have

$$\langle HK^{n+1}x, y\rangle = \langle HK^nKx, y\rangle = \langle Kx, HK^ny\rangle = \langle KHx, K^ny\rangle = \langle x, HK^{n+1}y\rangle$$

and the assertion is proved.

9.1.4 EXAMPLE (Garbe Operator) Let H be an integral operator with the kernel $H(x, y)$ such that

$$\int |H(x, y)||f(y)| \, d\mu(y) \in L^2(\mu)$$

for all $f \in L^2$ and

$$\int H(x, y)\bar{f}(x)f(y) \, d\mu \geq 0$$

If $A(x)$ is measurable and bounded almost everywhere with real values, then the operator defined by

$$(Af)(x) = f(x)A(x)$$

is bounded and hermitian. The operator $K = AH$ is symmetrizable relative to H. [It is also worth noting that the kernel of K is $A(x)H(x, y)$.]

9.1.5 EXAMPLE (Pell Operator) This is a generalization of Garbe operator assuming A to be of the form $A(x, y)$.

9.1.6 THEOREM If K is symmetrizable relative to H and λ and μ are two eigenvalues such that $\mu^* \neq \lambda$, then

$$M_\lambda = \{x \mid Kx = \lambda x\} \perp M_\mu = \{y \mid Kx = \mu x\}(H)$$

Proof. Let x and y be such that $Kx = \lambda x$ and $Ky = \mu y$. We have then

$$\langle Hx, y\rangle = \langle \lambda^{-1}HKx, y\rangle = \lambda^{-1}\langle x, HKy\rangle = \lambda^{-1}\mu^*\langle x, y\rangle$$

and since $\lambda \neq \mu^*$, we must have $\langle Hx, y \rangle = 0$.

We define now two notions connected with symmetrizability.

9.1.7 DEFINITION 1. A bounded linear operator K is *fully symmetrizable* relative to H if

a. K is symmetrizable relative to H.

b. $Hx \neq 0$ for every $x \in H$ such that $Kx = zx \neq 0$.

2. A bounded linear operator K is called *strongly symmetrizable* if $Hx = 0$ implies $Kx = 0$ and K is symmetrizable relative to H.

9.1.8 REMARK It is clear that every strongly symmetrizable operator is fully symmetrizable.

9.1.9 THEOREM If K is a fully metrizable and H is positive, then for any eigenvalue of K is real.

Proof. Let λ and $x \in X$ such that $Kx = \lambda x$. In this case we have

$$\langle HKx, x \rangle = \lambda \langle Hx, x \rangle = \langle x, Hx \rangle = \lambda^* \langle Hx, x \rangle$$

and this gives that $\lambda^* = \lambda$. The assertion is proved.

9.1.10 REMARK It is clear from the above proof that the assertion remains true if we assume only the positiveness of the operator HK.

A fundamental problem in the case of symmetrizable operators is the existence of nonzero eigenvalues. In what follows we prove a theorem which gives an answer to this problem in the case of special classes of operators, namely, for operators which are compact or in which an iterate is a compact operator. Of course, we suppose that H is a positive operator.

First we give a result which has its own interest.

9.1.11 THEOREM Let K be a symmetrizable operator relative to H which is positive. Then for any $x \in X$,

$$|\langle x, HKx \rangle| \leq \|K\| \langle x, Hx \rangle$$

Proof. First we remark that the above inequality is equivalent to the fact that the operator $\|K\|H \pm HK \geq 0$, or that

$$H \pm H \frac{K}{\|K\|} \geq 0$$

Now since H is positive, there exists a square root of H, denoted by $H^{1/2}$. For the proof we need the following result which is easy to prove: For any elements of a Banach algebra a and b, except possibly the point zero,

$$\sigma(ab) = \sigma(ba)$$

We consider now the operators

1. $a = H^{1/2}$
2. $b = H^{1/2}\left(I \pm \frac{K}{\|K\|}\right)$

and we remark that since the operator ba is hermitian, its spectrum is in the set of real numbers. From the above result, it follows that since ba is a positive operator, ab has also the spectrum on the positive real axis and this gives the assertion of the theorem.

9.1.12 REMARK The above inequality was proved by Reid; his arguments are more involved than those presented here.

We are now in a position to give the fundamental existence theorem.

9.1.13 THEOREM Let K be a symmetrizable operator relative to a positive operator H and suppose that K is completely continuous. Then there exists a nonzero real eigenvalue of K in the case when $HK \neq 0$.

Proof. Since HK is a hermitian operator and K is completely continuous, the operator HK is also completely continuous. Since $HK \neq 0$, according to Theorem 5.1.11, there exists a sequence $\{x_n\}$ such that

1. $\langle Hx_n, x_n\rangle = 1 \qquad n = 1, 2, 3, \ldots$
2. $\sup |\langle x, Hkx\rangle| = \lim \langle x_n, HKx_n\rangle = s$

Now, if s is not in the spectrum of K, then $R_s = (sI - K)^{-1}$ exists and we define $\{y_n\}$ by the relation $y_n = R_s y_n$. Now,

$$1 = \langle x_n, Hx_n\rangle = \langle x_n, H(sI - K)y_n\rangle = \langle y_n, (sI - K)Hx_n\rangle \leq \|R_s\|$$

If we set $z_n = x_n - 1/\|R_s\|y_n$, we get

$$\langle z_n, H(sI - K)z_n\rangle = \langle x_n, H(sI - K)x_n\rangle - \frac{2}{\|R_s\|}$$

$$+ \langle y_n, H(sI - K)y_n\rangle \frac{1}{\|R_s\|^2}$$

$$< \langle x_n, H(sI - K)x_n\rangle - \frac{1}{\|R_s\|}$$

Since $\langle x_n, H(sI - K)x_n\rangle \to 0$, we obtain that for large n,

$$\langle z_n, H(sI - K)z_n\rangle \leq 0$$

which is a contradiction since $H(sI - K) \geq 0$.

Thus s is an eigenvalue for K. The theorem is proved.

9.1.14 REMARK The assertion of the theorem remains valid if we suppose that for some p, K^p is completely continuous.

Indeed, we have the factorization

$$s^{2p+1} - K^{2p+1} = (s - K)(s^{2p} + s^{2p-1}K + \cdots + sK^{2p-1} + K^{2p})$$

which implies that $sI - K$ has the inverse equal to

$$(s^{2p} + \cdots + K^{2p})^{-1}(s^{2p+1} - K^{2p+1})^{-1}$$

and this gives the assertion.

The following theorem gives an example connected with Theorem 9.1.13.

9.1.15 THEOREM On a separable Hilbert space X there exists a fully symmetrizable operator K such that

1. K is symmetrizable relative to a positive operator H.
2. HK is compact.
3. K^p is not compact for any $p \geq 1$.

Proof. Let M and N be two infinite dimensional orthogonal subspaces such that

1. M has an orthonormal basis $\{\varphi_n\}$.
2. N has an orthonormal basis $\{\psi_n\}$.

We define on the space generated by these subspaces an operator by the formulas

$$H\varphi_n = a_n\varphi_n \qquad H\psi = 0$$

where $a_n > 0$ with $\lim a_n = 0$. The operator K is defined as follows:

$$K\varphi_n = H_n\varphi \qquad K\psi_n = \psi_{n+1}$$

It is clear that $HK\varphi_n = a_n^2\varphi_n$ and $HK\psi_n = 0$. Thus K is symmetrizable relative to H and since $a_n \to 0$, it is a compact operator. It is obvious that every power of K is not a compact operator.

For other results about symmetrizable operators on Hilbert spaces we refer to Zaanen (1953).

9.2 SYMMETRIZABLE ELEMENTS IN BANACH ALGEBRAS

The purpose of the present section is to consider two classes of elements in Banach algebras. The first class is considered in arbitrary Banach algebras and the second class in B*-algebras or, more generally, in Banach algebras with involution.

9.2.1 DEFINITION An element $k \in A$, where A is a Banach algebra with unit, is called *symmetrizable* if there exists a hermitian element h such that hk is a hermitian element of A.

Let X be a compact Hausdorff space and C(X) be the Banach algebra of all continuous complex-valued functions on X and let L(C(X)) be the Banach algebra of all bounded linear operators on C(X). Using Lumer's result about hermitian elements in L(C(X)), we are able to determine the form of all symmetrizable elements of L(C(X)).

9.2.2 THEOREM Every symmetrizable element k of L(C(X)) with respect to a hermitian element h of L(C(X)) is of the form

$$(kf)(x) = h(x)f(x)$$

where h is a real function in C(X).

Proof. Since h is hermitian, we have that

$$(hf)(x) = h(x)f(x)$$

and since hk is hermitian, we obtain that

$$(hk)(f)(x) = \varphi(x)f(x)$$

where φ is a real continuous function. Now we have

$$\varphi(x)f(x) = h(x)(kf)(x)$$

and thus

$$h(x)(kf)(x) = \varphi(x)f(x)$$

This gives that at any point x, $h(x) \neq 0$, we have

$$(kf)(x) = \frac{\varphi(x)}{h(x)} f(x)$$

An interesting problem is to determine all symmetrizable operators on the space C^n with nonhilbertian norms.

In the case when A is a C*-algebra with unit, using a result of Sinclair [see Bonsall and Duncan (1971, 1973)], we obtain the following theorem.

9.2.3 THEOREM Every symmetrizable element of $L(A)$ is of the form

$$Tx = kx + Dx$$

where k is a symmetrizable element of A and D is a star derivation, i.e.,

1. $D(xy) = (Dx)y + xDy$ for all $x,y \in A$.
2. $Dx^* = -(Dx)^*$ for all $x \in A$.

Proof. Since T is symmetrizable, we can find an element h of $L(A)$ which is also a hermitian element of A such that hT is hermitian.

From Sinclair's result we have that

$$(hT)x = k_1x + D_1x$$

where k_1 is a hermitian element and D_1 is a star derivation of A.

Since h is hermitian, we have that

$$(hT)x = h_1(Tx) + DTx$$

where h_1 is a hermitian element and D is a star derivation of A.

Further, we obtain

$$h_1 Tx + DTx = k_1 x + D_1 x$$

and this gives

$$h_1 Tx = k_1 x \qquad DTx = D_1 x$$

and the assertion of the theorem is proved.

One of the interesting theorems in the theory of Banach algebras obtained in the recent years is the Dye-Russo (1966) theorem, which asserts that the closed unit ball of a B*-algebra A is the closed convex hull of unitary elements of A. An interesting generalization of this result, obtained by Palmer, asserts that the closed unit ball of A is in fact the closed convex hull of the set E, where

$$E = \{e^{ih} \mid h \text{ hermitian}\}$$

Our question connected with these results is the following one: Does the Banach algebra of all bounded operators on a B*-algebra have the properties stated in the Dye-Russo or Palmer theorem?

We give now some results in the case of B*-algebras. The results are like those for operators on Hilbert spaces.

9.2.4 DEFINITION Let A be a unital C*-algebra and $h \in A$. We say that k is symmetrizable if there exists a hermitian element $h \in A$ such that hk is hermitian.

Exactly as in the case of operators on Hilbert spaces, we have the following.

9.2.5 THEOREM Let k be symmetrizable (relative to h), then for any integer n, k^n is symmetrizable relative to h.

9.3 INNER PRODUCTS ON BANACH SPACES: SYMMETRIZABLE OPERATORS AND SOME GENERALIZATIONS

In order to generalize some aspects of the theory of symmetrizable operators on Hilbert spaces to the case of elements of the Banach algebra of all bounded linear operators on a Banach space, we consider here some results of the theory of inner products on Banach spaces.

Let X be a complex Banach space and F be a continuous bilinear functional on X having the following properties:

1. $F(x, y) = F(y, x)^*$
2. $F(x, x) > 0$ for $x \neq 0$

Let

$$x \mapsto F(x, x)^{1/2}$$

and exactly as in the case of Hilbert spaces, we can show that this is a norm on X, which we denote by $|\cdot|$. The continuity of F implies that there is a constant $k > 0$ such that for all $x \in X$,

$$|x| \leq k\|x\|$$

where $\|\cdot\|$ denotes the original norm of X. Since the space $(X, |\cdot|)$ has many of the properties of Hilbert spaces, except possibly the completeness (of course, with respect to $|\cdot|$), we can define with respect to the bilinear form F the notions hermitian operator, normal and unitary operator, etc.

Thus we are led naturally to the problem of the existence of inner products on Banach spaces. We mention that the inner products on Banach spaces were used by Lax (1954) in his study of symmetrizable operators on Banach spaces. Also Olagunju [see Murphy and West (1972)] has given an example of a Banach space which does not possess an inner product $< \, , \, >$ satisfying the following axioms:

1. $\langle x, x\rangle > 0$ and $\langle x, x\rangle = 0$ if and only if $x = 0$
2. $\langle x, y\rangle = \langle y, x\rangle^*$
3. $\langle ax + by, z\rangle = a\langle x, z\rangle + b\langle y, z\rangle$
4. $\langle x, x\rangle \leq k^2\|x\|^2$

for all $x,y,z \in X$ and $a,b \in C$ and for some fixed constant k.

9.3.1 EXAMPLE Let K be a compact in R^m and C(K) be the set of all complex-valued continuous functions on K. With the sup norm this is a Banach space and the sup norm is denoted by $\|\cdot\|$. We can define, as usually, the inner product by

$$<f, g> = \int_K f(x)g(x)^* \, dx$$

where dx is the Lebesgue measure on R^m. It is not difficult to see that this inner product satisfies all the above axioms.

9.3.2 EXAMPLE Let B_I be the space of all bounded complex-valued functions on I = [0, 1], with the norm

$$f \mapsto \|f\| = \sup_t |f(t)|$$

For each $t_0 \in I$ we consider the function

$$x_{t_0}(t) = \begin{cases} 1 & t = t_0 \\ 0 & t \neq t_0 \end{cases}$$

which is obvious in B_I.

Suppose now that B_I has an inner product satisfying the above axioms. We show that in this case we obtain a contradiction. For $a > 0$ we define

$$T_a = \{t \in I \mid <x_t, x_t> \geq a\}$$

First we show that for at least one n, $T_{1/n}$ is an infinite set. Indeed since $<x, x> = 0$ implies $x = 0$, we have that $I = \bigcup_n T_{1/n}$, and if for all n, $T_{1/n}$ is a finite set, we obtain that I is countable and this is a contradiction. Let n_0 be the integer for which $T = T_{1/n}$ is an infinite set and if $t_1, t_2 \in I$, $t_1 \neq t_2$, then we set

$$y_1 = x_{t_1} \qquad y_2 = x_{t_2} e^{is_2}$$

where s_2 is an appropriate real number such that $<x_{t_1}, x_{t_2}>e^{-s_2} \geq 0$.

We define

$$z_1 = y_1 \qquad z_2 = y_1 + 2^{-1/2} y_2$$

Suppose now that $z_1, \ldots, z_{n-1}$ are given. Let $t_n \in T$, $t_n \neq t_i$ for $i = 1, 2, \ldots, n - 1$. We take s_n such that

$$y_n = e^{is_n} x_{t_n} \qquad \langle z_{n-1}, y_n \rangle \geq 0$$

and we define

$$z_n = z_{n-1} + n^{-1/2} y_n$$

Since t_i are distinct points, we have that

$$\|z_{n+1} - z_m\| = \|(n + 1)^{-1/2} e^{is_{(n+1)}} x_{t_{n+1}} + n^{-1/2} x_{t_n} e^{is_n} + \cdots + m^{-1/2} e^{is_m} x_{t_m}\| = (n + 1)^{-1/2}$$

for $m < n$.

From this it is clear that $\{z_n\}$ is a Cauchy sequence and thus we find an element $z \in B_I$ such that

$$\lim z_n = z$$

Axiom 4 of the inner product implies that

$$\lim \langle z_n, z_n \rangle = \langle z, z \rangle$$

Since we have the relation

$$\langle z_n, z_n \rangle \geq \langle y_1, y_1 \rangle + \cdots + \langle y_n, y_n \rangle$$

we obtain a contradiction. The above inequality can be proved by an induction argument. The assertion is true for $n = 1$. Suppose that it is true for all $i \leq n - 1$ and we show that it holds for $i = n$.

Since

$$\langle z_n, z_n\rangle = \langle z_{n-1} + n^{-1/2} y_n, z_{n-1} + n^{-1/2} y_n\rangle$$
$$= \langle z_{n-1}, z_{n-1}\rangle + \langle z_{n-1}, n^{-1/2} y_n\rangle + n^{-1/2}\langle y_n, z_{n-1}\rangle$$
$$+ \frac{1}{n}\langle y_n, y_n\rangle$$

and $\langle z_{n-1}, y_n\rangle > 0$, the assertion follows.

Now, for all t_i we have

$$\langle x_{t_i}, x_{t_i}\rangle \geq \frac{1}{n_0}$$

and this gives

$$\langle z_n, z_n\rangle \geq \frac{1}{n_0}\left(\sum_{i=1}^{n} \frac{1}{i}\right)$$

which implies that the norm of z_n tends to ∞ and this gives that the norm of z is ∞. This contradiction proves the assertion.

Thus we are led to the following problem: What are the classes of Banach spaces which possess an inner product satisfying the axioms 1 through 4? For the treatment of this problem, we recall some notations. Let X be a complex Banach space and X* its dual, Ω is the closed unit ball with the weak *-topology. We show further that this is a compact Hausdorff space and from the extension theorems it follows that X is isomorphic and isometric with a closed subspace of $C(\Omega)$ with the norm

$$f \mapsto \|f\| = \sup\{|f(t)| \mid t \in \Omega\}$$

Suppose now that there exists a Radon measure μ whose support is Ω (the support of a Radon measure is the complement of the largest open set on which μ vanishes). Clearly the support of μ is equal to the support of $|\mu|$, the total variation of μ. Thus we can restrict our consideration to the case of positive μ Radon measures. In this case, the formula

$$\langle x, y\rangle = \langle f_x, g_y\rangle = \int f_x(t) g_y^*(t)\, d\mu(t)$$

defines an inner product on X, and of course f_x and g_y denote the images of x and y in $C(\Omega)$.

We remark that all the properties of the inner product are clear except possibly 1.

Let $x \in C(\Omega)$ and suppose that

$$\int_\Omega |x(t)|^2\, d\mu(t) = 0$$

If for some $t_0 \in \Omega$, $x(t_0) \neq 0$, then from the continuity it follows that for some neighborhood V of t_0, $x(t) \neq 0$ for all $t \in V$, and further there exists $\varepsilon_0 > 0$ such that $|x(t)| \geq \varepsilon_0$ for $t \in V$. In this case we have

$$0 = \int_\Omega |x(t)|^2\, d\mu \geq \int_V |x(t)|^2\, d\mu \geq \varepsilon_0^2 \mu(V)$$

This contradicts the fact that the support of μ is Ω. This proves our assertion.

9.3.3 EXAMPLE Let Ω be a compact Hausdorff space and we suppose that it is separable, i.e., it contains a dense countable subset, say (t_i). Let $\{a_i\}$ be a sequence of positive numbers such that $\Sigma_{i=1}^\infty a_i = 1$. In this case

$$\mu(f) = \sum_{i=1}^\infty a_i f(t_i)$$

defines a Radon measure with the support equal to Ω.

Following Murphy and West (1972) we define the (ccc).

9.3.4 DEFINITION Let Ω be a topological space. We say that Ω satisfies the *countable chain condition* (ccc) if every disjoint family of nonempty sets in Ω is countable.

9.3.5 REMARK If Ω is separable, then it satisfies the (ccc).

9.3.6 THEOREM Let Ω be a compact Hausdorff space. Then a necessary condition that Ω supports a Radon measure is that it satisfy the (ccc).

Proof. Suppose that we have a Radon measure μ and supp $\mu = \Omega$. Suppose on the contrary that Ω fails to satisfy the (ccc). We find thus a family of disjoint sets (open) $\{V_i\}_{i \in I}$, where I is uncountable. Consider, for each n, the set

$$I_n = \left\{ i \in I \mid \mu(V_i) \geq \frac{1}{n} \right\}$$

Since the support of μ is Ω, the measure of every open set V_i is positive and thus $I = \bigcup I_n$. Since I is uncountable, we find at least one I_n which is uncountable. Since μ is countable additive, we find that $\mu(\Omega) = \infty$ which is false for every Radon measure.

The theorem is proved.

For the application to our problem, we recall some details about the weak*-topology. Let X be a Banach space and

$$S^* = \{x^* \in X^* \mid \|x^*\| \leq 1\}$$

The weak*-topology on X* is the topology for which the neighborhood of the origin are the sets

$$V(0, \varepsilon, x_1, \ldots, x_n) = \{x^* \mid |x^*(x_i)| < \varepsilon, i = 1, \ldots, n\}$$

where $\varepsilon > 0$ and $x_1, \ldots, x_n$ is a finite set in X.

The following is known as the Banach-Alaoglu theorem [Dunford and Schwartz (1967)].

9.3.7 THEOREM S* is compact in the weak*-topology.

Proof. For any $x \in X$ we define

$$a_x = \{a \mid |a| \leq \|x\|\}$$

and we consider the cartesian product

$$P = \prod_{x \in X} a_x$$

with the Tychonov topology (the product topology). Then P is a compact set and we consider the application

$$p: S^* \to P$$

defined by the relation

$$p(x^*) = (x^*(x))$$

Let $p(S^*)$ be the image of S^* in P and since the product topology is the induced topology on S^* when we consider the weak*-topology on this set.

Thus to prove our theorem it is sufficient to show that $p(S^*)$ is closed. Let p_0 be a point in the closure of $p(S^*)$ and $x,y \in X$ as well as $\varepsilon > 0$ such that

$$V(p_0) = \{p \mid |p(z) - p_0(z)| < \varepsilon,\ z = x,\ y,\ ax,\ x + y\}$$

We take in this neighborhood the point $p = (x^*(x))$ and we have

1. $|x^*(x) - p_0(x)| < \varepsilon$
2. $|x^*(y) - p_0(y)| < \varepsilon$
3. $|x^*(ax) - p_0(ax)| < \varepsilon$
4. $|x^*(x + y) - p_0(x + y)| < \varepsilon$

and since $x^* \in S^*$, we have that

$$|p_0(x + y) - p_0(x) - p_0(y)| < 3\varepsilon$$

$$|ap_0(x) - p_0(ax)| < \varepsilon(|a| + 1)$$

$$|p_0(x)| < \|x\| + \varepsilon$$

Since $\varepsilon > 0$ is arbitrary, it follows that p_0 is a linear functional and is a member of S^*. Thus $p_0 \in p(S^*)$ and the theorem is proved.

9.3.8 THEOREM For every Banach space X, X is isomorphic and isometric with a closed subspace of $C(S^*)$.

Proof. Here S^* is considered with the topology induced by the weak*-topology.

For each $x \in X$ we consider the element in X^{**} defined by the relation

$$f_x(x^*) = x^*(x)$$

From the extension theorems, it follows that this gives an isometric isomorphism between X and a closed subspace of $C(S^*)$, and from the definition of the weak*-topology we have that f_x are continuous functions.

If X is a Banach space which is separable, then the weak*-topology is metrizable and the metric can be given as follows:

$$d(x^*, y^*) = \sum_{i=1}^{\infty} \frac{1}{2^i} \frac{|(x^*(x_i)| + |y^*(x_i)|}{1 + |x^*(x_i)| + |y^*(x_i)|}$$

where $\{x_i\}$ is an arbitrary countable dense subset of X.

9.3.9 DEFINITION A Banach space X satisfies the condition (M - W) of (+) if it contains an uncountable subset $\{x_i\}_{i \in I}$ such that

1. $\|x_i\| = 1, \; i \in I$
2. $\|x_i + x_j\| \leq \delta < 2$

for some fixed $\delta > 0$ and all $i \neq j$.

9.3.10 EXAMPLE The following spaces satisfy the condition (+):

1. B_1 the space in Example 9.3.2 for $\delta = 1$
2. Any inseparable Hilbert space for $\delta = 2^{1/2}$
3. Any inseparable uniformly convex space

It is clear that the spaces in 1 and 2 satisfy the (+) condition. We show now that the space in the third example also satisfies the condition (+). For this it is sufficient to remark that for $\varepsilon > 0$ there exists $\delta(\varepsilon) > 0$ such that $\|x - y\| \geq \varepsilon$ implies that $\|x + y\| \leq 2(1 - \delta(\varepsilon))$.

The assertion follows immediately from the following.

9.3.11 THEOREM Let S be an inseparable metric space with the metric d. In this case there exists $s > 0$ and an uncountable set $\{x_i\}_{i \in I}$ in S such that $d(x_i, x_j) \geq s$.

Proof. Let $a > 0$ and consider the family F of all subsets of S such that for any distinct points x,y in the set, we have $d(x, y) \geq a$. We can introduce a partial order in the family; by inclusion and as an application of Zorn's lemma, we get a maximal set with this property. We denote this set as $H(a)$. Now, if the assertion of the theorem is false, then for each a, $H(a)$ is countable and the set $H = \bigcup_n H(1/n)$ is obviously countable and dense in S. But S is inseparable and this contradiction proves the theorem.

Using this theorem, we can show that the spaces in Example 9.3.11, part 3, satisfy the (ccc). Since

$$S = \{x \mid \|x\| \le 1\}$$

is an inseparable metric space by the above result contains an uncountable set, say $\{x_i\}_{i\in I}$ such that

$$\|x_i - x_j\| \ge \varepsilon$$

and thus by the fact that X is uniformly convex there exists $\delta(\varepsilon)$ such that

$$\|x_i + x_j\| \le 2[1 - \delta(\varepsilon)]$$

and the assertion is completely proved.

The basic fact about Banach spaces which possess an inner product satisfying axioms 1 through 4 is given in our next theorem.

9.3.12 THEOREM Let X be an inseparable Banach space satisfying the (+) condition. Then $\Omega = \{x^* \mid \|x^*\| \le 1\}$ fails to satisfy the (ccc).

Proof. Let $\{x_i\}_{i\in I}$ be an uncountable set in X satisfying (+) with some $\delta < 2$. We consider the sets

$$\sigma_i = \{f \mid \|f\| \le 1,\ |f(x_i) - 1| < \varepsilon\} \qquad i \in I$$

for some $\varepsilon > 0$. These sets are nonempty by the extension theorems and we show that if ε is sufficiently small, these are disjoint.

Indeed, suppose that $g \in \sigma_i \cap \sigma_j$, $i \ne j$, and thus we have

$$|g(x_i) - 1| < \varepsilon \qquad |g(x_j) - 1| < \varepsilon$$

Further we obtain

$$|g(x_i + x_j)| \le \|x_i + x_j\|\|g\| \le \delta\|g\|$$

and

$$|g(x_i) + g(x_j)| = |2 + g(x_i) - 1 + g(x_j) - 1| \geq 2 - 2\varepsilon$$

which implies

$$\|g\| \geq \delta^{-1}(2 - 2\varepsilon)$$

Since δ is fixed and less than 2, we may choose ε such that $0 < \varepsilon < 1 - \delta/2$ which gives $\|g\| > 1$ and thus the sets are disjoint. Since the family is uncountable, the theorem is proved.

In what follows we show that another class of Banach spaces satisfies the above theorem. First we consider a localization of the condition (+) as follows.

9.3.13 DEFINITION A Banach space X satisfies the (+) *condition locally* if it contains an uncountable set $\{x_i\}_{i \in I}$ such that

1. $\|x_i\| \leq 1$, $i \in I$
2. $\|x_i + x_j\| \leq \delta_i$, $i \neq j$, $\delta_i \in [0, 2)$

9.3.14 DEFINITION A Banach space X is said to be *locally uniformly convex* if for each $\varepsilon > 0$ and x, $\|x\| = 1$, and y, $\|y\| \leq 1$ $\|x - y\| \geq \varepsilon$ there exists $\delta_x(\varepsilon)$ such that $\|x + y\| \leq 2[1 - \delta_x(\varepsilon)]$.

We have the following generalization of Example 9.3.10, part 3.

9.3.15 THEOREM Every inseparable locally uniformly convex space satisfies (+) condition locally.

Proof. Let $S = \{x \mid \|x\| \leq 1\}$ and $\{x_i\}$ be an uncountable set in S such that $\|x_i - x_j\| \geq \varepsilon$. Since X is locally uniformly convex, we can find $\delta_i(\varepsilon)$ such that

$$\|x_i + x_j\| \leq 2[1 - \delta_i(\varepsilon)]$$

and the assertion of the theorem follows.

We give now a generalization of Theorem 9.3.12.

9.3.16 THEOREM Every inseparable locally uniformly convex space fails to satisfy the (ccc).

Proof. Let $\{x_i\}_{i\in I}$ be an uncountable set in X satisfying (+) locally and consider the sets

$$C_i = \{f \mid \|f\| \le 1,\ |f(x) - 1| < \varepsilon_i,\ \varepsilon_i < 1/2\}$$

and we show that if ε_i are sufficiently small, these sets are disjoint. First we remark that these are nonempty sets and suppose that $g \in C_i \cap C_j$. Thus we have

$$|g(x_i) - 1| < \varepsilon_i \qquad |g(x_j) - 1| < \varepsilon_j$$

and thus

$$\|x_i + x_j\| \cdot \|g\| > g(x_i + x_j) = |2 + g(x_i) + g(x_j) - 1 - 1|$$
$$> 2 - |g(x_i) - 1| - |g(x_j) - 1|$$

Thus we obtain

$$\delta_i\|g\| \ge 2 - \varepsilon_i - \varepsilon_j$$

and similarly

$$\delta_j\|g\| \ge 2 - \varepsilon_i - \varepsilon_j$$

which gives

$$(\delta_i + \delta_j)\|g\| \ge 4 - 2(\varepsilon_i + \varepsilon_j)$$

Now if we take

$$0 < \varepsilon_i < 1 - \frac{\delta_i}{2}$$

we have that

$$0 < \varepsilon_i + \varepsilon_j < 2 - (\delta_i + \delta_j)$$

which implies that

$$\|g\| \ge \frac{1}{\delta_i + \delta_j}\ [4 - 2(2 - \delta_i + \delta_j)] > 1$$

and this contradiction proves the assertion.

We are now in a position to consider symmetrizable operators on Banach spaces and also some generalizations such as normalizable, hyponormalizable, and quasi-normalizable operators.

In what follows we suppose that X is a complex Banach space which possesses an inner product satisfying axioms 1 through 4.

Also we denote, as is usual, by L(X) the set of all bounded linear operators on X.

9.3.17 DEFINITION An element $T \in L(X)$ is called *symmetrizable* if

$$\langle Tx, y\rangle = \langle x, Ty\rangle$$

holds for all $x,y \in X$, where $\langle\ ,\ \rangle$ denotes the inner product.

9.3.18 DEFINITION An element $T \in L(X)$ is said to be *normalizable* if there exists $T^* \in L(X)$ such that for all $x,y \in X$,

$$\langle T^*Tx, y\rangle = \langle TT^*x, y\rangle$$

9.3.19 DEFINITION An element $T \in L(X)$ is said to be *quasi-normalizable* if for all $x \in X$, $\langle x, x\rangle = 1$,

$$\langle Tx, Tx\rangle \le \langle T^2x, T^2x\rangle^{1/2}$$

9.3.20 DEFINITION An element $T \in L(X)$ is said to be *hyponormalizable* if there exists $T^* \in L(X)$ such that for all $x \in X$,

$$\langle (T^*T - TT^*)x, x\rangle \ge 0$$

It is not difficult to see that any hyponormalizable operator is quasi-normalizable, and using the Halmos example of a hyponormal operator whose square is not hyponormal, we find that the inclusion is strict.

Also, another feature of the class of quasi-normalizable operators is that it is not supposed that the adjoint exists.

The fundamental property of quasi-normalizable operators is given in the following.

9.3.21 THEOREM Let H be the Hilbert space obtained from X under completion with respect to the norm induced by the inner product on X. Suppose that T is a quasi-normalizable operator on X. In this case T is bounded on H.

First we show that for all x, $x \in X$, $\langle x, x\rangle = 1$, and all n,

$$\|T^n x\|^2 = \langle T^n x, T^n x\rangle \geq \langle Tx, Tx\rangle^n$$

We remark that for n = 2 this is just the definition of quasi-normalizable operators. Suppose the inequality valid for all $i \leq n$ and we prove it for n + 1. Indeed,

$$\|T^{n+1}x\| = \left\|T^n \frac{Tx}{\|Tx\|}\right\| \|Tx\| \geq \|T^2x\|^n \frac{1}{\|Tx\|^{n-1}}$$

$$\geq (\|Tx\|)^{2n} \frac{1}{\|Tx\|^{n-1}} = \|Tx\|^{n+1}$$

and the assertion follows.

If $|\cdot|$ denotes the norm of X, we have

$$\|Tx\|^n \leq \|T^n x\| \leq k|T^n x| \leq k|T|^n|x|$$

which gives

$$\|Tx\| \leq k^{1/n}|T|\,|x|^{1/n}$$

and for $n \to \infty$ we obtain that

$$\|Tx\| \leq |T|$$

We remark also that the above computation gives

$$\|Tx\| \leq r_T$$

for all x, $\langle x, x\rangle = 1$.

9.3.22 THEOREM Let $T \in L(X)$ and suppose that for all complex numbers z, $T_z = T - zI$ is a quasi-normalizable operator. Then $\sigma(T)$ with respect to H is a subset of $\sigma(T)$ with respect to X.

First we prove a theorem which has also an independent interest.

9.3.23 THEOREM Let $T \in L(X)$ be a quasi-normalizable operator and suppose that $T^{-1} \in L(X)$. In this case T^{-1} is quasi-normalizable.

Proof. Let $x \in X$, $\|x\| = 1$, and $y \in X$ such that $T^2y = x$. We have

$$\|T^{-1}x\|^2 = \|Ty\|^2 = \left\|T^2 \frac{y}{\|y\|}\right\|^2 \|y\|^2 \leq \|T^2y\|\|y\| = \|y\|\|x\| = \|y\| = \|T^{-2}x\|$$

and the theorem is proved.

Proof of Theorem 9.3.22. Suppose that z is not in the $\sigma(T)$ and thus $(T - z)^{-1}$ exists and by our result it is also quasi-normalizable. By Theorem 9.3.21 it is bounded on H and

$$(T - z)S = S(T - z) \qquad S = (T - z)^{-1}$$

on a dense set in H and thus the relation holds in H. This proves the theorem.

9.3.24 THEOREM Let $T \in L(X)$ be a hyponormalizable operator. Then the following assertions hold:

1. For every z, $T_z = T - zI$ is hyponormalizable.
2. If for some $x \in X$, $Tx = zx$, then $T^*x = z^*x$.
3. If T is a compact operator, then T is normalizable.

Proof. The first assertion is obvious and the second follows by computation,

$$\|T^*x - z^*x\|^2 = \langle T^*x - z^*x, T^*x - z^*x\rangle \leqq \|Tx - zx\|^2$$

For 3 we remark that from the properties of the inner product we have that T is a compact operator on H. Then the assertion follows from the result about hyponormal operators.

9.3.25 THEOREM Let $T \in L(X)$ be a hyponormalizable operator and $z \in \sigma_p(T)$, the point spectrum of T on X. Then z is in the point spectrum of T as an operator on H.

Proof. Let z in the point spectrum of T as an operator on X and thus $\dim N(T - z) < \infty$ and is equal to $\dim X/(T - z)X$. We remark also that every element in $(T - z)X$ is orthogonal on elements in $N(T - z)$. Indeed, let $y \in (T - z)X$, i.e., $y = (T - z)x$, and for $z \in N(T - z)$ we have that $z \in N(T^* - z^*)$. Now

$$\begin{aligned} \langle y, z\rangle &= \langle (T - z)x, z\rangle = \langle Tx, z\rangle - z\langle x, z\rangle = \langle x, T^*z\rangle - z\langle x, z\rangle \\ &= \langle x, z^*z\rangle - z\langle x, z\rangle = 0 \end{aligned}$$

and the assertion is proved. From this we obtain that $(T - z)X$ is exactly the set of all elements in X which are orthogonal on $N(T - z)$. Thus $T - z$ is a surjection if we consider it on the subspace $(T - z)X$. By the closed graph theorem (since it is also an injection) we obtain that it is bounded. It is also bounded in the hilbertian norm. Since $(T - z)S$ is an identity on a dense subset, we obtain that $(T - z)$ has a bounded inverse on the orthogonal complement of $N(T - z)$ with respect to H. This means that the point spectrum of T as an operator on H contains z. Also from these, it follows that $N(T - z)$ are the same space in X as well as in H.

The following result is related to the above, but in some sense, is more general.

9.3.26 THEOREM Let X and X' be two Banach spaces with the norms $\|\cdot\|$ and $\|\cdot\|'$, respectively, X is densely in X', and there exists there $k > 0$ such that

$$\|x\| \leq k\|x\|'$$

for all $x \in X$. Suppose that there exists a bounded linear operator $T: X' \to X'$ having X as an invariant subspace. Let Z in the point spectrum of T be an operator on X, then it is in the point spectrum of T as an operator on X'.

Proof. Let $N = N(T - z)$ when we consider T as an operator on X and $N' = N(T - z)$ when we consider T as an operator on X'. Similarly, let $J = (T - z)X$ and $J' = (T - z)X'$. Let z as in the theorem. In this case we have

$$n = \dim N = \dim \frac{X}{J}$$

$$n' = \dim N' = \dim \frac{X'}{J'}$$

and since $J \subset J'$ we obtain that

$$\dim \frac{X'}{J'} \leq \dim \frac{X}{J}$$

i.e., $n' \leq n$. Indeed, let $x_1, x_2, \ldots, x_{n+1}$ elements in X'. Since X is densely in X', let $x_1', x_2', \ldots, x_{n+1}'$ be elements in X nearly to above elements, respectively. Since $\dim X/J = n$, we find a linear combination

$$\sum_{k=1}^{n+1} a_k^m x_k$$

which is in J. We normalize the coefficients so that the largest of these is 1 and the others are less than 1 in absolute value.

By the Bolzano-Weierstrass theorem we can select a subsequence for each sequence $\{a_k^m\}$ such that they converge to $a_1, \ldots, a_{n+1}$.

Thus, we can find a nontrivial combination

$$\sum_{i=1}^{n+1} a_i x_i$$

in J and thus in J' (J' being closed). This proves that

$$n' = \dim \frac{X'}{J'} \leq n$$

Since $N \subset N'$, we have that $\dim N \leq \dim N'$ so that $n \leq n'$. Thus $n = n'$, and the theorem is proved.

It is easy to see that if T is symmetrizable operator, then every power T^n is also symmetrizable. Now we prove that this property holds also for quasi-normalizable operators.

9.3.27 THEOREM Let $T \in L(X)$ be a quasi-normalizable operator. Then for any n the operator T^n is also quasi-normalizable.

Proof. Since T is quasi-normalizable, we have for any $x \in X$,

$$\frac{\|Tx\|}{\|x\|} \leq \frac{\|T^2x\|}{\|Tx\|} \leq \cdots \leq \frac{\|T^nx\|}{\|T^{n-1}x\|} \leq \cdots \leq \frac{\|T^{2n}x\|}{\|T^{2n-1}x\|} \leq \cdots \leq \|T\|$$

and since we can assume, without loss of generality, that $\|T\| \leq |T| \leq 1$, we obtain

$$\|T^{2n}x\| \geq \|T^nx\|^2$$

and thus the theorem.

We present now a sufficient condition for an operator to be quasi-normalizable when it admits an adjoint T* with respect to the inner product on X and, of course, we suppose that $T^* \in L(X)$.

9.3.28 THEOREM Let $T \in L(X)$. Then T is quasi-normalizable if and only if for all $s > 0$,

$$T^{*2}T^2 - 2sT^*T + s^2 \geq 0$$

with respect to the inner product.

Proof. Since for any $x \in X$ we have

$$\|T^2x\|\|x\| = \inf_{s>0} \frac{1}{2} \{s^{-1}\|T^2x\|^2 + s\|x\|^2\}$$

$$= \inf \frac{1}{2s} \{<T^{*2}T^2x, x> + s^2\}$$

and thus we obtain that

$$\|T^2x\|\|x\| = \inf_{s>0} \frac{1}{2s} \{<T^{*2}T^2 + s^2)x, x> > <T^*Tx, x>$$

which is equivalent with

$$\inf_{s>0} \frac{1}{2s} <T^{*2}T^2 - 2sT^*T + s^2)x, x> \geq 0$$

and this clearly gives our assertion.

From this theorem we obtain, in the case of the existence of an adjoint, a simple proof of Theorem 9.3.23. We have for $x = T^{-2}y$,

$$<y, y> - 2s<T^{-1}y, T^{-1}y> + S^2<T^{-2}y, T^{-2}y> \geq 0$$

and this gives that T^{-1} is a quasi-normalizable operator.

9.3.29 THEOREM Let $\{T_n\}$ be a sequence of symmetrizable or quasi-normalizable operators which converges strongly to an operator $T \in L(X)$. Then T is symmetrizable or quasi-normalizable, respectively.

In connection with this the following problem appears to be of interest: Let T be a symmetrizable operator and $S \geq 0$, then there exists $S^{1/2} \in L(X)$ which is symmetrizable.

We close this section with another class of operators suggested by the class (N,k).

9.3.30 DEFINITION An element $T \in L(X)$ is said to be *quasi-normalizable* of type (N,k) if for all $x \in X$, $\|x\| = 1$, $\langle x, x\rangle = 1$, the following inequality holds:

$$\|Tx\|^k \leq \|T^k x\|$$

The proof of the following theorem is almost exactly as the proof of Theorem 9.3.21, and we omit it.

9.3.31 THEOREM Let T be a quasi-normalizable operator of type (N,k), then it is a bounded operator on H.

Another extension of the notion of quasi-normalizable operators is the following.

9.3.32 DEFINITION An operator on a Banach space X is said to be *double-positivable* if with respect to the norm induced by the inner product, for all $n > 1$,

$$\|T^n x\|^2 \leq \|T^{n-1}x\| \cdot \|T^{n+1}x\|$$

We remark that the class of quasi-normalizable operators is included in this class.

9.4 SOME APPLICATIONS OF SYMMETRIZABLE OPERATORS AND QUASI-NORMALIZABLE OPERATORS

We give now some applications of the above classes of operators. These applications refer to Hilbert quadratic form and to an interpolation theorem. The first application is due to P. Lax and the other uses a technique from the theory of quasi-normalizable operators due to Nieto (1974).

Hilbert has studied the following quadratic form:

$$\sum_{n,m=1}^{\infty} \frac{1}{n+m} x_n x_m$$

and this was generalized by Magnus, who considered the quadratic form

$$\sum_{n,m=1}^{\infty} \frac{1}{(n + m)^{1-a}} \frac{1}{(nm)^{a/2}} x_n x_m \qquad 0 \le a < 1$$

Also Magnus [see Lax (1954)] has proved that this quadratic form is bounded for all $a \in [0, 1)$ and for $a = 1$ it is unbounded. We give in what follows the proof of this result as an application of Theorem 9.3.21.

Let ℓ^2 be the Hilbert space of all sequences $x = (x_i)$ such that $\Sigma_{i=1}^{\infty} |x_i|^2 < \infty$ and with the coefficients of the quadratic form we define a linear operator on ℓ^2 by the formula

$$T(x) = y = (y_i)$$

where

$$y_i = \sum_{n=1}^{\infty} \frac{1}{(i + n)^{1-a}} \frac{1}{(in)^{a/2}} x_n \qquad x = (x_n)$$

It is easy to see that this operator is symmetric with respect to the inner product on ℓ^2. We show that it is bounded with respect to another norm, namely

$$|x| = \sup |x_m| m^s$$

where $s \in (a/2, 1 - a/2)$. Since for $s > 1/2$ the norm dominates the Hilbert norm, we have

$$\begin{aligned} \langle x, x\rangle^{1/2} &= (\Sigma |x_i|^2)^{1/2} \le \sup |x_m| (1 + 2^{-s} + \cdots + m^{-s})^{1/2} \\ &\le \sup |x_m| m^{1/2} \end{aligned}$$

Thus if we prove that T is bounded with respect to the above norm, then we obtain that T is bounded in the Hilbert norm and this implies the Magnus result.

For this we need the upper bound for $|y_i| i^s$. We have

$$|y_i| i^s \le \sum_{n=1}^{\infty} \frac{1}{(i + n)^{1-a}} \frac{i^s}{(in)^{a/2}} x_n \le \sum_{i=1}^{\infty} \frac{1}{(i + n)^{1-a}} \frac{1}{(in)^{a/2}}$$

and further our sum is computed as follows:

$$\Sigma = I_1 + I_2$$

where

$$I_1 = i^{s-a/2} \sum_{n=1}^{i} \left(\frac{1}{(i + n)}\right)^{1-a} \frac{1}{n^{s+a/2}}$$

$$I_2 = i^{s-a/2} \sum_{n=i+1}^{\infty} \left(\frac{1}{(i + n)}\right)^{1-a} \frac{1}{n^{s+a/2}}$$

Since $(1/(i + n))^{1-a} \le 1/i^{1-a}$, we get

$$I_1 \le (\text{constant}) n^{s-a/2} i^{1-s-a/2}$$

for $s < 1 - a/2$ and thus

$$I_1 \le (\text{constant}) n^{s-1/2-1+a+1-s-a/2} = \text{constant}$$

Since the constant is $O(1 - a/2 - s)^{-1}$, the first sum is computed.

Now for the second we have

$$I_2 \le \Sigma\, n^{a-1} n^{-a/2-s} \le (\text{constant}) i^{a/2-s}$$

if $s > a/2$ and the value of the constant is $O(s - a/2)$. Thus we have

$$I_2 \le (\text{constant}) i^{a/2-s} i^{s-a/2} = \text{constant}$$

Now $I_1 + I_2$ is less than a constant for $s \in [a/2, 1 - a/2)$. The value of the constant depends on $O(\max\{(s - a/2)^{-1}, (1 - s - a/2)^{-1}\})$ and for s close to $1/2$ the upper bound is $O((1 - a)^{-1})$. To prove that the order of magnitude of the quadratic form is in fact $(1 - a)^{-1}$ for $a \to 1$, we can choose $x = (x_n)$, $x_n = n^{1/2}$ for $n = 1, 2, \ldots, N$, $x_n = 0$ for $n > N$.

Before we go on to the interpolation theorem, we mention that P. Lax has given interesting applications of symmetrizable operators to the theory of hyperbolic systems of first order

$$U_t + A_i U_{x^i} + BU = 0$$

where U is a vector with an unknown function; the coefficient matrix A_i and B may depend on time as well as the space variables x. The system is supposed to be hyperbolic and using the results on symmetrizable operators, it proves the fundamental theorem of I. G. Petrovsky, which, in fact, generalizes a theorem of Friederichs [see Lax (1954)], for the case of symmetric hyperbolic systems.

We can also consider quadratic forms of a more general type such that the operator T defined as above to be quasi-normalizable, and thus we obtain the boundedness of T in the Hilbert space norm. It is probable that this can be applied to formally normal operators.

Now we give an application to interpolation theory. Let X be a reflexive Banach space with the norm $|\cdot|$ and suppose that there is an inner product $\langle\ ,\ \rangle$ on X and the induced norm is $\|\cdot\|$. The Hilbert space obtained is denoted by the letter H. We suppose that the following conditions hold:

1. X is densely in H.
2. The identity map of X into H is continuous.

Since we can define the map i: $H \to X^*$ (the *antidual* of X) by the relation

$$\langle i(y), x\rangle = \langle y, x\rangle \qquad x \in X$$

we can identify H with its image in X* such that

$$X \subset H \subset X^*$$

and each embedding is continuous.

The following theorem was given by J. L. Lions and J. Peetre, but their proof is valid only for reflexive Banach spaces.

9.4.1 THEOREM Let $T\colon X^* \to X^*$ be a bounded linear operator on X* having X as an invariant subspace. Then the following assertions hold:

1. $TH \subseteq H$.
2. The restriction of T to X and H are bounded and

$$\|T\|_H \le (\|T\|_X \|T\|_{X^*})^{1/2}$$

where $\|T\|_X$, $\|T\|_H$, $\|T\|_{X^*}$ are the norms of T with respect to the spaces X, H, and X*, respectively.

As is well known a linear form φ on a complex Banach space X is called *antilinear* if

$$\varphi(ax + by) = a^*\varphi(x) + b^*\varphi(y)$$

for all $x,y \in X$ and all a,b complex numbers.

The antidual of X which is denoted by X* is the set of all antilinear and continuous forms on X. If we define for any $\varphi \in X^*$,

$$\|\varphi\| = \sup_{\|x\|\le 1} |\varphi(x)|$$

then we have that X* is a Banach space. If we suppose that X is reflexive, then the canonical map $\lambda: X \to X^{**}$ (X** is the antidual of X*) defined by

$$\lambda(x)(\varphi)) = \varphi(x)$$

is onto and isometric. From the closed graph theorem we obtain the following assertion.

9.4.2 THEOREM Let X_1 and X_2 be two Banach spaces such that

1. $X_1 \subset X_2$.
2. $i: X_1 \to X_2$ by $i(x) = x$, is continuous.
3. $T: X_2 \to X_2$ has X as an invariant subspace.

Then T/X_1 is in $L(X_1)$.

We are now ready to prove Theorem 9.4.1. Let T* be the adjoint of T,

$$\langle T^*y, x\rangle = \langle x, Ty\rangle$$

for all $y \in X^{**}$ and $x \in X^*$, which has also the property that

$$\|T^*\| = \|T\|_{X^*}$$

Consider now the operator

$$Q = \lambda^{-1}T\lambda$$

and since λ is an isometric operator, it follows that Q is a bounded operator and

$$\|Q\| = \|T^*\| = \|T\|_{X^*}$$

In this case the operator $A = QT_0$, T_0 being the restriction of T to X is bounded. From Theorem 9.4.2 it follows that it is bounded on X and

$$\|A\| = \|Q\| \cdot \|T_0\| = \|T\|_{X^*} \cdot \|T\|_X$$

Since

$$\langle Ax, y\rangle = \langle x, Ay\rangle = \langle T_0x, T_0y\rangle$$

we see that A is quasi-normalizable, and since

$$\|T_0x\|_H = \langle Ax, x\rangle$$

we have by Theorem 9.3.21 that

$$\|Tx\|_H \le (\|T\|_X \cdot \|T\|_{X^*})^{1/2}\|x\|$$

for all $x \in X$. From this it is clear that $T: H \to H$ and the restriction of T to H is a bounded operator with the norm less than

$$\|T\|_H \le (\|T\|_X \cdot \|T\|_{X^*})^{1/2}$$

9.5 FURTHER RESULTS ON SYMMETRIZABLE OPERATORS ON HILBERT SPACES

First we recall that a bounded linear operator on a Hilbert space H is called *quasi-hermitian* if

1. There exists a hermitian operator $H > 0$.
2. $HT = T^*H$.

Obviously this class contains the class of hermitian operators and if we consider the inner product

$$\langle\langle x, y\rangle\rangle = \langle Hx, y\rangle$$

then T is quasi-hermitian if and only if it is symmetric with respect to this inner product.

If the Hilbert space is finite dimensional, the condition $H > 0$ implies that H is invertible and the class of quasi-hermitian operators is exactly the class of diagonalizable operators with real eigenvalues.

Suppose now that the Hilbert space is infinite dimensional and there exists an invertible operator $H^{1/2}$ such that $(H^{1/2})^2 = H$. Thus we have

$$(H^{1/2})^2 T = T^*(H^{1/2})^2$$

which gives that

$$H^{1/2}TH^{-1/2} = H^{-1/2}T^*H^{1/2}$$

and if we define

$$L = H^{1/2}TH^{1/2}$$

then

$$TH^{1/2} = LH^{1/2}$$

and L is hermitian.

Thus we are led to extend this decomposition to general quasi-hermitian operators and this is given in Theorem 9.5.1. This result was obtained by J. Dieudonné (1960) and, as was noted by P. Lax, it is a consequence of the results on quasi-normalizable operators (in fact, symmetrizable operators).

Since we think the Dieudonné's proof is interesting in itself, we give it here.

9.5.1 THEOREM Let T be a quasi-hermitian operator on a Hilbert space. Then there exists a unique bounded hermitian operator L such that

$$H^{1/2}T = LH^{1/2}$$

The proof is a consequence of some results which also have an independent interest.

9.5.2 THEOREM Let A be an invertible positive hermitian operator, B a hermitian operator. Then $\|A^2B\| \le M$ implies that $\|ABA\| \le M$.

Proof. Let p be the largest positive element in the spectrum of ABA (if there exists), and thus for any $\varepsilon > 0$ there exists x such that $\|x\| = 1$ and

$$\|ABAx - px\| \le \varepsilon$$

which gives

$$\|A^2BAx - pAx\| \le \varepsilon\|A\|$$

Now since

$$1 \le \|A^{-1}\| \cdot \|Ax\|$$

we have

$$\|A^2Bx - pAx\| \le (\varepsilon\|A\| \cdot \|A^{-1}\|)\|Ax\|$$

and for $b = \varepsilon\|A\|\|A^{-1}\|$ this gives

$$\|A^2BAx\| \ge (p - b)\|Ax\|$$

(if we suppose that $Ax \ne 0$). Thus $p - b \le M$. Since b is arbitrary (because ε is arbitrary), we have that $p \le M$. If we replace B by -B we see that if q is the smallest element in the spectrum of ABA, then $-q \le M$ and this proves the theorem.

Now Theorem 9.5.1 can be proved as follows: Let P_n be the orthogonal projection on the invariant subspace X_n which corresponds to the intersection of $[1/n, \infty)$ with the spectrum of $H^{1/2}$. It is clear that $\{P_n\}$ converges strongly to the identity operator and clearly commutes with $H^{1/2}$.

Let $H_n = P_nH^{1/2}P_n$ and $T_n = P_nTP_n$; thus $H_n^2T_n = T_n^*H_n^2$. If we consider T_n and H_n as operators in X_n, then H_n is positive and invertible. Since $\|T\| \le M$ and if we set $S_n = H_n^2T_n$, S_n is a hermitian operator on X_n and $\|H_n^{-2}S_n\| \le M$. By Theorem 9.5.2 the operator $L_n = H_n^{-1}S_nH_n^{-1}$ is with the property $\|L_n\| \le M$.

Let L_n' be the operator defined on X_n as L_n and 0 on the orthogonal complement of X_n. Obviously $\|L_n'\| \leq M$ and since, in the case of Hilbert spaces, the closed ball of bounded radius in the space of all bounded linear operators is compact for the weak topology, $\{L_n'\}$ converges weakly (or some subsequence, but for convenience we suppose without loss of generality that just $\{L_n'\}$ is with this property) to L, which is bounded. Since for any integer m and any $x \in X_m$, for $n > m$ we have

$$X_n \supset X_m$$

and thus $H_n x = H^{1/2} x$. Hence, we get

$$H_n T_n x = L_n H_n x = L_n H^{1/2} = L_n' H^{1/2} x$$

But $\{H_n T_n x\}$ converges in the weak topology to $H^{1/2}Tx$ and $\{L_n' H^{1/2} x\}$ converges weakly to $LH^{1/2}x$. Thus we obtain the equality $H^{1/2}Tx = LH^{1/2}x$, and this is true for x in the union of all X_n which is dense in the Hilbert space and thus the above relation is valid for all x. Since $H^{1/2}X$ is dense in X, X being our Hilbert space, it follows that L is unique.

The following example of Dieudonné (1960) of a quasi-hermitian operator with nonreal spectrum shows another difference within the class of hermitian operators.

Let X be a Hilbert space and $\{X_n\}$ be a sequence of finite dimensional subspaces, dim $X_n = n$, and X is the sum of these spaces. On each X_n we consider the operator P_n with the matrix

$$\begin{bmatrix} a_1 & 0 & \cdot\cdot & \cdot\cdot & \cdot\cdot & 0 \\ 1 & a_2 & 0 & \cdot\cdot & \cdot\cdot & 0 \\ \vdots & & & & & \\ \vdots & & & & & \\ \vdots & \cdot\cdot & \cdot\cdot & \cdot\cdot & 1 & a_n \end{bmatrix}$$

with respect to an orthogonal basis (e_i) of X_n.

We suppose that $\{a_i\}$ are distinct real numbers and $\Sigma_{i=1}^n a_i^2 = 1$. This gives that $\|P_n\| \le 2$. Let ξ be a complex number with the property $|\xi| = 1$ and consider the element

$$x = e_1 - \xi^{-1}e_2 + \cdots + (-1)^{n-1}\xi^{-n+1}e_n$$

and clearly $\|x\| = n^{1/2}$. Also,

$$(\xi I_n + P_n)x = (\xi + a_1)e_1 - \xi^{-1}a_2e_2 + \cdots + (-1)^{n-1}\xi^{-n+1}a_ne_n$$

and thus

$$\|\xi I_n + P_n\| \le 2$$

which implies that

$$\|(\xi I_n + P_n)^{-1}\| \ge \frac{n^{1/2}}{2}$$

We define an operator T on X as being equal to P_n on each X_n and since the eigenvalues of P_n are real, we can find a positive operator H_n such that $H_nP_n = P_n^*H_n$. We can assume, without loss of generality, that $\|H_n\| \le 1$ and we define H on X to be equal to H_n on each X_n. Clearly $H > 0$ and $HT = T^*H$; thus T is quasi-hermitian.

But there exists, for any ξ, $|\xi| = 1$ an element $x \in X$ such that $(\xi + T)x$ has an arbitrarily small norm. Thus for ξ, $|\xi| = 1$ is in the spectrum of T.

Dieudonné (1960) gives the structure of Riesz operators which are quasi-hermitian. The following theorem shows that all the Riesz operators which are quasi-hermitian are in fact compact operators.

9.5.3 THEOREM Let T be an operator such that

1. T is quasi-hermitian.
2. T^p is a Riesz operator for some $p \ge 1$.

Then T is a compact operator.

Proof. We consider the Calkin algebra L(X)/K(X) and clearly $T^\sim$, the image of T in this algebra, is also quasi-hermitian and for p, $T^{\sim p} = 0$. This implies that $T^\sim = 0$, and thus T is a compact operator.

The above result is in fact valid for all quasi-normalizable operators.

9.5.4 THEOREM Let T be a quasi-normalizable operator and suppose that for some $p \geq 1$, T^p is a Riesz operator. In this case T is a compact operator.

Proof. The proof is exactly as the proof of Theorem 9.5.3.

An interesting class of elements in Hilbert spaces is the class of subnormal operators, i.e., the class of operators which are the restrictions of normal operators to invariant subspaces.

We can define, using this suggestion, a new class of operators as follows.

9.5.5 DEFINITION An element $T \in L(X)$ is called *subsymmetrizable* if there exists a Hilbert space $Y \supset X$ and an operator $T^{\sim}: Y \to Y$ such that

1. $T^{\sim}$ is quasi-hermitian (with respect to $H^{\sim}$).
2. X is invariant for $T^{\sim}$ and for $H^{\sim}$ and $T^{\sim}/X = T$.

For results about this class of operators, see Istrăţescu.

Chapter 10

INVARIANT SUBSPACES AND SOME STRUCTURE THEOREMS

10.1 INVARIANT SUBSPACES: SOME EXISTENCE THEOREMS

First we shall give some historical background about the problem of invariant subspaces for bounded operators on Hilbert and Banach spaces. As is mentioned in the fundamental paper of Aronszajn and Smith (1954), the first result about the existence of invariant subspaces was proved by J. von Neumann: If T is a bounded linear and compact operator on a Hilbert space H, then there exists a closed subspace $H_1 \subset H$ with the property that if $x \in H_1$, then $Tx \in H_1$. In the above assertion we suppose that H_1 has the property

$$\{0\} \neq H_1 \neq H$$

Any closed subspace with the above property is called an *invariant subspace* for T. Independently, but later, Aronszajn proved the same result (with the same proof!) and learned from von Neumann about the above result. With K. T. Smith, Aronszajn has proved that every compact operator on a (separable) Banach space has an invariant subspace. The method of the proof in the paper of Aronszajn and Smith was refined and developed by many authors and we note here the contributions of Arveson (1968), Feldman (1968), Halmos, etc.

It is worth noting a very interesting contribution to the invariant subspace problem made by Bernstein and Robinson (1966), which used the nonstandard analysis to prove the following result:
If T is a bounded operator on a Banach space and for some nonzero

polynomial p, p(T) is compact, then T has an invariant subspace. W. Arveson and J. Feldman have proved the following result: If T is quasi-nilpotent [$\sigma(T) = \{0\}$] and the closed algebra generated by T and I contains a nonzero compact operator, then T has an invariant subspace.

Godement (1947) has proved that for a certain class of operators T there exist invariant subspaces having an additional property: these subspaces are invariant for all operators commuting with T.

Also it is worth mentioning the beautiful work of Beurling (1947) on the structure of all invariant subspaces of the unilateral shift operator.

We recall that the unilateral shift operator is the operator defined on ℓ^2 as follows:

$$x = (x_0, x_1, x_2, \ldots) \to (0, x_0, x_1, x_2, \ldots)$$

An interesting theorem about the existence of invariant subspaces for a class of operator for which $\|T^n\|$ has a special growth was obtained by Wermer (1952).

In 1973 Lomonosov (1973) introduced a new method of proving the existence of invariant subspaces for a class of operators. The main result of Lomonosov is the following: If T is a nonzero operator commuting with a nonzero compact operator, then there exists an invariant subspace of T which is invariant for all operators commuting with T.

It is worth mentioning that Lomonosov's method involves an ingenious application of the famous Schauder fixed point theorem (for this theorem and related fixed points, see V. Istrăţescu, 1981).

In what follows we present the results mentioned above and at the same time we give some applications to the structure of operators. We mention here the following nice result of Ando (1963): Let T be a bounded operator on a Hilbert space, having the following properties:

1. T is compact.
2. If H_1 is an invariant subspace of T, then H_1 is an invariant subspace of T* (in other words, H_1 reduces T).

Then T is normal.

We give also further refinements and results related to the method of Lomonosov.

Let X be a complex Banach space and L(X) be the set of all linear and bounded operators on X.

10.1.1 DEFINITION A closed subspace X_1 of X is called *invariant* for T if for any $x \in X_1$, $Tx \in X_1$. An invariant subspace is called *nontrivial* if it is different from X and {0}.

10.1.2 DEFINITION If $T \in L(X)$ and X_1 is a nontrivial invariant subspace of T, then X_1 is called a *hyperinvariant* subspace for T if X_1 is invariant for all operators commuting with T.

The invariant subspace problem has the following form.

10.1.3 INVARIANT SUBSPACE PROBLEM If $T \in L(X)$ is a nonzero element and $T \neq I$, then T has nontrivial invariant subspaces.

A similar assertion is concerned with the case of hyperinvariant subspaces.

10.1.4 DEFINITION If F is a set in L(X) and X_1 a closed subspace of X, then we say that X_1 is invariant for F if X_1 is invariant for all operators in F.

10.1.5 DEFINITION A closed subalgebra A of L(X) is called *transitive* if the only subspaces invariant for A are X and {0}.

The following is basic in proving the existence of nontrivial invariant subspaces for operators related to compact operators.

10.1.6 LEMMA If A is a transitive subalgebra of L(X) and K is an arbitrary nonzero compact operator in L(X), then there exists an element of A, say S, such that SK has $z = 1$ as an eigenvalue and similarly, there exists S_1 in A such that KS_1 has $z = 1$ as an eigenvalue.

The proof of this lemma uses in an ingenious way the Schauder fixed point theorem, which, for the reader's convenience, we give here. For a detailed account as well as for other applications, see Pearcy (1974) and Sadovski (1967, 1972).

10.1.7 THEOREM (Schauder Fixed Point Theorem) If C is a convex and bounded set in a Banach space X and $f: C \to C$ is a continuous (in general nonlinear) mapping, then there exists a point $x_0 \in C$ such that $f(x_0) = x_0$.

For the proof we need also another result which is a consequence of the Schauder fixed point theorem.

10.1.8 LEMMA Let $f: C \to C$, where C is a convex closed set in a Banach space X. If f(C) is contained in a compact subset of C, then there exists a point $x_0 \in C$ such that $f(x_0) = x_0$.

Proof. If we consider the set $\overline{\text{conv } f(C)}$, then this set is compact, and clearly it is in C. Since $f(C) \subseteq K$, where K is the compact set which exists, we have

$$f(\text{conv } K) \subseteq \text{conv } K$$

and the Schauder fixed point theorem applied to the set conv K gives the assertion.

Now we are in a position to prove Lemma 10.1.6.

Of course we can suppose, without loss of generality, that $\|K\| = 1$. Let x_0 be any element of X such that $\|Kx_0\| > 1$. Clearly this implies that $\|x_0\| > 1$. Consider now the set

$$S = \{x \mid \|x - x_0\| < 1\}$$

and let S_K be the closure of the set KS. It is obvious that S_K is compact and convex, and

1. $0 \notin S$
2. $0 \notin S_K$

Consider now the set U_T defined as follows:

$$U_T = \{y \in X \mid \|Ty - x_0\| < 1\}$$

Clearly U_T is an open set, and we show now that

$$\bigcup_{T \in A} U_T = X \setminus \{0\}$$

Let y be fixed and consider the subspace $\{Ty \mid T \in A\}$; it is clear from the definition that this set is invariant for A.

Since A is a transitive algebra, it follows that this subspace is X. Thus we can find $T \in A$ such that $\|Ty - x_0\| < 1$. This gives us that the set S_K is contained in $X \setminus \{0\} = \bigcup U_T$.

Now the set S_K is compact, and thus we can find a finite number of operators, say, $T_1, T_2, \ldots, T_n$ in A such that

$$S_K \subset U_{T_1} \cup U_{T_2} \cup U_{T_3} \cup \cdots \cup U_{T_n}$$

We define now the following functions on the set S_K. For each i, $1 < i < n$,

$$\varphi_i(y) = \max\{0, 1 - \|T_i y - x_0\|\}$$

for all $y \in S_K$. From the definition it follows that these are continuous functions and

1. $0 < \varphi_i < 1$
2. $\Sigma\, \varphi_i(y) > 0$

In this case we may define the functions ψ_i by the relations,

$$\psi_i(y) = \frac{1}{(\Sigma\, \varphi_j(y))}\, \varphi_i(y)$$

Clearly these are also continuous functions and $\Sigma\, \psi_j = 1$. We define now a function on S_K with values in X as follows:

$$f: S_K \to X \qquad f(y) = \Sigma\, \psi_j(y) T_j y$$

Since $x_0 = \Sigma\, \psi_j(y) x_0$, we have that the values of f are in S. Indeed,

$$\|f(y) - x_0\| = \|\Sigma\, (T_j y - x_0)\psi_j(y)\| \leq \Sigma\, \psi_j(y) = 1$$

and the assertion is proved. We define now a function on S, say f_1, as $f_1(y) = f(Ky)$ and this function maps S into itself.

Since f is continuous, f_1 is continuous, and since K is a compact operator, the range of f_1 is contained in a compact set. In this case, by the above result, there exists a nonzero element in S, say x_1, such that $f_1(x_1) = x_1$. If we consider the linear operator

$$S = \Sigma\ \psi_j(Kx_1)T_j$$

then obviously $SKx_1 = x_1$.

Thus the first part of the lemma is proved. For the second part we define the function f_2 on S_K by the formula

$$f_2(y) = Kf(y)$$

which is also continuous and maps S_K unto itself. From the Schauder fixed point theorem we find an element z_1 in S_K such that $Kf(z_1) = z_1$. Now if we set

$$S_1 = \Sigma\ \psi_j(z_1)T_j$$

we find that S_1 satisfies the last part of the lemma.

From this lemma we obtain easily the Lomonosov result.

10.1.9 THEOREM Let $T \in L(X)$ be any operator which is nonscalar and T commutes with a nonzero compact operator K. In this case T has a nontrivial hyperinvariant subspace.

Proof. For any algebra A in L(X),

$$A' = \{S \mid S \in L(X),\ ST = TS \text{ for all } T \in A\}$$

A' is called the *commutant* of A and (A')' = A" is called the *bicommutant* of A. Similarly, we can define $(A)'^n$.

Now, in our case we consider the algebra A of all polynomials in T and the closure in L(X) of this algebra, which is a commutative Banach algebra. The assertion of the theorem is in fact that A' is a nontransitive algebra. Suppose that this is not so. Thus A' is transitive, and by Lemma 10.1.6 we can find an operator in A' such that $z = 1$ is an eigenvalue of the operator SK. Now, since T commutes with SK and SK has a finite dimensional eigenspace which is obviously invariant for T, it follows that T has an eigenvalue. Since T is not scalar, the associated eigenspace is not X; thus since this eigenspace is invariant for A', we have a contradiction. The theorem is proved.

The following refinement of the Lomonosov result about the existence of invariant subspaces was proved by Daughtry (1975a).

10.1.10 THEOREM If $T \in L(X)$ and there exists a nonzero compact operator K such that

$$C = TK - KT$$

is of rank 1, then T has a nontrivial invariant subspace.

The proof of this theorem consists in an ingenious application of the Lomonosov theorem.

We recall that by the rank of an operator we mean the dimension of the range.

For the proof of Theorem 10.1.10 we need the following lemma attributed to Luenberger (1965).

10.1.11 LEMMA Let A, B, and T be in L(X) and

$$C = TA - BT$$

of rank 1. If the largest invariant subspace of A contained in the kernel of C is $\{0\}$ and the smallest invariant subspace of B containing the range of C is X, then T is one-to-one or T is with dense range.

Proof. Suppose that the assertion of the lemma is not true. Thus T has a nontrivial kernel and a nondense range. Let $x \in \ker T$ and $x \neq 0$. In this case we have

$$Cx = (TA - BT)x = TAx$$

and clearly this gives

1. $Cx \neq 0$
2. $Cx \in \ker T$

Now, if 2 is true and since C is of rank 1, we obtain that

$$TA^2x = CAx$$

and thus $CAx \neq 0$ or $A^2x \in \ker T$.

We can continue in this way and since $\{A^nx\}$ cannot be an invariant subspace for A contained in the kernel of C, we find $y \in \ker T$ such that $Cy \neq 0$. Since $Cy = TAy$ and thus the kernel of A^*T^*

annihilates the range of C, which is one-dimensional, or ker A^*T^* is in ker C^*. In this case we obtain

$$\ker A^*T^* \subset \ker T^*B^*$$

since

$$A^*T^* - T^*B^* = C^*$$

Now if $x^* \in \ker A^*T^*$, then

$$(A^*T^*)B^*x^* = A^*(T^*B^*)x^* = 0$$

and thus ker A^*T^* is an invariant subspace for B^* contained in the null space of C^*. This implies that the closure of the range of TA is an invariant subspace for B containing the range of C, and this gives the assertion of the lemma.

We are now in a position to prove Theorem 10.1.10. Suppose that T has no invariant subspaces. By Lemma 10.1.6 there exists an operator B commuting with T such that $BKy = y$ for some $y \in X$, $y \neq 0$. Since we have

$$BC = B(TK - KT) = BTK - BKT = T(BK - I) - (BK - I)T$$

and BC has the rank 1, $BK - I$ has a nontrivial kernel and thus has a nondense range. This contradiction proves the theorem.

An interesting way to obtain invariant and hyperinvariant subspaces for operators is connected with the so-called quasi-similarity.

10.1.12 DEFINITION The operators $A,B \in L(X)$ are called *quasi-similar* if there exist injective operators $P,Q \in L(X)$ with dense ranges and such that

1. $AP = PB$
2. $QA = BQ$

The importance of this notion in connection with the hyperinvariant subspaces is given here.

10.1.13 THEOREM If $A,B \in L(X)$ are quasi-similar operators and A has a nontrivial hyperinvariant subspace, then B has a nontrivial invariant subspace.

Proof. Let P and Q be the operators in the definition of quasi-similarity of A and B. Let X_1 be a nontrivial hyperinvariant subspace of A and consider the following subspace:

$$X_2 = \overline{\{SQx \mid x \in X_1 \text{ and } SB = BS\}}$$

which is hyperinvariant for B. Of course, from the definition it is clear that $X_2 \neq \{0\}$. We prove now that X_2 is not equal to X.

Indeed, since SB = BS, we have that

$$(PSQ)A = PSQB = PBSQ = A(PSQ)$$

and thus $PSQX_1$ is in X_2. Hence, $PX_2 \subset X_1$ and this clearly implies that X_2 is not X.

As an application of this theorem, we give the following result of Nagy and Foias (1970).

10.1.14 *THEOREM* If T is a contraction on a Hilbert space H and

1. $\|T^n x\| \not\to 0$
2. $\|T^{*n} x\| \not\to 0$

for all $x \in H$, then T is quasi-similar to a unitary operator.

Proof. Consider the sequence of hermitian positive operators $\{T^{*n}T^n\}$ which converges strongly to a positive hermitian operator and let A be the square root of $\lim T^{*n}T^n$. We consider the following operator V defined on the closure of AH. The definition of V is VA = AT. Since $\|Ax\| = \lim \|T^n x\|$ and from the conditions imposed on T, it is clear that A is injective and thus $(\overline{AH}) = H$.

We show now that in fact V is unitary and that there exists B an injective operator with dense range such that BV = TB.

Since z = 0 is not an eigenvalue of T*, it follows that T has a dense range and this implies that V has a dense range. Since V is an isometry, this implies that V is unitary. Let A_1 be the square root of the limit $\lim T^n T^{*n}$ and V_1 be the operator defined by $V_1 A_1 = A_1 T^*$. As above we can show that in fact V_1 is a unitary operator and this gives that $A_1 = V_1^* A_1 T^*$. Furthermore

$$A_1 AVA = A_1 A^2 T = V_1^* A_1 T^* A^2 T = V_1^* A_1 A^2$$

As we know A has a dense range and this implies that

$$A_1 AV = V_1^* A_1 A$$

and the operator $B = A_1^2 A$ is injective and with dense range. Since

$$TB = TA_1^2 A = (A_1 T^*)^* A_1 A = A_1 V_1^* A_1 A = A_1^2 AV = BV$$

we have that T is quasi-similar to the unitary operator V.

10.1.15 COROLLARY If T is a contraction on a Hilbert space and there exists elements in H, say x_0 and y_0, such that

1. $\|T^n x_0\| \not\to 0$
2. $\|T^{*n} y_0\| \not\to 0$

then T has a nontrivial hyperinvariant subspace.

Proof. We consider the subspaces

$$X_1 = \{x \mid T^n x \to 0\}^\perp$$

and

$$X_2 = \{y \mid T^{*n} y \to 0\}^\perp$$

which are hyperinvariant subspaces. If both are proper, then T is quasi-similar to a unitary operator, and the existence of hyperinvariant subspaces follows from Theorem 10.1.13. If the unitary operator V is scalar, then T is also scalar and the theorem is proved.

An interesting and important class of operators with the set of all invariant subspaces with a special property is the so-called class of unicellulary operators.

10.1.16 DEFINITION An operator $T \in L(X)$ is called *unicellular* if for any invariant subspaces X_1 and X_2, either $X_1 \subset X_2$ or $X_2 \subset X_1$.

First we give a general result.

10.1.17 THEOREM If $T \in L(X)$ and F is a countable family of invariant subspaces with the property that for any invariant subspaces $m \in F$ and $n \notin F$, either $m \subset n$ or $n \subset m$. In this case every invariant subspace in F is hyperinvariant.

Proof. Let z such that $|z| > \|T\|$ and in this case we have that all invariant subspaces of T and $(T - z)^{-1}$ are the same. Indeed, the operator $(T - z)^{-1}$ is, in this case, of the form

$$(T - z)^{-1} = \sum \frac{T^{n+1}}{z^n}$$

and this gives immediately that the invariant subspaces of T are invariant subspaces for $(T - z)^{-1}$.

Let X_1 be an invariant subspace for $(T - z)^{-1}$ and thus

$$X_1 \subset (T - z)X_1$$

If the above inclusion is proper, we can find an element $x \in X_1$ of norm 1 such that

$$\langle x, (T - z)x\rangle = 0$$

and thus we obtain

$$0 = \langle x, (T - z)x\rangle \geq |z| - |\langle Tx, x\rangle| \geq |z| - \|T\| > 0$$

which is a contradiction. Thus $X_1 = (T - z)X_1$ and this implies that $TX_1 \subset X_1$ and the assertion is proved. Let S be an operator in L(X) and commuting with T. Let m be an element in F. By the above remark if $|z| > \|S\|$, the invariant subspaces for S and for (S - z) are the same. Now since (S - z)m is invariant, for T we have two possibilities: if (T - z)m is not in F for some z then by the hypothesis

$$m \subset (S - z)m \qquad \text{or} \qquad (S - z)m \subset m$$

From this we have

$$(S - z)^{-1}m \subset m \qquad \text{or} \qquad m \subset (S - z)^{-1}m$$

and thus $Sm \subset m$.

Now if $(S - z)m \in F$ for all z, $|z| > \|S\|$, then we obtain that

$$(S - z_1)m = (S - z_2)m$$

for some $z_1 \neq z_2$ since F is countable. This gives

$$m = (S - z_2)^{-1}(S - z_1)m = (I + (z_2 - z_1)(S - z_2)^{-1})m$$

and this gives $Sm \subset m$ and the theorem is proved.

For the case of unicellulary operators we obtain the following interesting property.

10.1.18 THEOREM If $T \in L(X)$ is a unicellulary operator, then every invariant subspace of T is hyperinvariant.

Proof. Follows from Theorem 10.1.17.

In what follows we give some examples of unicellular operators.

One of the most famous operators is the so-called Volterra operator V defined on $L^2[0, 1]$ by

$$Vf(x) = \int_0^x f(t)\, dt$$

If $0 \leq a \leq 1$, then the closed subspace

$$L_a = \{f \in L^2[0, 1] \mid f(t) = 0 \text{ almost everywhere on } [0, a]\}$$

of $L^2[0, 1]$ is invariant under V. The fact that V is a unicellulary operator was first proved by Dixmier (1949). Interesting proofs of the unicellularity of V have been given by Brodski (1965), Donoghue (1957), and Kalisch (1962).

In what follows we give a proof of the unicellularity of V following Kalisch and Schreiber (1963).

10.1.19 THEOREM If m is an invariant subspace of V, then $m = L_a$ for some $a \in [0, 1]$.

First we recall the notion of a support for a measurable function defined on [0, 1]: the support of f, denoted by S(f), is the complement of the largest open set in which f = 0 almost everywhere.

If f and g are functions in L^1, then the product (f*g), called the *convolution*, can be defined as follows:

$$(f*g)(x) = \int_0^1 f(x - s)g(s)\, ds$$

and it is easy to see that with this product L^1 is a Banach algebra. An induction argument shows that the following representation for T is true:

$$(V^n f)(x) = \frac{1}{(n-1)!} \int_0^1 (x-s)^{n-1} f(s)\, ds$$

This implies the following estimates for the norm of V^n,

$$\|V^n\| \le \frac{1}{n!}$$

and thus V is a quasi-nilpotent operator.

10.1.20 DEFINITION If $T \in L(X)$ and $x \in X$, then x is called a *cyclic vector* for T if the vector subspace $\{T^n x\}$ is dense in X.

For the proof of Theorem 10.1.19 we need the form of cyclic vectors for V restricted to some subspaces.

First we note that the above form for $V^n f$ can be written using the convolution:

$$V^n f = e_n^* f$$

where e_n represents the n-fold convolution of the function

$$e(x) = 1 \qquad \text{for all } x \in [0, 1]$$

Also we need the following famous theorem of Titchmarsh: If $f*g = 0$ (a.e.) in $[0, b]$ and $0 \in S(f)$, then $g = 0$ (a.e.) in $[0, b]$.

10.1.21 LEMMA $f \in L^2[a, 1]$ is a cyclic vector for the restriction of V to this space if and only if $a \in S(f)$.

Proof. Let $g \in L^2[0, 1]$ such that $<(V^n f), g> = 0$ for all $n \ge 0$.

If we set $h(x) = \overline{g(1 - x)}$, then we have

$$<(V^n f), g> = \int_0^1 (V^n f)(s)\overline{g(s)}\, ds = \int_0^1 (V^n f)(s) h(1-s)\, ds$$

$$= (V^n f)*h)(1) = e_n^* f^* h)(1) = (V^n (f*h)(1)$$

$$= \frac{1}{(n-1)!}\int_0^1 (1-s)^{n-1}(f*h)(s)\,ds$$

$$= \frac{1}{(n-1)!}\int_0^1 t^{n-1}(f*h)(1-t)\,dt$$

and this implies that $f*g = 0$ (a.e.) in [0, 1]. If we assume that $f = 0$ (a.e.) in [0, a] and that $a \in S(f)$, we define the function

$$f_1(t) = f(t + a)$$

for all $t \in [0, 1 - a]$, then we have the following relation

$$(f*h)(t + a) = (f_1{}^*h)(t)$$

in [0, 1 - a]. Since $0 \in S(f_1)$, we can apply the Titchmarsh theorem for $f_1^*h = 0$, and thus we obtain that $h = 0$ (a.e.) in [0, 1 - a].

This clearly gives that $g = 0$ (a.e.) in [a, 1] and the lemma is proved.

The proof of Theorem 10.1.19 is now as follows. Let $f \in L^2[0, 1]$ and define $a(f) = \inf \{s \mid s \in S(f)\}$. If m is an invariant subspace, we have that $L^2[a(f), 1]$ is in m for all $f \in m$. From this we have that $L^2[a, 1]$ is in m, where $a = \inf \{a(f) \mid f \in m\}$.

Since the inclusion in the opposite direction is obvious, the assertion of the theorem is proved.

Another example of a unicellulary operator on Hilbert spaces are some classes of weighted shifts. As we know, if $H = \ell^2$, then the weighted shift with the weight sequence $\{w_n\}_{n=0}^{\infty}$ is defined as follows:

$$S(x_0, x_1, x_2, \ldots) = (0, w_0x_0, w_1x_1, w_2x_2, \ldots)$$

It is obvious that the closed subspaces

$$m_n = \{x = (x_i) \mid x_j = 0 \text{ for all } j < n\}$$

are invariant subspaces for S. We show that S has only this set of subspaces as invariant subspaces if the sequence has some special properties.

10.1.22 PROPOSITION If $\lim w_n = 0$, then S is quasi-nilpotent.

Proof. Let $x = (x_i)$ be an arbitrary element in ℓ^2 and set $S^n x = y$, and we have immediately that if $y = (y_i)$, then $y_j = 0$ for all $j < n$.

Also a simple computation shows that

$$y_{n+k} = w_k w_{k+1} \cdots w_{k+n-1} x_k$$

for all $k > 0$ and this gives

$$\|S^n\| \leq \sup |w_k w_{k+1} \cdots w_{k+n-1}|$$

which implies that

$$\|S^n\| = a_0 a_1 \cdots a_{n-1}$$

where

$$a_k = \sup \{|w_{k+1}|, \ldots\}$$

But $\lim w_n \to 0$ and thus $\lim a_k = 0$. This implies that

$$\lim (a_0 a_1 \cdots a_{n-1})^{1/n} = 0$$

and the proposition is proved.

The following lemma gives a sufficient condition for finite dimensionality of subspaces of ℓ^2.

10.1.23 LEMMA If m is a subspace of ℓ^2 and there exists y_0 in ℓ^2 such that for all $x \in m$, $x = (x_i)$,

$$|x_n| \leq y_n \|x\|$$

where $y_0 = (y_n)$, then m is finite dimensional.

Proof. If $e_1, e_2, \ldots, e_n$ is an orthonormal set in m and let

$$f^p = (e_p^{-1}, \ldots, e_p^{-n})$$

for all $p > 0$. If $x = \Sigma\, a_i e_i \in m$, then

$$|\langle x, f^p\rangle| = |\Sigma\, a_i e_p^i| = |x_n| \le y_n\|x\| = y_n\|a\|$$

where $a = (a_1, \ldots, a_n)$.

From this relation we obtain that

$$\sum |e_p^i|^2 = \|f^p\|^2 \le y_n^2$$

and summing over p, we obtain that $n \le \|y\|^2$, and this assertion of the lemma is proved.

We give now a result of Nikolskii about the unicellularity of a certain class of weighted shifts.

10.1.24 THEOREM Suppose the weighted shift S has the following properties:

1. $\{w_n\}$ is a nonincreasing sequence.
2. $w_n \ne 0$ for all n.
3. $\{w_n\}$ is in ℓ^p for some $p \in (0, \infty)$.

Then for any invariant subspace m of S there exists n such that $m = m_n$.

Proof. Let n be the last integer such that $x_n \ne 0$, where x_n is the nth component of some $x \in m$. We have that m_n contains m.

We show now that in fact we have an equality. We may assume that $n = 0$ and in this case we show that m is ℓ^2. Let N be an integer and $p \le 2N$. In this case $\{|w_n|^N\}$ is in ℓ^2 since $\Sigma\, |w_n|^{2N} < \infty$.

Let $x \in m$ and suppose that $x = (x_i)$ with $x_0 \ne 0$. Since the elements $x, Sx, \ldots, S^N x$ are linearly independent, we can find in the subspace generated by these elements a vector $y = (y_i)$ with the following property: $y_0 = 1$ and $y_1 = \cdots y_N = 0$.

Consider now an element z which is orthogonal to m. The relations $\langle z, S^n y\rangle = 0$ for all $n > 0$ implies

$$\sum w_k w_{k+1} \cdots w_{k+n-1} y_k \bar{z}_{n+k} = 0$$

$$\bar{z} = -(w_0 w_1 \cdots w_{n-1})^{-1} \sum_{k=N+1} w_k \cdots w_{k+n-1} y_k \bar{z}_{n+k}$$

and since for n, k > N + 1 we have the relations

$$\left|\frac{w_k \cdots w_{k+n-1}}{w_0 \cdots w_{n-1}}\right| \le \frac{|w_n|}{w_0 \cdots w_N}$$

we obtain that

$$|z_n| \le C|w_n|^N \|z\|$$

for all $n \ge N + 1$, where $C = |w_0 \cdots w_N|^{-1}\|y\|$.

We can apply Lemma 10.1.23 to obtain that the orthogonal complement of m is finite dimensional. Since the orthogonal complement is finite dimensional and is an invariant subspace for S*, it follows that S* has x as an eigenvector. Since S* is a quasi-nilpotent operator (since S is quasi-nilpotent), it follows that $S^*z_0 = 0$, where z_0 is the eigenvector. If $z_0 = (a_0, a_1, \ldots)$, it follows that $a_i = 0$ for $i \ge 1$. Now since $\langle z_0, y\rangle = 0$ and the special form of y implies that $a_0 = 0$. Thus the orthogonal complement of m is equal to {0} and the theorem is proved.

In connection with the Volterra operator we mention the following result; for the proof, see Ringrose (1962).

10.1.25 THEOREM Suppose T is a bounded operator on a Hilbert space and has the following properties:

1. T is quasi-nilpotent.
2. There is a unit element in H, say e, such that $(T - T^*)x = i\langle x, e\rangle e$ for all $x \in H$.
3. $\overline{\{T^n e\}} = H$.

Then T is unitarily equivalent to V and the subspaces of T (we consider $H = L^2$) are L_a only.

Suzuki (1971) has proved the following theorem for the case of Hilbert space operators.

10.1.26 THEOREM If A and B are two positive operators on a Hilbert space, then the operator T = AB has a proper invariant subspace.

Istrăţescu (1972b) extended this theorem to Banach space operators, obtaining a slight generalization in the sense that only A is supposed to be positive.

10.1.27 THEOREM If A and B are hermitian operators on a Banach space and $A^{1/2}$ exists such that $A^{1/2}BA^{1/2}$ is a hermitian operator, then T = AB has a nontrivial invariant subspace.

Proof. Let m be a nontrivial invariant subspace for $A^{1/2}BA^{1/2}$ and we define the following subspace:

$$a(m) = \overline{\{A^{1/2}x \mid x \in m\}}$$

We show that this satisfies the theorem.

Indeed, if $y = A^{1/2}x$, then we have

$$ABy = (A^{1/2}A^{1/2}B)A^{1/2}x = A^{1/2}(A^{1/2}BA^{1/2}x)$$

which proves that $Ay \in a(m)$. From the definition of a(m) and this remark it follows that a(m) is an invariant subspace.

Since m is nontrivial, it follows that a(m) is not {0}.

We show now that a(m) is not X. Indeed, in the contrary case $\{A^{1/2}x \mid x \in m\}$ is dense in X. In this case $A^{1/2}\ \{BA^{1/2}x \mid x \in m\}$ is dense in X which is impossible. The theorem is proved.

With the same method we can prove the following general result.

10.1.28 THEOREM If A and B are operators on a Banach space and ABA has a nontrivial invariant subspace, then A^2B has a nontrivial invariant subspace.

The above result gives rise to the following problem.

10.1.29 PROBLEM If A and B are operators in L(X) and AB has an invariant subspace, then BA has a nontrivial invariant subspace.

For results in this direction, see Istrăţescu (1979).

For the following invariant subspace theorem we need the notion of maximal vector for a bounded operator.

10.1.30 DEFINITION An element $x \in X$ is called a *maximal vector* for the operator $T \in L(X)$ if $\|Tx\| = \|T\| \cdot \|x\|$.

10.1.31 THEOREM If $T \in L(H)$, H is a Hilbert space and T is of class (N,k), then the set of all maximal vectors is an invariant subspace.

Proof. We can assume, without loss of generality, that $\|T\| = 1$. Since $\|Tx\| = \|x\|$, we obtain that

$$\|T^*Tx - x\|^2 = \|T^*Tx\| - 2 \operatorname{Re} \langle T^*Tx, x\rangle + \|x\|^2 \leq 0$$

and thus $T^*Tx = x$ and this gives that the set of maximal vectors is a closed subspace. Now if x is a maximal vector and since T is of class (N,k), we may assume that $\|x\| = 1$. Now from

$$\|T^k x\| \geq \|Tx\|^k = \|T\|^k$$

we obtain that

$$\|T(Tx)\| \leq \|Tx\| \|T\| \leq \|T\|^2 = \frac{1}{\|T\|^{k-2}} \|T\|^k \leq \frac{1}{\|T\|^{k-2}} \|T^k x\|$$
$$= \frac{1}{\|T\|^{k-2}} \|T^{k-2}(T^2 x)\| \leq \|T(Tx)\|$$

and thus

$$\|T(Tx)\| = \|Tx\|$$

and the theorem is proved.

10.1.32 COROLLARY If T is an operator of class (N,k) on a Hilbert space and has maximal vectors, then T has invariant subspaces.

Proof. From the above result we can consider only the case when the set of all maximal vectors is H. In this case T is a multiple of an isometry, and clearly there exists invariant subspaces.

10.2 REDUCING INVARIANT SUBSPACES

First we recall the definition of reducing subspaces.

10.2.1 DEFINITION An invariant subspace H_1 of an operator $T \in L(H)$, H is a Hilbert space, is called a *reducing subspace* for T if it is invariant for T^*.

10.2.2 DEFINITION An operator $T \in L(H)$ is called *irreducible* if the only reducing subspaces are $\{0\}$ and H.

Simple examples show that there exist operators on a Hilbert space and invariant subspaces which are not reducing. We present here one.

Let ℓ^2 be the space of all sequence $x = (x_i)_{-\infty}^{\infty}$ with the property

$$x \mapsto \|x\|^2 = \sum_{-\infty}^{\infty} |x_i|^2 < \infty$$

We consider the unitary operator U defined as follows:

$$Ux = (x_{i+1})$$

It is clear that the closed subspace

$$\ell_+ = \{x \in \ell^2 \mid x = (x_i),\ x_i = 0 \text{ for } i < 0\}$$

is invariant for U and is not reducing.

As an example of a irreducible operator, we can take the weighted shift S, where the weight sequence $\{w_i\}$ has the property described in Nikolskii's theorem.

We give now two results which show the position of the set of irreducible operators in L(H). The first is given by Radjavi and asserts that any operator is the sum of two irreducible operators, and the next is given by Halmos and asserts that the set of irreducible operators is dense in L(H) in the norm topology.

For the proof we need some results which present also an independent interest.

10.2.3 LEMMA If $\{e_i\}_{i=1}^{\infty}$ is an orthonormal basis for H and $A \in L(H)$ is not a multiple of I and n,m fixed integers, then for any $\varepsilon > 0$ there exists a unitary operator U such that $\|I - U\| < \varepsilon$ and $\langle U^*AUe_n, e_m\rangle \neq 0$.

Proof. Suppose that $\langle Ae_n, e_m\rangle = 0$. If Ae_n is not a multiple of e_n, we define $U_1 = I$. If $Ae_n = ze_n$, we choose k such that $Ae_k \neq ze_k$ and for d small, $(1 - d^2)^{1/2}e_n + de_k$ is not an eigenvector of A.

Let d be such that $(1 - (1 - d^2)^{1/2})^2 + d^2 < \varepsilon/8$ and set $d' = (1 - d^2)^{1/2}$. We define

$$U_1e_n = d'e_n + de_k$$

$$U_1e_k = -de_n + d'e_k$$

$$U_1e_i = e_i \qquad \text{if } i \neq n,k$$

It is clear that $\|I - U_1\| < \varepsilon/2$ and if $<AU_1e_n, U_1e_m> \neq 0$, U_1 satisfies the lemma. Suppose that this is not so. Then for some n,m we have $<AU_1e_n, U_1e_m> = 0$. Since A is not a multiple of I, for some $i \neq n$, AU_1e_n is not a multiple of U_1e_n. This clearly implies that $i \neq m$. We have two cases to discuss.

Case 1. $n \neq m$. In this case we define the unitary operator U_2 as follows:

$$U_2U_1e_m = d'U_1e_m + dU_1e_m$$

$$U_2U_1e_i = -dU_1e_m + d'U_1e_i$$

$$U_2U_1e_p = U_1e_p \qquad \text{for } p \neq i,m$$

The operator $U = U_2U_1$ is clearly unitary and since

$$<U^*AUe_n, e_m> = <AU_2U_1e_n, U_2U_1e_m> = <AU_1e_n, d'U_1e_m + dU_1e_i>$$
$$= 0 + <AU_1e_n, U_1e_i> \neq 0$$

Case 2. $n = m$. If we consider the fourth roots of the unity, we define the elements f_j by the relations

$$f_j = d'U_1e_n + w_jdU_1e_i \qquad w_j^4 = 1,\ j = 1, 2, 3, 4$$

and we can find a j such that $<Af, f> \neq 0$ since

$$4dd'<AU_1e_p, U_1e_i> = \Sigma\, w_j<Af_j, f_j>$$

For this j we define the unitary operator U_2 as follows:

$$U_2U_1e_n = f_j$$

$$U_2U_1e_i = -w_j^*de_n + d'U_1e_i$$

$$U_2 U_1 e_p = U e_p \qquad \text{for } p \neq i,n$$

It is clear that $U_2U_1 = U$ satisfies the relation $\|I - U\| < \varepsilon$ and satisfies also the lemma.

10.2.4 LEMMA If $\{A_i\}$ is a countable set of elements in L(H) none of which is a multiple of I, then there exists an orthonormal basis of H such that $\langle A_i e_n, e_m\rangle \neq 0$.

Proof. For a fixed orthonormal basis of H, say $\{f_j\}$, we consider the following unitary operators:

$$U_{i,n,m} = \{U \mid U \text{ unitary and } \langle U^*A_iUf_n, f_m\rangle = 0\}$$

and it is obvious that these are closed sets in L(H).

We show now that these sets are nowhere dense in the set of unitary operators considered as a metric subspace of L(H). Indeed, let U in $U_{i,n,m}$ and by the above lemma applied to the orthonormal basis $\{Uf_i\}$, for each $\varepsilon > 0$ we can find a unitary operator V such that $\|I - V\| < \varepsilon$ and $\langle V^*A_iVUF_n, UF_m\rangle \neq 0$.

Since

$$0 \neq \langle U^*V^*A_iVUf_n, f_m\rangle = \langle (VU)^*A_iVUf_n, f_m\rangle$$

the assertion is proved since $\|U - VU\| < \varepsilon$. We can apply the Baire category theorem and we obtain that there exists a unitary operator $\bar{U}$ which is not in the union of the sets $U_{i,n,m}$. We define now $e_i = \bar{U}f_i$, and this satisfies the lemma.

10.2.5 LEMMA If M is a finite or countable set of nonscalar operators on a Hilbert space H, then there exists a hermitian operator $K \in L(H)$ such that no member of M leaves invariant a nontrivial invariant subspace of K.

Proof. By the above lemma there exists a basis $\{e_i\}$ of H such that every entry in the matrix of $A \in M$ with respect to $\{e_i\}$ is nonzero. If K is the hermitian operator defined on H whose matrix with respect to $\{e_i\}$ is diag $\{1, 1/2, 1/3, \ldots\}$, then clearly K is a compact operator and since the subspaces of K are of the form

$$\bigcup_{j \in B} \{e_j\}$$

where B is a subset of natural numbers, for any $A \in M$, A and K have no common invariant subspaces since no entry in the matrix of A is zero.

10.2.6 THEOREM Every operator $T \in L(H)$ is the sum of two irreducible operators.

Proof. If $T \in L(H)$ is scalar, then we can take an arbitrary irreducible operator C, and

$$T = (T - C) + C$$

gives the desired decomposition.

Now if T is not scalar, $T = \text{Re } T + i \text{ Im } T$ and we have two cases.

Case 1. The operators Re T and Im T are nonscalar operators. In this case we can apply the above lemma for $M = \{\text{Re } T, \text{Im } T\}$ and we find the compact operator K (hermitian) and

$$T_1 = (\text{Re } T - K) - iK$$

$$T_2 = K + i(\text{Im } T + K)$$

satisfies the theorem.

Case 2. If either Re T or Im T is a scalar, we may suppose that $\text{Re } T = cI$ and we can apply the above lemma for $M = \{\text{Im } T\}$, and we find the compact operator K. The operators

$$T_1 = K + cI + \frac{i}{2} \text{Im } T$$

$$T_2 = -K + \frac{i}{2} \text{Im } T$$

satisfy the theorem.

We give now the result about the density.

10.2.7 THEOREM The set of all irreducible operators in L(H) is dense in L(H).

Proof. Let $T \in L(H)$ be arbitrary. By the spectral theorem we can find a hermitian operator K whose matrix is diagonal with respect to an orthonormal basis $\{e_i\}$ and K satisfying the inequality

$$\left\| K - \frac{T^* + T}{2} \right\| < \frac{\varepsilon}{2}$$

Now we can find another hermitian operator K_1, diagonal with respect to $\{e_i\}$, all eigenvalues of K_1 are distinct, and

$$\|K - K_1\| < \frac{\varepsilon}{4}$$

If K_2 is any hermitian operator within $\varepsilon/2$ of Im T with all entries nonzero (with respect to $\{e_i\}$), the operator $K_1 + \varepsilon K_2$ is irreducible and $\|T - (K_1 + \varepsilon K_2)\| < \varepsilon$.

An interesting class of operators on Hilbert spaces connected with the reducing invariant subspaces is the class of the so-called completely reducible operators. Their definition is as follows.

10.2.8 DEFINITION An operator $T \in L(H)$ is called *completely reducible* if the restriction of T to any reducing subspace has a reducing subspace.

As an example of a completely reducible operator, we can take every normal operator.

The following result gives the structure of completely reducible operators on finite dimensional spaces.

10.2.9 THEOREM If T is a completely reducible operator on a finite dimensional space, then T is normal.

Proof. Since T is compact, we can find the invariant subspaces

$$M_1 \subset M_2 \subset \cdots \subset M_n = H$$

In this case dim $M_i = i$ and M_1 is a reducing subspace. In this case $M_2 = M_1 \oplus M_1^{\perp}$, where the orthogonal complement is taken with respect to M_2. Clearly $M_1^{\perp}$ is a reducing subspace. We can continue in this way to obtain that T is normal.

The following example shows that there exists completely reducible operators which are not normal.

Let $H = L^2[0, 1]$ and we consider the operator $T: H \to H$ defined as follows:

$$Tf(x) = xf(x)$$

which is clearly hermitian. On the space $H \times H$ with the inner product

$$\langle\langle (x, y), (x', y') \rangle\rangle = \langle x, x' \rangle + \langle y, y' \rangle$$

we define the operator with the matrix

$$T^{\sim} = \begin{bmatrix} 0 & T \\ 0 & I \end{bmatrix}$$

and it is easy to see that

$$\tilde{T^*} = \begin{bmatrix} 0 & 0 \\ T^* & 0 \end{bmatrix} \qquad \tilde{T^*}T^{\sim} = \begin{bmatrix} 0 & 0 \\ 0 & T^2 + I \end{bmatrix}$$

We show that every reducing subspace of $T^{\sim}$ is of the form $N \oplus N$, where N is an invariant subspace of T. Let M be any invariant subspace which is reducing for $T^{\sim}$ and let N be the invariant subspace of T generated by the elements f of L^2 for which exist $g \in L^2$ such that $(f, g) \in M$ or $(g, f) \in M$.

From the definition of N it is clear that $M \subset N \oplus N$. We show now the inclusion in the opposite direction. Let (h, k) be an element of M. Since for any integer n, $(\tilde{T^*}\tilde{T})^n(h, k)$ is in M, it follows that $\{0\} \oplus V(T^2 + I)k\}$ is in M. We consider the algebra of polynomials without constant term having as variable $z = x^2 + 1$ which satisfies the conditions of the Stone-Weierstrass approximation theorem and thus is dense in $C_{[0,1]}$. From this we obtain that (0, k) is in M and thus also (h, 0) is in M.

In this case

$$\{0\} \oplus V\{T^{2^n}k\} \qquad \{0\} \oplus V((T^{2^n+1})h) \qquad V\{T^{2^n}h\} \oplus \{0\}$$

and

$$V\{T^{2^n+1}h\} \oplus \{0\}$$

are in M by applying $T^{\sim}$ and $\tilde{T^*}$ to the elements (h, 0) and (0, k).

Since (h, k) is in M, it follows that (h, 0) and (0, k) are in M and the assertion is proved.

The results given above show the importance of the reducing subspaces for an operator defined on a Hilbert space. We give now a nice theorem of Ando's which gives a structure theorem for operators having only reducing subspaces and being in a special class of operators.

First we give an interesting property of normal operators.

10.2.10 THEOREM Suppose T is a normal operator on a Hilbert space and has one of the following properties:

1. T is compact.
2. Re T or Im T is a compact operator.

Then every invariant subspace is reducing.

For the proof we need the following lemma.

10.2.11 LEMMA If T is a hyponormal operator and H_1 is an invariant subspace such that $T_{H_1} = TP_{H_1}$ is a normal operator, then H_1 is reducing.

Since the proof is not difficult, it is omitted.

Proof of Theorem 10.3.10. Let H_1 be an invariant subspace for T and we consider the operator T_{H_1} which is hyponormal and satisfies conditions 1 or 2 for T. In this case we know that T_{H_1} is a normal operator and Lemma 10.3.11 implies the assertion.

The following theorem was proved for p(z) = z by Ando and for $p(z) = z^n$ by Saito. The case of polynomials was considered by the present author; in the subsequent discussion we follow the exposition in Saito.

10.2.12 THEOREM Suppose T is an operator in L(H) and

1. T is polynomially compact.
2. T has only reducing subspaces.

Then T is a normal operator.

For the proof we need the following lemma.

10.2.13 LEMMA If K is a compact operator on a Hilbert space and $\{M_\alpha\}$ is a family of invariant subspaces which is totally ordered and

$$\|P_{M_\alpha} K\| = \|K\|$$

then we have that

$$\|P_{M_0} K\| = \|K\|$$

where $M_0 = M_\alpha$.

Proof. Let $\varepsilon > 0$. For each α there exists x_α of norm 1 such that

$$\|(P_{M_0} - P_{M_\alpha})Kx_\alpha\| > \|(P_{M_0} - P_{M_\alpha})K\| - \frac{\varepsilon}{2}$$

Since $\{x_\alpha\}$ is a bounded set and K is a compact operator, we can find a sequence in $\{Kx_\alpha\}$ which is convergent, say to x. Since P_{M_0} is the strong limit of $\{P_{M_\alpha}\}$, there exists ν such that

$$\|(P_{\alpha_\nu} - P_{M_0})x\| < \frac{\varepsilon}{4} \mid \|Kx_{\alpha_\nu} - x\| < \frac{\varepsilon}{4}$$

and thus we obtain

$$\begin{aligned}\|(P_{M_0} - P_{M_\alpha})K\| &< \|(P_{M_0} - P_{M_\alpha})Kx_{\alpha_\nu}\| + \frac{\varepsilon}{4}\\ &< \|(P_{M_0} - P_{M_\alpha})(Kx_{\alpha_\nu} - x)\| + \|(P_{M_0} - P_{M_{\alpha_\nu}})x\| + \frac{\varepsilon}{4}\\ &< 2\|Kx_{\alpha_\nu} - x\| + \|(P_{M_0} - P_{M_\alpha})x\| + \frac{\varepsilon}{4} < \varepsilon\end{aligned}$$

Since $\varepsilon > 0$ is arbitrary, we have

$$\|P_{M_0} K\| = \|K\| = \lim \|P_{M_{\alpha_\nu}} K\|$$

and the lemma is proved.

Proof of Theorem 10.2.12. It is obvious that for every point in the point spectrum of T, say z_0, the subspace

$$E_{z_0}(T) = \{x \mid Tx = z_0 x\}$$

is reducing. Also it is easy to see that the family of spaces $\{E_z(T) \mid z$ in the point spectrum of $T\}$ is an orthogonal family of subspaces, and if we consider the space

$$H_0 = \Sigma \oplus \{E_z(T) \mid z \text{ in the point spectrum of } T\}$$

then this is a reducing subspace and the restriction of T to this subspace is a normal operator. We denote this restriction by T_0.

Now, on the orthogonal complement of H_0, which is also a reducing subspace, the operator TQ, where Q is the orthogonal projection of H on the orthogonal complement of H_0, has the form

$$p(TQ) = p(T)Q$$

and so it is a polynomially compact operator. Of course, $\|p(T)Q\|$ is not zero (in the contrary case there is nothing to prove).

The operator TQ being polynomially compact has a nontrivial invariant subspace which is also reducing for T. We consider now the family F of all invariant subspaces of T for which

$$\|P_M p(T)\| = \|p(T_1)\|$$

for all $M \in F$, where T_1 is the restriction of T to M. Using Zorn's lemma, we can see that the family F contains a minimal element, say M_0.

From the condition about norms it is clear that M_0 is not $\{0\}$.

Now if the dim $M_0 \geq 2$, the restriction of T has again an invariant subspace which is also reducing. Let N be this invariant subspace. In this case we have

$$\|p(T_1)\| = \max \{\|P_N p(T_1)\|, \|(P_{M_0} - P_N)p(T_1)\|\}$$

and this gives that either N or $M_0 \setminus N^-$ is in F, contradicting the minimality of M_0. In this case dim $M_0 = 1$, the operator T has an eigenvalue, and this contradicts the definition of H_0. Thus on H_0^- the operator T is zero and the theorem is proved.

10.3 SOME STRUCTURE THEOREMS

In what follows we give some structure theorems for operators using invariant subspaces. First we give the structure of polynomially compact operators on Banach spaces, and next we apply this to operators on Hilbert spaces which are normal.

First we recall that an operator $T \in L(X)$ is called *polynomially compact* if there exists a nonzero polynomial $p(z)$ such that $p(T)$ is a compact operator. Of course, we can suppose that $p(z) = z^n + a_1 z^{n-1} + \cdots + a_{n-1} z + a_n$ and in this case the polynomial is unique, as is easy to see. We call this polynomial the *minimal polynomial* for T.

The following theorem gives the structure of polynomially compact operators.

10.3.1 THEOREM Let T be a polynomially compact operator and $p(z) = (z - z_1)^{n_1}(z - z_2)^{n_2} \cdots (z - z_k)^{n_k}$ its minimal polynomial. In this case the Banach space X has a decomposition in the direct sum of invariant subspaces for T

$$X = X_1 \oplus X_2 \oplus \cdots \oplus X_k$$

and T has the form

$$T = T_1 \oplus T_2 \oplus \cdots \oplus T_2$$

such that for each $i \in [1, k]$, T_i is the restriction of T to X_i and for all operators $(T - z_i)^{n_i}$ are compact. The spectrum of T consists of a countable set of points having as possible limits points $z_1, \ldots, z_k$ and every z_i is either a limit point or $(T - z_i)^{n_i}$ is a quasi-nilpotent operator.

Proof. First we show that every root z_i is in $\sigma(T)$. Suppose that this is not so. Thus $(T - z_i)^{-1}$ exists and

$$p(T)(T - z_i)^{-n_i}$$

is a compact operator and a polynomial in T. Clearly this contradicts the minimality of $p(T)$. Since by the spectral mapping theorem

$$\sigma(p(T)) = p(\sigma(T))$$

it follows that $\sigma(T)$ is a finite or countable set.

Now since the only limit point of $\sigma(p(T))$ is possibly zero, it follows that the possible limits points of $\sigma(T)$ are $z_1, \ldots, z_k$.

Let z be a root of p which is an isolated point in $\sigma(T)$ and $p(z) \neq 0$. We prove that z is an eigenvalue and

$$E_z(T) = \{x \mid Tx = zx\}$$

is a finite dimensional subspace.

Let τ be a domain in the complex plane such that $z \in \tau$ and $\bar{\tau} \cap \sigma(T) = \{z\}$. We define the projection

$$P_\tau = \frac{1}{2\pi i} \int_{\partial\tau} (T - z)^{-1}\, dz$$

and set $X_\tau = P_\tau X$, which is an invariant subspace for T, and $\sigma(T/X_\tau) = \{z\}$. Since $p(z) \neq 0$, it follows that $p(T/X_\tau)$ is an invertible operator and also a compact operator. This is possible if and only if X_τ is a finite dimensional space by the Riesz lemma. But $\{z\} = \sigma(T/X_\tau)$ and it follows that

$$X_\tau = \{x \mid (T - z)^n x = 0\}$$

for a sufficiently large integer.

We consider now the case of a point $z \in \sigma(T)$ and $p(z) = 0$, and also an isolated point in $\sigma(T)$. Let τ_1 and τ_2 be two domains in the complex plane such that

$$\bar{\tau}_1 \cap \bar{\tau}_2 \neq \emptyset$$

and also

1. $\tau_1 \cup \tau_2 \supset \sigma(T)$
2. $z \in \tau_1$
3. $\bar{\tau}_1 \cap \sigma(T) = \{z\}$

We define

$$X_{\tau_1} = P_{\tau_1} X \qquad X_{\tau_2} = P_{\tau_2} X$$

which have the property that

$$X = X_{\tau_1} \oplus X_{\tau_2} \qquad T = T_{\tau_1} \oplus T_{\tau_2}$$

where T_{τ_i} are the corresponding restrictions of T. Since z is exactly the spectrum of T_{τ_1} and $z \notin \sigma(T_{\tau_2})$, we obtain that

$$(T_{\tau_2} - z)^{-1} p(T_{\tau_2})$$

is a compact operator, and since the above polynomial is minimal, we obtain that X_{τ_1} is an infinite dimensional subspace and $(T_{\tau_1} - z)$ is a quasi-nilpotent operator.

Now for the roots of p, let τ_i be domains in the complex plane such that

1. $z_i \in \tau_i$, $i = 1, \ldots, k$.
2. $\partial\tau_i$ is a rectifiable Jordan curve.
3. $\tau_i \cap \tau_j = \emptyset$.
4. If z_i is an isolated point in $\sigma(T)$, then $\tau_i \cap \sigma(T) = \{z_i\}$.
5. $\sigma(T) \subset \bigcup_{i=1}^{k} \tau_i$.

For each j we define the projections P_j by

$$P_j = \frac{1}{2\pi i} \int_{\partial\tau_j} (T - z)^{-1}\, dz$$

and $X_i = P_i X$, which are invariant subspaces for T. If T_i denotes the restriction of T to X_i, from properties 1 to 5 of τ_i we have

1. $X = \Sigma_{i=1}^{k} \oplus X_i$
2. $T = \Sigma_{i=1}^{k} \oplus T_i$
3. $\sigma(T_i) = \sigma(T) \cap \tau_i$

and also the operators $(T_i - z_i)^{n_i}$ are compact. The theorem is proved.

From this theorem we obtain in the case of normal operators on Hilbert spaces the following result.

10.3.2 THEOREM If T is a normal operator on a Hilbert space and is also a polynomially compact operator and p(z) is $(z - z_1)^{n_1} \cdots$

$(z - z_k)^{n_k}$ is its minimal polynomial, then the following assertions hold:

1. $\sigma(T)$ is a countable set.
2. Some z_i are eigenvalues of finite multiplicity.
3. If z_i is not as in 2, then they are either isolated eigenvalues of infinite multiplicity or limits of eigenvalues.
4. $H = \Sigma H_i$, direct orthogonal sum, $T = \Sigma T_i$ and each H_i is reducing and $T - z_i$ are compact operators.

Proof. All the assertions follow from the above theorem except the last assertion. Since $(T - z_i)^{n_i}$ are compact operators and T is normal, we have for any x, $\|x\| = 1$,

$$\|(T - z_i)^{n_i}x\| \geq \|(T - z_i)x\|^{n_i}$$

and this implies the assertion.

Chapter 11

THE WEYL SPECTRUM OF AN OPERATOR

11.1 PRELIMINARIES AND SOME GENERAL RESULTS

In his famous paper H. Weyl proved that for a hermitian operator defined on a Hilbert space the set of limits points in the spectrum (a point in the spectrum of a hermitian operator is called a *limit point* if it is an eigenvalue of infinite multiplicity) remains invariant under compact hermitian perturbations. This very elegant result was extended to more general classes of operators. The first result in this direction was obtained by Coburn (1965), for the case of hyponormal operators and for a class of Toeplitz operators. The extension to certain classes of operators satisfying a growth condition was obtained in Istrăţescu (1968). In Schechter (1970) and Berberian (1969) these results are extended further to larger classes of operators. All of these results were obtained using some results of Schechter (1970). In his paper Schechter (1970) extends the results to more general operators and also results for the case of Banach space operators. Related results for some sorts of essential spectra were obtained by several authors.

In what follows we give results about the Weyl spectrum for the case of Hilbert space operators and also for the case of Banach space operators. Also we consider this notion in the setting of von Neumann algebras.

We introduce now some notations which will be used below.

For any operator T on a Banach space X, $T \in L(X)$, we denote by N(T) and R(T) the null space and the range of T, respectively;

$\sigma(T)$ is the spectrum of T, $\pi_0(T)$ is the set of eigenvalues of T, $\pi_{0f}(T)$ is the set of eigenvalues of finite multiplicity, π_{0i} is the set of eigenvalues of infinite multiplicity, and $\pi_{00}(T)$ is the set of isolated points in $\sigma(T)$ which are eigenvalues of finite multiplicity. If L(X) is the Banach algebra of all bounded operators on X, then $\mathcal{F}$ denotes the set of all Fredholm operators in L(X).

For each operator $T \in L(X)$ we can define, as we know, the *Fredholm region* of T, that is, the set of all complex numbers z for which $T - zI = T - z$ is not a Fredholm operator. This set in what follows is denoted by

$$\sigma_{\mathcal{F}}(T) = \{z \mid T - z \notin \mathcal{F}\}$$

and also we recall that the quotient Banach algebra $\mathcal{C} = L(X)/K(X)$, where K(X) is the two-sided, closed ideal of compact operators, is called the *Calkin algebra* of X (J. W. Calkin has provided deep results about this algebra in the case of Hilbert spaces). As we know an operator is Fredholm if and only if it is invertible in the Calkin algebra of X. In this case the set $\sigma_{\mathcal{F}}(T)$ is exactly the spectrum of the image of T in this algebra. As a notation, we use for the image of T in $\mathcal{C}$, $\hat{T}$.

Also if T is a Fredholm operator, we can define the *index* of T by the formula

$$\text{ind }(T) = \dim N(T) - \dim R(T)^{-}$$

and this in the case of Hilbert spaces admits also the form

$$\text{ind }(T) = \dim N(T) - \dim N(T^*)$$

In what follows $\mathcal{F}_0$ is the set of all Fredholm operators of index zero. Obviously every invertible operator is with the index zero.

11.1.1 DEFINITION The Weyl spectrum of an operator in L(X) is the set

$$\omega(T) = \{z \mid T - z \notin \mathcal{F}_0\}$$

From the definition and the above discussion about the Calkin algebra we have the following assertion.

11.1.2 THEOREM For any $T \in L(X)$ the set $\omega(T)$ is not empty and is contained in $\sigma(T)$.

For a useful characterization of the Weyl spectrum of an operator we need an interesting result about the Fredholm region of an operator.

11.1.3 THEOREM For any operator $T \in L(X)$, the Fredholm region $\sigma_F(T)$ is an open set and the function

$$z \mapsto i(T - z)$$

is constant on each connected component of $\sigma_F(T)$.

Proof. If T is a Fredholm operator, then we find $\varepsilon > 0$ such that for all $S \in L(X)$, $\|S\| < \varepsilon$, the operator $T + S$ is also Fredholm.

Indeed, since T is a Fredholm operator, we find the compact operators K_1 and K_2 such that

$$S_1T = I - K_1 \qquad TS_1 = I - K_2$$

where $S_1 \in L(X)$. Thus we have

$$S_1(T + S) = I - K_1 + S_1S \qquad (T + S)S_1 = I - K_1 + SS_1$$

and if we take $\varepsilon = 1/\|S_1\|$, we obtain that

$$\|SS_1\| < 1 \qquad \|S_1S\| < 1$$

and thus the operators $I + SS_1$ and $I + S_1S$ have inverse. This implies that

$$(I + S_1S)^{-1}S_1(T + S) = I - (I + S_1S)^{-1}K_1$$

$$(T + S)(I + SS_1)^{-1} = I - K_2(I + SS_1)^{-1}$$

and thus $T + S$ is a Fredholm operator.

Using this, the assertion about the set $\sigma_F(T)$ follows easily.

Indeed, let $z_0 \in \sigma_F(T)$. From the above assertion we find $\varepsilon > 0$ such that if s has the property $|s| < \varepsilon$, the operator $T - z_0 - s$ is also a Fredholm operator. From the relations about the index, which follows from the formulas given above, we have

$$\text{ind } (T - z_0 - s) = \text{ind } (T - z_0)$$

$$\dim N(T - z_0 - s) < \dim N(T - z_0)$$

Thus for all complex numbers z such that $|z - z_0| < \varepsilon$ the operator $T - z$ is a Fredholm operator and

$$\text{ind } (T - z) = \text{ind } (T - z_0)$$

$$\dim N(T - z) < \dim N(T - z_0)$$

which implies that $\sigma_{\mathcal{F}}(T)$ is an open set.

We show now that the index is constant on each connected component of $\sigma_{\mathcal{F}}(T)$. Indeed, if z_1 and z_2 are two points in the same component, we find a smooth curve C through these points.

Thus for each $z \in C$ there exists $\varepsilon > 0$ such that

$$\text{ind } (T - z) = \text{ind } (T - s)$$

for all s such that $|z - s| < \varepsilon$. By the Borel-Lebesgue lemma we find a finite number of points, say $s_1, s_2, \ldots, s_n$, such that the circles with the centers at these points covers C, and we can introduce an order such that

$$\{z \mid |z - s_i| < \varepsilon\} \cap \{z \mid |z - s_{i+1}| < \varepsilon\} \neq \phi$$

and this implies obviously that the index is the same at z_1 and z_2.

11.1.4 COROLLARY If $T \in L(X)$, then for any $K \in K(X)$

1. $\text{ind } (T - z + K) = \text{ind } (T - z) \qquad z \in \sigma_{\mathcal{F}}(T)$
2. $\sigma_{\mathcal{F}}(T + K) = \sigma_{\mathcal{F}}(T)$

Now we are in a position to prove the following characterization of the Weyl spectrum.

11.1.5 THEOREM If $T \in L(X)$, then a point z is in $\omega(T)$ if and only if

$$z \in \bigcap \sigma(T + K)$$

where the intersection is taken over all compact operators in L(X).

Proof. First we remark that z is not in $\cap\, \sigma(T + K)$ if and only if

1. $z \in \sigma_F(T)$
2. $\text{ind}\,(T - z) = 0$

Indeed, if z is not in the set $\cap\, \sigma(T + K)$, then there exists a compact operator K such that $z \in \rho(T + K)$ and thus $z \in \sigma_F(T + K)$ and $\text{ind}\,(T + K - z) = 0$. Now, if we add T + K to the compact operator K we obtain that $z \in \sigma_F(T)$ and $\text{ind}\,(T - z) = 0$.

Conversely, if $z \in \sigma(T)$ and $\text{ind}\,(T - z) = 0$, we show that z is not in $\cap_K\, \sigma(T + K)$. We may assume, without loss of generality, that z = 0. Let $x_1, x_2, \ldots, x_n$ be a basis for N(T) and $y_1', y_2', \ldots, y_n'$ be a basis for the dual of $R(T)^-$, and we may suppose that the following relations hold:

1. $x_i'(x_j) = \delta_{ij} \qquad 1 \le i,j \le n$
2. $y_i'(y_j) = \delta_{ij} \qquad 1 \le i,j \le n$

For each $x \in X$ we define

$$F(x) = \sum_{i=1}^{n} x_i'(x) y_i$$

and clearly this is an operator of finite rank and we have the following properties:

1. $N(T) \cap N(F) = \{0\}$
2. $R(T) \cap R(F) = \{0\}$

Indeed, if $x \in N(T)$, then $x = \sum a_i x_i$ and thus $a_i = x_i'(x)$. If x is in N(F), then $x_i'(x) = 0$ and thus x = 0. Similarly, we can prove the assertion about ranges.

Since F is a compact operator $0 \in \sigma(T + F) = 0$. Now if $x \in N(T + F)$, it follows that Tx is in $R(T) \cap R(F)$. Thus Tx = 0, which implies that $\dim N(T + F) = 0$ and $R(T + F) = X$. Thus $0 \in \rho(T + F)$ and the assertion is proved.

From this assertion it follows that we have the equality of the sets $\omega(T)$ and $\cap_K\, \sigma(T + K)$.

We give now a result about the set $\sigma(T) - \omega(T)$, which in the case of Hilbert spaces was obtained by Baxley (1971) and extended to Banach spaces in Istratescu (1972).

First we give the following definition for the notion of isolated eigenvalues in $\sigma(T)$.

11.1.6 DEFINITION The point z_0 in the spectrum of a bounded operator $T \in L(X)$ is called an *isolated point* if there exists an $\varepsilon > 0$ such that $z \in \rho(T)$ for all $0 < |z - z_0| < \varepsilon$ and z_0 is called an *eigenvalue of finite multiplicity* if it is an isolated point in $\sigma(T)$ and there exists a finite projection (i.e., PX has a finite dimension) P such that

1. $PT = TP$.
2. $T/R(I - P) = T_1$ has the property that $\sigma(T_1) = \{z_0\}$.

The notations $\pi_{oo}(T)$, $\pi_{of}(T)$, etc. are defined similarly, and we remark that at the beginning of this section these notations have been considered only for the case of Hilbert spaces.

11.1.7 DEFINITION For an operator $T \in L(X)$ we say that Weyl's theorem holds if the following equality holds

$$\omega(T) = \sigma(T) - \pi_{oo}(T)$$

Following Baxley we consider the following conditions:

C_1: If $\{z_n\}$ is an infinite sequence of distinct points in π_{of} and $\{x_n\}$ is any sequence of corresponding normalized eigenvectors, then the sequence $\{x_n\}$ does not converge.

C_2: If $z \in \pi_{oo}(T)$, then $T - z$ has a closed range and the index is zero.

11.1.8 THEOREM If $T \in L(X)$ and conditions C_1 and C_2 are fulfilled, then the Weyl theorem holds for T.

The proof will be a consequence of some lemmas which also have their own interest.

11.1.9 LEMMA For any $T \in L(X)$, $\sigma(T) - \pi_{of}(T) \subseteq \omega(T)$.

Proof. Suppose that z is an eigenvalue of infinite multiplicity and K is an arbitrary compact operator. We show that z is in $\sigma(T + K)$. If $\{x_n\}$ are eigenvectors corresponding to eigenvalue z, then from the hypothesis and using Riesz's lemma, we find a sequence of elements, say $\{y_n\}$, in the space generated by $\{x_n\}$ such that

1. $\|y_n\| = 1$
2. $\|y_n - y_m\| \geq 1/2 \quad (n \neq m)$

Also we can suppose, without loss of generality, that $\lim Ky_n$ exists. In this case we have that

$$\lim z_n = \lim (T + K - z)y_n = \lim Ky_n$$

which is a contradiction if z is not in $\sigma(T + K)$.

11.1.10 LEMMA If $T \in L(X)$ satisfies C_1, then

$$\sigma(T) - \pi_{oo}(T) \subset \omega(T)$$

Proof. Since we have the identity

$$\sigma(T) - \pi_{oo}(T) = [\sigma(T) - \pi_{of}(T)] \cup [\pi_{of}(T) - \pi_{oo}(T)]$$

from the above lemma it will be sufficient to prove that if z is in $\pi_{of}(T) - \pi_{oo}(T)$ and not in $Cl(\sigma(T) - \pi_{of}(T))$, then z is in $\sigma(T + K)$ for every compact operator K. Since $z = \lim z_n$, with z_n in $\pi_{of}(T)$, we can consider a sequence of unit elements in X such that $Tx_n = z_n x_n$ for every n. Of course, we can suppose, without loss of generality, that $\lim Kx_n$ exists. In this case we obtain

$$\begin{aligned}\lim y_n &= \lim (T + K - z)x_n = \lim (z_n - z)x_n + \lim Kx_n \\ &= \lim Kx_n = y\end{aligned}$$

Now if we suppose that $(T + K - z)^{-1}$ is in $L(X)$, we obtain

$$\lim x_n = \lim (T + K - z)^{-1}y_n$$

exists and thus we have a contradiction with C_1.

11.1.11 LEMMA If $T \in L(X)$ satisfies C_2, then $\sigma(T) - \pi_{of}(T) \subset \omega(T)$.

Proof. Since condition C_2 holds, the operator $T - z$ has a closed range and index zero. We can decompose the space X as follows

$$X = R(T - z) \oplus W$$
$$X = N(T - z) \oplus W_1$$

and we define an operator K_1: $N(T - z) \to W$ as follows: if $\{e_i\}_{i=1}^m$ is a basis for $N(T - z)$ and $\{f_i\}_{i=1}^m$ is any basis for W (here we use the fact that the index is zero), then $Ke_i = f_i$, $i = 1, 2, \ldots, m$. We define an operator compact on X such that it coincides with K_1 on $N(T - z)$ and is zero on W. From the definition it follows that the operator $(T + K - z)$ is one-to-one and onto. Thus $(T + K - z)^{-1}$ exists and is in $L(X)$. Thus z is not in $\sigma(T + K)$ and the lemma follows.

Proof of Theorem 11.1.8. Obvious from these lemmas.

From Theorem 11.1.8 we obtain easily the following results.

We give now further properties of the Weyl spectrum in the case of Hilbert space operators. We remark that some of these results extend to Banach space operators.

First we remark that if N is a normal operator in $L(H)$, then

$$\omega(N) = \sigma(\hat{N})$$

Indeed, this follows from the definition of the Weyl spectrum and the relation

$$\omega(T) = \sigma(T) \cup \{z \mid T - z \in L(H) \text{ and } i(T - z) \neq 0\}$$

since the second part is empty.

11.1.12 THEOREM If $T \in L(H)$ is a normal operator, then

$$\omega(p(T)) = p(\omega(T))$$

for every complex-valued function on $\sigma(T)$.

Proof. This result follows from the above remark about Weyl's spectrum of a normal operator and the fact that the image of any normal operator in the Calkin algebra is also a normal operator.

Connected with this we can prove the following result.

11.1.13 THEOREM If $T \in L(X)$, X is a Banach space, then for any polynomial $p(\cdot)$ the relation

$$\omega(p(T)) \subseteq p(\omega(T))$$

holds.

Proof. Let p be a nonconstant polynomial and suppose that

$$p(z) - Z_0 = a(z - z_1)(z - z_2)\cdots(z - z_n)$$

which gives that

$$p(T) - Z_0 = a(T - z_1)(T - z_2)\cdots(T - z_n)$$

Clearly for each j, $p(z_j) = Z_0$ and if we suppose that $Z_0 \notin p(\omega(T))$, then we have that z_j are not in $\omega(T)$. Thus $(T - z_j) \in F_0$. From the form of $p(T) - Z_0$ it follows that it is in F_0. Thus the assertion of the theorem is proved.

We give now an example to show that the inclusion in the above theorem may be strict.

11.1.14 EXAMPLE Consider the Hilbert space ℓ^2 and U be the unilateral shift and on the Hilbert space $\ell^2 \oplus \ell^2$ consider the operator $T = U \oplus (U^* + 2I)$. In this case, for some polynomial $p(\cdot)$, $0 \in p(\omega(T))$ and not in $\omega(p(T))$.

Proof. We consider the polynomial $p(z) = z(z - 2)$. The unilateral shift is a Fredholm operator of index -1 and since U - 2 and U* + 2 are invertible operators, it follows that T and T - 2 are Fredholm operators of indices -1 and 1, respectively. Thus we have that $p(T) \in F_0$. This gives that $0 \notin \omega(p(T))$. Since the Weyl spectrum of the unilateral shift is the closed unit disk we have that $0 \in p(\omega(T))$.

Since the spectral mapping theorem holds for the spectrum defined in Banach algebras, the above example shows that the Weyl's spectrum is not derived in this way.

11.2 WEYL'S THEOREM

In what follows we give several conditions which assure that for an operator $T \in L(X)$ Weyl's theorem holds. As we have seen, such a condition was formulated in Theorem 11.1.8. Following Schechter (1970) we give now a general necessary and sufficient condition for the fact that for an operator T in L(X) Weyl's theorem holds.

11.2.1 THEOREM Weyl's theorem holds for $T \in L(X)$ if and only if

1. $z_0 \in \pi_{00}(T)$ implies that $R(T - z_0$ is closed.
2. If $z_0 \in \sigma_F(T)$ and $\mathrm{ind}(T - z_0) = 0$, then z_0 is not an interior point of $\sigma(T)$.

For the proof we need the following notion, which may be useful also in other settings.

11.2.2 DEFINITION A pair of closed subspaces X_1 and X_2 is said to *completely reduce* T if there is a bounded projection P on X such that $R(P) = X_1$ and $N(P) = X_2$ and X_i, $i = 1, 2$, are invariant subspaces of T. A closed subspace X_1 of X is said to be *reductive* for T if there exists X_2, also a closed subspace of X, such that the pair X_1, X_2 completely reduce T.

For the following results we need the next lemma.

11.2.3 LEMMA Let $T \in L(X)$ and z be an isolated point in $\sigma(T)$ and $\varepsilon > 0$ such that $s \in (T)$ is $0 < |s - z| < 2\varepsilon$. If P_z is the projection defined by the formula

$$P_z = -\frac{1}{2\pi i}\int_{|s-z|=\varepsilon} (T - s)^{-1}\, ds$$

then the following assertions hold:

1. P_z is a projection and $R(P_z)$ and $N(P_z)$ completely reduce T.
2. $x \in R(P_z)$ if and only if $\|(T - z)^n x\|^{1/n} \to 0$.
3. If $\dim N(T - z) < \infty$ and $R(T - z)$ is closed, then P_z is of finite rank.
4. If P_z is of finite rank, then $z \in \sigma\ (T)$ and $\mathrm{ind}\ (T - z) = 0$.

Proof. The first assertion is clear from the definition of P_z and since the spectrum of the restriction of T to P_zX is $\{z\}$, $\sigma(T/P_zX) = \{z\}$, the operator $T/P_zX - I_{P_zX}$ is quasi-nilpotent and this obviously implies the second assertion. Assertions 3 and 4 follow from the results given in the section on Fredholm operators.

For any operator $T \in L(X)$ let

$$r(T) = \lim\ (\dim N(T^n)) \qquad r^*(T) = \lim \operatorname{codim} R(T^n)$$

11.2.4 LEMMA If $T \in L(X)$ is a Fredholm operator and $r(T)$, $r^*(T)$ are finite, then there exists $\varepsilon > 0$ such that $z \in \rho(T)$ if $0 < |z| < \varepsilon$.

Proof. First we remark that from the definition of the numbers $\dim N(T^n)$ and $\operatorname{codim} R(T^n)$ it follows that the sequences

$$\{\dim N(T^n)\} \qquad \{\operatorname{codim} R(T^n)\}$$

are nondecreasing sequences of integers, and if $r(T - z) < \infty$, then z is an isolated eigenvalue of finite multiplicity.

Now by the result proved in Theorem 11.1.3, the Fredholm region of T contains a neighborhood of the origin, and since we have $\operatorname{ind}(T) = 0$, it follows that $\operatorname{ind}(T - z) = 0$ for z in the neighborhood. As a consequence of the above discussion about the numbers $\dim N(T^n)$ and $\operatorname{codim} R(T^n)$, we can find the integer n_1 such that

$$X = N(T^{n_1}) \oplus R(T^{n_1}) \qquad N(T^{n_1}) = N(T^k) \qquad R(T^{n_1}) = R(T^k)$$

for all $k \geq n_1$. It is also clear that $N(T^{n_1})$ and $R(T^{n_1})$ are invariant subspaces for T. For the proof of the lemma it will be sufficient to show that $(T - z)u = 0$ implies $u = 0$ for $u \in N(T^{n_1})$ and $(T - z)v = 0$ implies $v = 0$ for $v \in R(T^{n_1})$. We write now $n = n_1$. If $u \in N(T^n)$ and $(T - z)u = 0$, then we have $T^{n-1}u = z^{-1}T^n u = 0$, $T^{n-2}u = z^{-1}T^{n-1}u = 0, \ldots, u = z^{-1}Tu = 0$; similarly for v. We show now that T has a bounded inverse on $R(T^n)$; clearly this property holds for z small, and this implies the assertion of the lemma.

Indeed, if $w \in R(T^n)$ and $Tw = 0$, then $w = T^n y$ for some $y \in X$. Also $T^{n+1}y = 0$ and since $N(T^n) = N(T^k)$ for all $k \geq n$, it follows that $w = T^n y = 0$. Thus T is one-to-one on $R(T^n)$. We show now that this is also onto. Indeed, if $y = R(T^n)$ then $y = T^n x$ for some $x \in X$. From the above decomposition of X we can suppose that $x \in R(T^n)$, and thus $y = T^{2n}x$. If we define $v = T^{2n-1}x$, we have obviously that $Tv = y$. From these properties, the invertibility as well as the continuity of the inverse follows.

11.2.5 LEMMA If $z \in \sigma_F(T)$ and $\operatorname{ind}(T - z) = 0$, $r(T - z) < \infty$, then z is an isolated eigenvalue of finite multiplicity.

Proof. Since ind (T - z) = 0, it follows that

$$r(T - z) = r^*(T - z)$$

and Lemma 11.2.4 gives the assertion.

11.2.6 LEMMA If z is in the Fredholm region of T and is also an isolated eigenvalue of finite multiplicity, then $r(T - z) < \infty$.

Proof. This assertion follows from Lemma 11.2.3.

The following theorem gives the necessary and sufficient conditions for an operator $T \in L(X)$ to have the property that Weyl's theorem holds for T.

11.2.7 THEOREM If $T \in L(X)$, then Weyl's theorem holds for T if and only if

1. $z \in \pi_{oo}$ implies that R(T - z) is closed.
2. If $z \in \sigma_F(T)$ and ind (T - z) = 0, then z is not an interior point of $\sigma(T)$.

Proof. Suppose that Weyl's theorem holds for T. Let $z \in \pi_{oo}(T)$ and since in this case $z \in \sigma_F(T)$, R(T - z) is closed. Now if Z is in the Fredholm region of T and ind (T - z) = 0, then $z \in \pi_{oo}(T) \cup \rho(T)$ and thus it is not an interior point of $\sigma(T)$.

Suppose now that for $T \in L(X)$ the above two conditions are fulfilled. Let z not be in the Weyl spectrum of T. Then $z \in \sigma_F(T)$ and ind (T - z) = 0. We find a neighborhood of z such that ind (T - Z) is constant on this neighborhood as well as dim N(T - z). By condition 2 this neighborhood contains points of (T). Thus we have that this neighborhood contains only point from $\rho(T)$. This gives that z is in $\pi_{oo}(T)$. Now if $z \in \pi_{oo}(T)$, then condition 1 implies that R(T - z) is closed and dim $N(T - z) < \infty$. Since ind (T - z) is constant in a neighborhood of z, and this neighborhood contains points from $\rho(T)$; this implies that z is in the Fredholm region of T and ind (T - z) = 0. Thus z is not in Weyl's spectrum of T and the theorem is proved.

In applications it is useful to have also equivalent statements for conditions 1 and 2 of Theorem 11.2.7. The following theorems are concerned with such problems.

11.2.8 THEOREM Condition 1 in Theorem 11.2.7 is equivalent to the following assertion: If $z \in \pi_{oo}(T)$, then $T - z$ is not quasi-nilpotent on any infinite dimensional invariant subspace.

Proof. Let $z \in \pi_{oo}(T)$ and if 1 holds, then the projection P_z defined as in Lemma 11.2.3 is of finite rank. This implies that $T - z$ cannot be quasi-nilpotent on any infinite dimensional subspace. Conversely, if the above condition holds, then the projection P_z is of finite rank and thus $R(T - z)$ is closed.

Thus Theorem 11.2.8 is proved.

We give now a theorem about the second condition of Theorem 12.2.7.

11.2.9 THEOREM Condition 2 in Theorem 11.2.7 is equivalent to the following assertion: If $z \in \pi_{oo}(T)$ and ind $(T - z) = 0$, then $r(T - z) < \infty$.

Proof. If $z \in \pi_{oo}(T)$, ind $(T - z) = 0$, and z is not an interior point of $\sigma(T)$, then every neighborhood of z contains points of $\sigma(T)$.

As we know the functions

$$z \mapsto \text{ind } (T - z)$$

and

$$z \mapsto \dim N(T - z)$$

are constant in some neighborhood of z. This gives that $z \in \pi_{oo}(T)$.

From the above lemma we obtain that $r(T - z) < \infty$.

Conversely, if $r(T - z) < \infty$, then from the above results we have that $z \in \pi_{oo}(T)$ and thus is not an interior point of $\sigma(T)$.

The following theorem is useful when we consider the restriction of T to invariant subspaces.

11.2.10 THEOREM Suppose that $T \in L(X)$ satisfies condition 1 of Theorem 11.2.7. In this case if $z \in \pi_{oo}(T)$ and X_1 is an invariant subspace of T, then z is an eigenvalue of the restriction of T to X_1 (denoted T_1).

Proof. We define the projection

$$P_{1,z} = \frac{1}{2\pi i} \int_{|s-z|=\varepsilon} (s - T)^{-1}/X_1 \, ds$$

where $\varepsilon > 0$ is such that all z with the property $0 < |z| < 2\varepsilon$ are in the resolvent of T_1. Since $(T - z)/X_1$ cannot be quasi-nilpotent on an infinite dimensional invariant subspace, we have that $P_{1,z}$ is of finite rank. This shows that $z \in \sigma(T_1)$ and ind $(T - z)/X_1 = 0$ implies that z is an eigenvalue for T_1.

The following theorem gives another sufficient condition for an operator $T \in L(X)$ to have the property that Weyl's theorem holds for T.

11.2.11 THEOREM If $T \in L(X)$ and if for any complex number z, dim $N(T - z) < \infty$ implies that $T - z$ is not quasi-nilpotent on any infinite dimensional invariant subspace, then Weyl's theorem holds for T.

Proof. The condition of the theorem implies the condition of Theorem 11.2.8 and also the condition of Theorem 11.2.9 since $T - z$ is quasi-nilpotent on

$$\bigcup_{n=1}^{\infty} N(T - z)^n$$

and these conditions imply that for T, Weyl's theorem holds.

For the next result we need the following.

11.2.12 LEMMA If $T \in L(X)$ and N(T) is reductive, then $N(T^2) = N(T)$.

Proof. By definition there exists an invariant subspace X_1 such that $X = N(T) \oplus X_1$. If $x \in N(T^2)$, then $x = x_0 + x_1$, where $x_0 \in N(T)$ and $x_1 \in X_1$. Thus we have $Tx = Tx_1 \in X_1$. Since $Tx \in N(T)$, we have that $Tx = 0$ and thus $x \in N(T)$.

From this lemma we obtain the further result.

11.2.13 THEOREM Let $T \in L(X)$ and suppose that T has the following properties:

1. $z \in \pi_{oo}(T)$ implies that $N(T - z)$ is reductive for T.
2. If X_1 is a reductive subspace for T and $z \in \pi_{oo}(T)$ is an isolated point of $\sigma(T/X_1)$, then z is an eigenvalue of T/X_1; thus T has property 1 of Theorem 11.2.7.

Proof. Let $z \in \pi_{oo}(T)$ and X_1 be an invariant subspace such that

$$X = N(T - z) \oplus X_1$$

and by definition X_1 is reductive for T. If z is not in $\sigma(T/X_1)$, then it is an isolated point in $\sigma(T/X_1)$. In this case z is an eigenvalue for T/X_1, which contradicts the fact that $T - z$ is one-to-one on X_1. Thus $R(T - z) = X_1$ and condition 1 holds.

Another form of condition 1 of Theorem 11.2.8 is in our next result.

11.2.14 THEOREM If $T \in L(X)$, then condition 1 of Theorem 11.2.8 is equivalent to the following property: $z \in \pi_{oo}$ implies that $T - z$ is not quasi-nilpotent on any infinite dimensional reductive subspace.

Proof. Clearly the condition of Theorem 11.2.8 implies the condition of this theorem. Conversely, if $z \in \pi_{oo}(T)$, we can define the projection P_z and the subspace $R(P_z)$ is reductive for T; this gives that P_z is of finite rank. Thus $T - z$ is not quasi-nilpotent on any infinite dimensional invariant subspace, and the result follows from Theorem 11.2.9.

11.3 WEYL'S THEOREM FOR SOME CLASSES OF OPERATORS

As an application of the above results, we obtain that Weyl's theorem holds for certain classes of operators including the class of normal operators.

11.3.1 THEOREM Weyl's theorem holds for hyponormalizable operators.

For the reader's convenience we recall that an operator $T \in L(H)$, H a Hilbert space, is called *hyponormal* if $T^*T - TT^* > 0$. Clearly, if T is hyponormal, then for any z, T - z is also hyponormal.

If X is a Banach space and < , > be a semi-inner product on X which satisfies the relation

$$\|x\|^2 < k\langle x, x\rangle$$

for all $x \in X$ and some fixed constant $k > 0$, then an operator $T \in L(X)$ is called *hyponormalizable* if there exists $T^* \in L(X)$ such that

$$\langle (T^*T - TT^*)x, x\rangle > 0$$

for all $x \in H$.

The proof of Theorem 11.3.1 is an application of Theorem 11.1.8. First we remark that if z is an eigenvalue for T, then z* is an eigenvalue for T* and this gives immediately that T satisfies the condition C_1.

Now, if $z \in \pi_{oo}(T)$, we know that the null space of T - z on X is the same as the null space on the completion with respect to the inner product. Since the image of X under T - z is precisely the set of elements in X which are orthogonal to N(T - z), we have that

$$\dim N(T - z) = \dim \frac{X}{R(T - z)}$$

and thus ind (T - z) = 0. Also R(T - z) is closed. From these it is clear that C_2 is satisfied.

11.3.2 COROLLARY If $T \in L(H)$ is a hyponormal operator, then Weyl's theorem holds for T.

11.3.3 THEOREM Weyl's theorem holds for normal operators on Banach spaces.

Proof. The result is also an application of conditions C_1 and C_2. See Baxler (1971).

We show now that Weyl's theorem holds for a class of operators defined using some growth conditions.

First we recall that an operator satisfies the growth condition G_1, or that T has G_1 property, if the following inequality holds:

$$\|(T - z)^{-1}\| = \frac{1}{d(z, \sigma(T))}$$

for all $z \in (T)$.

A point in the complex plane is a bare point for the closed set F if it lies on the circumference of some closed disk that contains the set F. A point in the complex plane is called a *semibare point* of a closed set F if it lies on the circumference of some closed disk that contains no other points of F.

Following Berberian (1969), who has systematized different conditions, we set

G_1': Every isolated point in $\sigma(T)$ is an eigenvalue for T.

α: The restriction of T to any invariant subspace satisfies G_1.

α': The restriction of T to any reducing subspace satisfies G_1.

α'': The restriction of T to any invariant subspace is convexoid.

α''': The restriction of T to any reducing subspace satisfies G_1'.

The condition says that every isolated point in the spectrum of a direct summand of T is necessarily an eigenvalue of the summand.

We consider now conditions related to the position of the points of the spectrum of T.

β: Every point of $\sigma(T)$ is a bare point of $\sigma(T)$.

β': Every eigenvalue of finite multiplicity is a semibare point of $\sigma(T)$.

To give the relations between these conditions, we need some results which also have their own interest.

11.3.4 LEMMA For any $T \in L(H)$, the G_1 property implies G_1'.

Proof. Let z_0 be an isolated point in $\sigma(T)$ and let $\varepsilon > 0$ such that the set $\{z \mid |z - z_0| \le \varepsilon\}$ intersects $\sigma(T)$ only in z_0. If we define

$$P_{z_0} = \frac{1}{2\pi i}\int_{|z-z_0|=\varepsilon} (T - z)^{-1}\, dz$$

then clearly this is a projection. The fact that T has G_1 property implies that P_{z_0} is hermitian. Indeed,

$$\|P_{z_0}\| \le \frac{1}{2\pi}\,\varepsilon\,\frac{1}{\varepsilon}\,2\pi = 1$$

and thus it is hermitian. Now if $x \in P_{z_0}H$, we have

$$\|(T - z_0)x\| = \frac{1}{2\pi}\left\|\int_{|z-z_0|=\varepsilon} (z - z_0)(T - z_0)x\right\| \le \varepsilon$$

and for $\varepsilon \to 0$ we obtain that z_0 is an eigenvalue.

From this we have the following corollary.

11.3.5 COROLLARY The condition α implies α'''.

Also since every operator with G_1 property is convexoid, we have another immediate corollary.

11.3.6 COROLLARY The condition α implies α''.

11.3.7 LEMMA For any $T \in L(H)$, α'' implies α'''.

Proof. Let $\{z \mid |z - z_0| \le \varepsilon\}$ be as above and define also the projection P_{z_0}. We have that $\sigma(T/P_{z_0}H) = \{z_0\}$, and since T/P_{z_0} is a convexoid operator, it follows that $T/P_{z_0} = z_0 I$.

Summarizing all the above implications as well as the obvious ones about β and β', we can state the following.

11.3.8 LEMMA The following implications hold: α implies α' implies α''', α implies α'' implies α''', and β implies β'.

The connection between these conditions and Weyl's theorem is given in our next result.

11.3.9 THEOREM Suppose $T \in L(H)$ and has the following properties:

1. α''' holds.
2. T is reduced by each of its finite dimensional eigenspaces.

Then Weyl's theorem holds for T.

First we remark that the conditions of the theorem are translation invariant, and thus we can consider the operator $T - z$ with z

arbitrary. Since Weyl's theorem holds for an operator if and only if

$$\sigma(T) - \omega(T) = \pi_{oo}(T)$$

to prove the theorem it suffices to prove the following assertion: $0 \in \pi_{oo}(T)$ implies that T is a Fredholm operator of index zero.

Also we remark that each side of the assertion implies that 0 is an eigenvalue of T. From the hypothesis of the theorem N(T) is a reducing subspace of T

$$N(T) \subset N(T^*) = R(T)^{\perp}$$

This implies that the operator T can be written in the form

$$T = 0 \oplus T_1$$

where 0 denotes the zero operator on N(T) and T_1 is the restriction of T to $N(T)^{\perp}$. Clearly T_1 is injective and

$$\sigma(T) = \{0\} \cup \sigma(T_1)$$

Now we are in a position to give the proof of Theorem 11.3.9.

Proof. Suppose that $0 \in \pi_{oo}(T)$. We show that T is a Fredholm operator of index zero.

First we prove that T_1 is invertible. Indeed, in the contrary case, 0 is an isolated point in $\sigma(T) = \sigma(T_1)$ and by α''' it is an eigenvalue of T_1. This contradiction proves the assertion.

Thus T_1 is surjective and $T(N(T)^{\perp}) = N(T)^{\perp}$ which implies that $N(T)^{\perp} \subset R(T)$. The relation $N(T) \subset N(T^*) = R(T)^{\perp}$ implies that $R(T) \subset R(T)^{\perp\perp} \subset N(T)^{\perp}$. This gives that $R(T) = N(T)^{\perp}$. Now it is clear that this implies that T is a Fredholm operator of index zero.

For the proof in the opposite direction, suppose that T is a Fredholm operator of index zero. As we know, 0 is an eigenvalue of T of finite multiplicity and we show now that an isolated point in $\sigma(T)$. From the above decomposition of T it will be sufficient to show that $0 \in \sigma(T_1)$. As we know, T_1 is an injective operator and we show that it is also onto. This is equivalent to $N(T)^{\perp} = T(N(T)^{\perp})$.

Since T is a Fredholm operator, R(T) is a closed subspace, and also,

$$\dim N(T) = \dim R(T)^{\perp} < \infty$$

and this gives that $N(T) = R(T)^{\perp}$. If we prove that $R(T) = T(R(T))$, the theorem is completely proved.

Clearly, R(T) contains T(R(T)). Now if $y \in R(T)$, y is of the form $y = Tx$ for some $x \in H$. But x can be written in the form $x = Tu + z$ with $z \in R(T)^{\perp} = N(T)$. In this case we have

$$y = Tx = T(Tu + z) = T^2u$$

and the assertion is proved.

11.3.10 REMARK From this theorem we can obtain that Weyl's theorem holds for hyponormal operators. Indeed, as we know, every hyponormal operator satisfies α (and thus α'''), and from Lemma 11.3.4 and the definition of hyponormal operators we conclude that any hyponormal operator is reduced by each of its eigenspaces.

From Theorem 11.3.9 we obtain the following.

13.3.11 COROLLARY If $T \in L(H)$ has the properties α' and β', then Weyl's theorem holds for T.

For the proof we need the following.

11.3.12 LEMMA If $T \in L(H)$ has the G_1 property and z_0 is a semi-bare point of σ(T), then

$$N(T - z_0) = N(T^* - z_0^*)$$

This lemma says that z_0 is in fact a normal eigenvalue.

Proof. Since T - z satisfies also the G_1 property, we may suppose, without loss of generality, that $z_0 = 0$.

Let s_0 be such that the disk $\{z \mid |z - s_0| \le |s_0|\}$ meets σ(T) only at zero. Since the distance of s_0 to σ(T) is exactly $|s_0|$ we have that

$$\|(T - s_0)^{-1}\| \le \frac{1}{|s_0|}$$

and if we set

$$S = -s_0(T - s_0)^{-1}$$

S is a contraction operator. If $Tx = 0$, then we have $(T - s_0)^{-1}x = -s_0^{-1}x$ and $Sx = x$. Since S is a contraction,

$$\{x \mid Sx = x\} = \{y \mid S^*y = y\}$$

and thus $T^*x = 0$.

The proof of Corollary 11.3.11 is now as follows: Clearly the condition α''' holds and by the lemma each eigenspace is reducing.

We mention here an application of Weyl's spectrum to the structure of operators. If $T \in L(H)$ is a polynomially compact operator which is also normal, the following theorem characterizes this class of operators.

11.3.13 THEOREM If $T \in L(H)$ is a normal operator, then the following assertions are equivalent:

1. T is polynomially compact.
2. There exists a complex-valued function defined and continuous on $\sigma(T)$ such that $f(T)$ is compact and f has at most finitely many zeros on $\omega(T)$.
3. $\omega(T)$ is finite.

Proof. Clearly 1 implies 2. If 2 holds, then we have that

$$f(\omega(T)) = \omega(f(T)) = \{0\}$$

Now since f has finitely many zeros on $\omega(T)$, we have that 2 implies 3. Conversely, if 3 holds, let p be any polynomial such that p is zero on $\omega(T)$. Since $p(T)$ is a normal operator, it follows that $p(T)$ is compact. Thus 3 implies 1.

11.3.14 REMARK From the previous results it follows that Theorem 11.3.13 is true for the case of normal operators on Banach spaces.

The following example shows that there exists an operator $T \in L(H)$ with only reducing eigenspaces for which Weyl's theorem fails and α''' is not satisfied.

Indeed, let T be an operator defined as follows: $T = T_1 \oplus T_2$, where T_1 is the one-dimensional zero operator and T_2 is any injective quasi-nilpotent operator. Clearly T is a compact operator and $\omega(T) = \{0\}$. Thus $\sigma(T) - \omega(T) = \emptyset \neq \pi_{oo}(T)$.

We mention that Weyl's theorem holds for the following class of operators: $T \in L(H)$ and T^*T commutes with $T + T^*$. We conjecture that Weyl's theorem holds for the following class of operators containing the above class: $T \in L(H)$ and $4T^*T - (T + T^*)^2 \geq 0$.

For some results about these classes, see Istratescu (1979), Fong and Istratescu (1979).

11.4 THE WEYL SPECTRUM OF AN ELEMENT IN A VON NEUMANN ALGEBRA

In what follows we present the notion of the Weyl spectrum for an element in a von Neumann algebra. For more details see Dixmier (1964).

Let H be a Hilbert space and L(H) be the set of all bounded operators on H. For any subset S of L(H), S' denotes the set of all elements in L(H) commuting with all elements in S.

11.4.1 DEFINITION A subalgebra A of L(H) is called a *von Neumann algebra* if and only if $A'' = (A')' = A$.

If A is a von Neumann algebra, then the set of all hermitian projections is denoted by P(A). P(A) is a lattice under the order defined as follows:

$$P \underset{\sim}{<} Q \text{ if and only if } PQ = P$$

On P(A) we can define another order relation as follows: $P \leq Q$ if and only if there exists $R \in P(A)$ such that $P \sim R \underset{\sim}{<} Q$ and where "$\sim$" means that there exists $U \in A$ such that $P = U^*U$ and $R = UU^*$.

It is not difficult to prove that "$\sim$" is an equivalence relation.

11.4.2 DEFINITION An element $E \in P(A)$ is called *finite relative to* A if for every $F \in P(A)$ the relations $E \sim F$ and $E \underset{\sim}{>} F$ imply $E \geq F$. A von Neumann algebra A is *finite* if its unit is a finite element.

Let A be a von Neumann algebra and $T \in A$. The *null projection* of T is defined as

$$N_T = \sup \{E \in P(A) \mid TE = 0\}$$

and the range projection of T as

$$R_T = \inf \{E \in P(A) \mid ET = T\}$$

From these definitions it is clear that

$$TN_T = 0 \qquad R_T T = T$$

and that

$$I - R_{T*} = N_T \qquad N_{T*} = I - R_T$$

11.4.3 DEFINITION An element T is called *finite* (or of finite rank) relative to A if R_T is finite.

For any von Neumann algebra A let m_0 be the set of all finite elements and m be the norm closure of m_0.

11.4.4 DEFINITION An element $T \in A$ is called *compact* if $T \in m$.

The notion of a Fredholm element is introduced as follows.

11.4.5 DEFINITION An element $T \in A$ is called a *Fredholm element* if

1. N_T is finite.
2. There is a finite projection E of A such that $(I - E)H = TH$.

If we consider an involutive algebra and σ is a representation of the algebra A in L(H) for some Hilbert space H, then we say that H is the representation space and we write this as $H_\sigma = H$.

If we have two representations σ_1 and σ_2, respectively, then we say that these representations are equivalent if there exists an isomorphism J such that

$$J\sigma_1(x) = \sigma_2(x)J$$

for all $x \in A^\sim$. We write this shortly as $\sigma_1 \sim \sigma_2$. It is easy to see that this is an equivalence relation in the set of all representations of $A^\sim$. The equivalence class for the representation σ

is denoted by $[\sigma]$. The set of all equivalence classes forms a commutative monoid when the "+" is defined as follows: If we have two representations, we can define the sum as the direct sum of the representations. Next we set

$$[\sigma_1] + [\sigma_2] = [\sigma_1 \oplus \sigma_2]$$

and it is easy to see that this is commutative.

Now suppose M(A) is the submonoid generated by the set of all $[\pi_E]$ with $E \in P(A)$, where

$$\pi_E(T) = TE$$

on the subspace H_E, the range of E, and also we suppose that E is finite relative to A.

We can consider the Grothendieck group of the commutative monoid M(A') and thus we can find a homomorphism

$$\beta\colon M(A') \to K(A', A)$$

having the universal property with respect to the homomorphism of M(A') into commutative groups.

11.4.6 DEFINITION For any finite projection $E \in P(A)$ we define

$$\dim E = \beta(\pi_E)$$

11.4.7 REMARK If the von Neumann algebra A is a factor of type I (a *factor* is a von Neumann algebra whose center contains only the scalars; a factor is of *type I* if it is isomorphic to a von Neumann algebra with the commutant abelian, and it is called a factor of *type II* if there exists a nonzero projection of the center such that A_E = {the restriction of ET to H} is discrete) (respectively a factor of type II) then the Grothendieck group is isomorphic to the integers (respectively, the real numbers).

This gives us the possibility to define the index of a Fredholm element in a von Neumann algebra.

11.4.8 DEFINITION For any Fredholm element T of a von Neumann algebra A the *index of T*, denoted ind T, is defined as follows:

$$\text{ind } T = \dim N_T - \dim N_{T^*}$$

For any von Neumann algebra A, F_0 denotes the set of all Fredholm elements of index zero.

Now we are in a position to define the Weyl spectrum of an element in a von Neumann algebra A.

11.4.9 DEFINITION For any T ∈ A the *Weyl spectrum* is the set

$$\omega(T) = \{z \mid T - zI \notin F_0\}$$

As in the case of A = L(H), we can consider the Calkin algebra of a von Neumann algebra and this is the following algebra:

$$C_A = \frac{A}{m}$$

Using the result that the Atkinson characterization of Fredholm operators extends to the case of von Neumann algebras, we have the following result.

11.4.10 LEMMA For any element T ∈ A,

$$\sigma(\hat{T}) \subset \omega(T)$$

where $\hat{T}$ denotes the image of T in C_A.

We give now another characterization of the Weyl spectrum in a von Neumann algebra.

11.4.11 THEOREM For any T ∈ A the following relation holds:

$$\omega(T) = \bigcap \sigma(T + K)$$

where K varies on the set m.

The proof of this theorem is a consequence of some lemmas.

11.4.12 LEMMA If K ∈ m, then I + K is a Fredholm element.

Proof. See Breuer (1969), Theorem 1.

11.4.13 LEMMA If S ∈ A is invertible and K ∈ m, then S + K is a Fredholm element of index zero.

Proof. Since we have the identity

$$S + K = S(I + S^{-1}K)$$

the assertion follows from Theorem 2.2 in Breuer (1969).

11.4.14 LEMMA If T is a Fredholm element of index zero, then there exists an element of finite rank such that T + K is invertible.

Proof. We choose a cofinite projection E of A such that EHTH and the range of ET is closed. If F is the range of the projection T*E and since ET is also a Fredholm element of index zero, there is a partial isometry $U \in A$ such that

$$I - E = UU^* \qquad I - F = U^*U$$

Since the range of ET is closed, ET + U is invertible. Now the element R = U + (E - I)T is of finite rank and satisfies the relation T + R = ET + U, and the lemma is proved.

Proof of Theorem 11.4.11. If K is any compact element of A and T + K - zI is invertible, then

$$T - zI = [T + K - zI] - K$$

is a Fredholm element of index zero. This gives

$$\omega(T) \subset \sigma(T + K)$$

Now, if z is not in $\omega(T)$, we can find an element of finite rank K such that T + K - zI is invertible. This gives that z is not in $\sigma(T + K)$.

As in the case when A = L(H), we obtain from this characterization some information about the Weyl spectrum of an element. We shall give some of these.

11.4.15 COROLLARY For any element $T \in A$, $\omega(T)$ is a nonempty compact set in the complex plane.

11.4.16 COROLLARY For any element $T \in K$ and for any $K \in m$,

$$\omega(T + K) = \omega(T)$$

11.4.17 COROLLARY For any $T \in A$ the following assertions are equivalent:

1. T is a Fredholm element.
2. $T = S + F$, where S is invertible and F is of finite rank.
3. $T = S^{\sim} + K^{\sim}$, where $S^{\sim}$ is invertible and $K^{\sim}$ is compact.
4. For any $T \in A$, $\omega(T) = \omega(T^*)$.

The following theorem gives a sort of "spectral mapping theorem" for the Weyl spectrum of an element in a von Neumann algebra.

11.4.18 THEOREM If $T \in A$, then for any polynomial $p(z)$ we have

$$\omega(p(T)) \subseteq p(\omega(T))$$

Proof. Since the proof using Corollary 11.4.17 does not differ essentially from the case $A = L(H)$, we omit it.

If $T \in A$ and z is a complex number, then we say that $z \in \sigma(T)$ is an eigenvalue for T if N_{T-zI} is nonzero. We say that the eigenprojection N_{T-z} is reducing if

$$N_{T-zI}T^*N_{T-zI} = T^*N_{T-zI}$$

We define now the case when Weyl's theorem holds for an element T in a von Neumann algebra A. Of course we define in a similar way the sets $\pi_{oo}(T)$, $\pi_{of}(T)$, etc.

11.4.19 DEFINITION If $T \in A$ we say that for T, Weyl's theorem holds if

$$\sigma(T) - \pi_{oo}(T) = \omega(T)$$

Using the cofinite projections, we can prove the following result.

11.4.20 THEOREM For any $T \in A$, $\sigma(T) - \omega(T)$ is empty or consists of eigenprojections of finite type.

As in the case of $A = L(H)$, we can define several classes of elements and the problem is whether Weyl's theorem holds for elements in these classes.

Also, there is the problem of characterization of polynomially compact elements in von Neumann algebras and more particularly for the case of polynomially compact elements which are also normal (i.e., $T^*T = TT^*$).

11.5 THE VON NEUMANN THEOREM

Suppose that A and B are bounded hermitian operators on a Hilbert space such that A - B is compact. In this case the image of A - B in the Calkin algebra of H is the zero operator. This gives that the Weyl spectrum of A and the Weyl spectrum of B are the same.

Von Neumann (1935) proved the following remarkable theorem in the case of separable Hilbert spaces.

11.5.1 THEOREM If A and B are two hermitian operators such that $\omega(A) = \omega(B)$, then there exists a unitary operator U such that $UAU^* - B$ is a compact operator.

Following Berberian (1970) we call two operators T and S in L(H) essentially similar if there exists a unitary operator U such that $UTU^* - S$ is a compact operator. It is obvious that if T and S in L(H) are essentially similar, then they are similar in the Calkin algebra of H. The following example shows that there exists unitary operators in the Calkin algebra having no unitary lifting to L(H), or even normal lifting.

11.5.2 EXAMPLE Let U be the unilateral shift and U be its image in the Calkin algebra $\mathcal{C}$ of H. In this case U is unitary in $\mathcal{C}$, but does not have unitary lifting nor a normal lifting. Indeed, in the contrary case, for K compact U + K is a Fredholm operator and thus

$$i(U + K) = i(U) = -1$$

Thus U + K cannot be normal.

11.5.3 REMARK If T is hermitian in $\mathcal{C}$, then it has a hermitian lifting to L(H).

For the proof we remark that if $T^{\sim} \in L(H)$ is such that $\hat{T}^{\sim} = T$, then $T_1^{\sim} = (T^{\sim} + \tilde{T}^*)/2$ is the required lifting.

11.5.4 DEFINITION Two elements in the Calkin algebra $\mathcal{C}$ of H are called *strongly equivalent* if they have unitary equivalent liftings to L(H).

11.5.5 THEOREM If a and b are two hermitian elements in $\mathcal{C}$, then the following assertions are equivalent:

1. a and b are strongly equivalent.
2. a and b are similar.
3. $\sigma(a) = \sigma(b)$.

Proof. It is obvious that 1 implies 2 implies 3. Suppose now that 3 holds. Let A and B in L(H) such that $\hat{A} = a$ and $\hat{B} = b$. From the results about the Weyl spectrum we have that $\omega(A) = \omega(B)$, and thus by the von Neumann theorem enunciated above, A and B are essentially equivalent.

11.5.6 REMARK The assertions of the above theorem do not remain true if we consider unitary operators.

As an example we can consider the unilateral shift and the bilateral shift. Then their image in the Calkin algebra is of unitary operators having the spectrum equal to the closed unit disk. A computation of the index in the case when the assertion of the remark is not true gives immediately a contradiction.

Now we give the converse of Weyl's theorem for the case of normal operators. For this we need the following lemma whose proof is given in Berg (1971) or Sikonia (1971).

11.5.7 LEMMA If $A \in L(H)$ is a normal operator on a separable Hilbert space H, then for any $\varepsilon > 0$ there exists a compact operator K, $\|K\| < \varepsilon$, such that A + K has only pure point spectrum and is also a normal operator.

11.5.8 REMARK In the case of hermitian operators K may be chosen to be of the Hilbert-Schmidt type.

The following lemma was given by von Neumann (1935).

11.5.9 LEMMA If A and B are normal operators in L(H), A and B with $\omega(A) = \omega(B)$, then there exists a unitary operator U such that $UAU^* - B$ is a compact operator.

Proof. Since A and B are normal operators, Weyl's theorem holds for A and B. From the fact that A and B have pure point spectrum for each $z \in \omega(A)$, we choose a unit vector x_z such that

$$Ax_z = zx_z$$

and similarly for B. The corresponding vectors for B are denoted by y_z. We define an operator

$$U_1 x_z = y_z$$

and extend this to a unitary operator on the orthogonal complement of the spaces generated by $\{x_z\}$ and $\{y_z\}$, respectively. We remark that this is possible because of the separability and $\omega(A) = \omega(B)$.

From the construction it follows that the unitary operator satisfies the lemma.

In the proof of this lemma we have used the following very interesting result of von Neumann about sequences having the same cluster set. We give the proof of Yorke (1969).

11.5.10 LEMMA If $\{a_n\}$ and $\{b_n\}$ are two sequences in a compact metric space with the metric d and having the same cluster set C, then there exists a permutation π of the set of positive integers Z such that $d(a_n, b_{\pi n}) \to 0$.

Proof. We define the permutation π by induction. First we set $\pi 1 = 1$. Now if $\pi 1, \pi 2, \ldots, \pi(n - 1)$ are given, if

$$Z_1 = Z \qquad Z_n = Z - \{\pi 1, \pi 2, \ldots, \pi(n - 1)\}$$

we define $\sigma(n) = \min Z_n$. In this case n is the smallest integer in Z_n such that

$$d(b_n, a_n) \leq d(a_n, C) + \frac{1}{\sigma(n)} + d(\sigma(n), C) \tag{*}$$

We show now that such an element exists. For this let c_n in the compact space such that $d(a_n, c_n) = d(a_n, C)$, and thus for some $m \in Z_n$, $d(b_m, c_n) < 1/\sigma(n)$ since c_n is a cluster point of b. We set

$\pi n = m$. To prove the assertion of the lemma, it is necessary to show that π is onto. Suppose that this is not so. Let q in $Z - \{\pi n\}$.

For all n, $\sigma(n) < q$, and N is defined as $\max \{n \mid q \neq \sigma(n)\}$ and is a finite number. Consider now $c^q \in C$ such that $d(b_q, c^q) = d(b_q, C)$ and since c^q is a cluster point of a, we can choose $n > N$ such that $1/q \geq d(a_n, c^q)$. Since

$$d(a_n, C) \geq \frac{1}{q} > d(a_n, c^q) \qquad n \geq N, \quad n \geq q$$

we obtain that

$$d(b_q, a_n) > d(a_n, C) + \frac{1}{q} + d(b_q, c^q) > d(a_n, c^q)$$

which contradicts the triangle inequality.

To complete the proof of the lemma it remains to show that

$$d(a_n, b_{\pi n}) \to 0$$

and we remark that since $\pi n \to \infty$ and $d(a_n, C) \to 0$, $d(b_n, C) \to 0$ from (*), the assertion follows.

Now we are in a position to prove the extension of the von Neumann theorem for the case of normal operators on separable Hilbert spaces.

11.5.11 THEOREM If A and B are two normal operators on a separable Hilbert space H and $\omega(A) = \omega(B)$, then there exists a unitary operator U such that $UAU^* - B$ is a compact operator.

Proof. From Lemma 11.5.7 there exist compact operators K_1 and K_2 such that $A + K_1$ and $B + K_2$ are normal operators with pure point spectrum. It is clear that

$$\omega(A + K_1) = \omega(B + K_2)$$

and thus there exists a unitary operator U such that

$$U(A + K_1)U^* - B - K_2 = K_3$$

where K_3 is a compact operator. From this we obtain obviously the assertion of the theorem.

11.5.12 REMARK The following example shows that Theorem 11.5.11 cannot be extended to quasi-normal operators. Let us consider the Hilbert space with the basis $\{e_i\}$ and define

1. $Se_n = e_{n+1}$ (the unilateral shift)
2. $Te_n = e_{n+2}$

and it is easy to see that

$\omega(T) = \omega(S) =$ the closed unit disk

The operators S and T are quasi-normal since $S^*S = T^*T = I$.

Suppose now that there exists a unitary operator U such that $UTU^* - S$ is a compact operator. In this case by the computation of the index we have

$$\text{ind } (UTU^*) = \text{ind } (S + K) = \text{ind } S$$

since K is a compact operator. This is a contradiction since $\text{ind } T = -2$ and $\text{ind } S = -1$.

We close by noting some problems which appear to be of some interest.

11.5.13 PROBLEM Is there an analogue of the von Neumann theorem for the case of hermitian operators on separable Banach spaces?

11.5.14 PROBLEM For any hermitian element in L(X), X is a separable Banach space, does there exist a hermitian operator which is of the Hilbert-Schmidt type [in the sense of Pietsch (1972)] such that $\|K\| < \varepsilon$ and T + K has only a pure point spectrum?

11.5.15 PROBLEM Does the von Neumann theorem hold for normal operators on separable Banach spaces?

11.5.16 PROBLEM For what class of operators $T \in L(H)$, where H is a Hilbert space, does the following spectral mapping theorem hold: $\text{Re } \omega(T) = \omega(\text{Re } T)$?

Chapter 12

ANALYTIC AND QUASI-ANALYTIC VECTORS

12.0 INTRODUCTION

One of the most interesting classes of functions in analysis is the class of analytic functions. Thus, interest in the classes of functions having some properties in common with the analytic functions appears very naturally. An interesting property of the analytic functions is the following one: If f is zero with all of its derivatives at a point, say x_0, then $f \equiv 0$. It was a famous problem of Hadamard to find conditions on the sequence $\{M_n\}$ such that a C^∞-function satisfying the estimates

$$|f^{(n)}x| \leq k^n M_n \qquad n = 1, 2, 3, \ldots$$

having the property that f and all of its derivatives are zero at some point, then necessarily $f \equiv 0$.

The class of functions having this property is called the class of *quasi-analytic* functions. Fundamental results about this class of functions were obtained by Denjoy, Carleman, Ostrovski, Bang, etc. For a detailed study of quasi-analytic functions we refer to the excellent book of Mandelbrojt (1952).

For the reader's convenience as well as for motivation of what follows, we quote some results from the theory of quasi-analytic functions.

Let $\{M_n\}$ be a sequence of positive numbers.

12.0.1 DEFINITION We denote by $C\{M_n\}$ the set of all C^∞ functions defined on R having the property that

$$\sup |f^{(n)}(x)| = \|f^{(n)}\| \leq k_f K_f^n M_n$$

for all n = 0, 1, 2, 3, ..., and where $f^{(0)} = f$ and k_f, K_f are constants depending only on f.

In what follows we suppose that the sequence $\{M_n\}$ is with the following properties:

1. $M_0 = 1$
2. $M_n^2 \leq M_{n-1}M_{n+1}$ $\qquad n = 1, 2, 3, \ldots$

It is known that every class $C\{M_n\}$ is equal to a class $C\{M_n'\}$ with $\{M_n'\}$ satisfying the above two conditions. Also it is not difficult to prove that every class $C\{M_n\}$ is an algebra of functions.

We define now the quasi-analytic classes $C\{M_n\}$.

12.0.2 DEFINITION The class $C\{M_n\}$ is called *quasi-analytic* if $f \in C\{M_n\}$ and $f^{(n)}(x_0) = 0$, $n = 0, 1, 2, \ldots$, implies that $f \equiv 0$.

The class of analytic functions is exactly the class with $M_n = n!$

The following theorem represents the basic information about the quasi-analytic classes.

12.0.3 THEOREM If for a sequence $\{M_n\}$ satisfying the above two conditions we define

$$Q(x) = \sum_{n=0}^{\infty} \frac{x^n}{M_n} \qquad q(x) = \sup_n \frac{x^n}{M_n}$$

for x > 0, then the following conditions are equivalent:

1. $C\{M_n\}$ is not a quasi-analytic class.
2. $\int_0^\infty \log Q(x)\, dx/1 + x^2 < \infty$.
3. $\int_0^\infty \log q(x)\, dx/1 + x^2 < \infty$.
4. $\Sigma_{n=1}^\infty (1/M_n)^{1/n} < \infty$.
5. $\Sigma_{n=1}^\infty M_{n-1}/M_n < \infty$.

We mention that we can organize some classes of quasi-analytic functions as Banach algebras. An interesting problem appears if we consider the problem of functional calculus for these algebras, more precisely, if there exists a functional calculus as was constructed by Dunford and Taylor.

Since they are one of the main applications of the analytic vectors and for their own interest, we give now some notions related to self-adjoint operators, although in this book the main theme is bounded operators.

12.1 SELF-ADJOINT OPERATORS

Let H be a Hilbert space over the complex numbers.

12.1.1 DEFINITION A function T defined on a part of H, say D_T, is called an *operator* if the following conditions are satisfied:

1. D_T is a linear subspace of H.
2. For all $x,y \in D_T$ and a,b complex numbers

$$T(ax + by) = aTx + bTy$$

In the above definition the set D_T is called the *domain* of T.

12.1.2 DEFINITION If T and S are two operators, we write $S \subset T$ and we say that T is an *extension* of S if

1. $D_S \subset D_T$
2. $Sx = Tx$ for all $x \in D_S$

We define now the sum and product of two operators.

12.1.3 DEFINITION If T and S are two operators, then the *sum* T + S of the operators is called the *operator with the domain* $D_T \cap D_S$ and defined by the relation

$$(T + S)(x) = Tx + Sx$$

For any operator T, the *range* R(T) is defined as the set

$$R(T) = \{y \mid y = Tx \text{ for } x \in D_T\}$$

Now the product of two operators is defined as follows.

12.1.4 DEFINITION The *product* of two operators T and S is the operator TS with the domain

$$S^{-1}(D_T) = \{z \mid Sz \in D_T\}$$

and the formula

$$(TS)(x) = T(Sx)$$

The product of an operator with a complex number is defined as in the case of bounded operators.

We define now the notion of the inverse of an operator.

12.1.5 DEFINITION If T is an operator, then the *inverse* of T, denoted by T^{-1}, is an operator with the domain equal to the range of T and satisfying the following relations:

$$T^{-1}(Tx) = x \qquad T(T^{-1}y) = y$$

for all $x \in D_T$ and for all $y \in R(T)$.

The following definition considers an important class of operators.

12.1.6 DEFINITION An operator T is *closed* if for any sequence $\{x_n\} \subset D_T$, the conditions

$$\lim x_n = x \qquad \lim Tx_n = y$$

imply

$$y = Tx$$

Using the Riesz theorem about the continuous linear functionals on Hilbert spaces, it is not difficult to prove the following result (see also the notion of adjoint operator for the case of bounded operators).

12.1.7 PROPOSITION If T is an operator and D_T is dense in H, then there exists an operator denoted T* and called the *adjoint* of T satisfying the following relation:

$$\langle Tx, y\rangle = \langle x, T^*y\rangle$$

for all $x \in D_T$ and $y \in D_{T^*}$.

The adjoint of an operator is an example of a closed operator. Indeed, if $\lim x_n = \bar{x}$ and $\lim T^*x_n = y$, then we have for each $x \in D_T$,

$$\langle Tx, x_n\rangle = \langle x, T^*x_n\rangle$$

and for $n \to \infty$ we obtain

$$\langle Tx, \bar{x}\rangle = \langle x, y\rangle$$

and thus $x \in D_T$ and $y = T^*x$.

We define now the notion of hermitian operator.

12.1.8 DEFINITION An operator T is called *hermitian* if

1. D_T is dense in H.
2. For any $x,y \in D_T$, $\langle Tx, y\rangle = \langle x, Ty\rangle$.

It is easy to see that T is hermitian if and only if

$$T \subset T^*$$

and thus for any operator of this type

$$D_T \subset D_{T^*}$$

Also it follows easily that if T is hermitian, the operator T^* is hermitian.

The very important notion of the self-adjoint operator is introduced in the following definition.

12.1.9 DEFINITION An operator T is called *self-adjoint* if

1. D_T is dense in H.
2. $T = T^*$.

Using the notion of hermitian operators, we can give the following characterization of self-adjoint operators.

12.1.10 THEOREM An operator T is self-adjoint if and only if

1. T is closed.
2. T^* is hermitian.

Proof. From the above results it is clear that the conditions are necessary. We show now that these are also sufficient.

Indeed, since T and T* are hermitian operators, we have

$$T \subset T^* \subset (T^*)^* = T$$

and thus $T^* = T$.

Another class of operators is the class of normal operators defined as follows.

12.1.11 DEFINITION An operator T is called *normal* if and only if

1. T is closed.
2. $D_T = D_{T^*}$ and D_T is dense in H.
3. $T^*T = TT^*$.

In the study of hermitian operators the following operator, called the *Cayley transform*, is of fundamental importance and is defined as follows:

$$T \mapsto V = (T - i)(T + i)^{-1}$$

with the domain D_{T+i}.

The following basic property of V can be established without much difficulty.

12.1.12 PROPOSITION V is a unitary operator if and only if T is self-adjoint.

In this case, as we know, V has the following spectral representation

$$\langle Vx, y\rangle = \int_0^{2\pi} e^{it} \, d\langle F_t x, y\rangle$$

and this gives [for details we refer to the book of Nagy and Riesz (1952)]

$$\langle Tx, y\rangle = - \int_0^{2\pi} \cot \frac{t}{2} \, d\langle F_t x, y\rangle$$

From this we can define a spectral measure $\{E_s\}$ on the real axis by the formula

$$E_s = F_t$$

and where

$$s = -\cot\frac{t}{2}$$

Thus we have

$$\langle Tx, y\rangle = \int_{-\infty}^{\infty} s \, d\langle E_s x, y\rangle$$

and the domain of T is exactly the set of all $x \in H$ for which

$$\int_{-\infty}^{\infty} s^2 \, d\langle E_s x, x\rangle < \infty$$

12.1.13 DEFINITION A hermitian operator is called *semi-bounded inferior* if

$$m(T) = \inf \frac{\langle Tx, x\rangle}{\langle x, x\rangle}$$

is not $-\infty$.

The existence of a self-adjoint extension of a semi-bounded self-adjoint operator was conjectured by J. von Neumann and was proved independently by Stone (1951) and Friederichs (1973). For these refer to the book of Nagy and Riesz and to the paper by Krein (1947).

If T is an operator on a Hilbert space, then it is possible that T is not closed. Thus the problem appears to be whether there exists an extension of T which is closed. The following result gives a condition for the existence of such an extension.

First we remark that the condition for an operator to be closed can be expressed in the following form.

12.1.14 PROPOSITION An operator T is closed if and only if

$$\{(x, Tx) \mid x \in D_T\}$$

is a closed subset of $H \times H$.

Now the condition for the existence of a closed extension of T, called the *closure* of T, is the following:

12.1.15 THEOREM If T is an operator on a Hilbert space, then its closure exists if and only if D_{T^*} is a dense subset of H.

Proof. If H is a Hilbert space, then a closed subspace G in H × H is not the graph of a closed operator means that for some $x \in Pr_1G$ there exist two distinct points, say (x, y_1) and (x, y_2), in G and that $(0, y_1 - y_2) \in G$. Now if the domain D_{T^*} is not dense in H, we can find an element $z \in H$, $z^\sim \neq 0$ such that $\langle x, z^\sim \rangle = 0$ for all $z^\sim$ in the closure of D_T. This gives that the element $(z^\sim, 0)$ is orthogonal to the graph of T*. This implies that $(0, -z^\sim)$ is in the closure of the graph of T.

Conversely, if D_{T^*} is dense in H, then T** has the property that its graph is the orthogonal complement of J (graph T*), where J denotes the operator defined by the relation

$$J(x, y) = (y, -x)$$

and since this orthogonal complement is equal to

$$J(\overline{J(\text{graph } T)}) = \overline{\text{graph } T}$$

it follows that T** is the closure of T.

12.1.16 REMARK If T is a closed operator and D_T is dense, then T** = T.

12.1.17 DEFINITION If T is an operator with D_T a dense set, then T is called essentially self-adjoint if the closure of T is a self-adjoint operator.

We need the following important theorem of Naimark about the existence of self-adjoint extension of operators.

12.1.18 THEOREM For any hermitian operator with D_T dense there exists an extension which is self-adjoint (and this may be defined on a larger space $K \supset H$).

For the proof we refer to the appendix of the book by Nagy and Riesz (1952).

12.2 CLASSES OF VECTORS FOR AN OPERATOR

If we consider the Hilbert space L^2 on the real axis and the operator

$$T = \frac{d}{dx}$$

then the set of C^∞ functions is exactly the set $\bigcap_{n=1}^{\infty} D_{T^n}$. This very simple remark suggests the introduction of some classes of vectors for operators.

12.2.1 DEFINITION An element x is called a C^∞ *vector* for the operator T if

$$x \in \bigcap_{n=1}^{\infty} D_{T^n}$$

The notion of analytic vector for an operator was introduced by Nelson (1959) and is as follows.

12.2.2 DEFINITION A C^∞ vector for an operator is called an *analytic vector* for T if the series

$$\sum \frac{z^n}{n!} \|T^n x\|$$

has a positive radius of convergence.

12.2.3 REMARK The notion of analytic vectors is important only in the case of unbounded operators since in the case of bounded operators any vector $x \in H$ is analytic.

The set of all analytic vectors of an operator T is denoted by $D^a(T)$.

As an important generalization of analytic vectors, Nussbaum (1965) has introduced the notion of a *quasi-analytic vector* and also the notion of a *Stieltjes vector* for an operator. Independently the notion of the Stieltjes vector was introduced by Masson and McClary (1972).

For the reader's convenience as well as for a precise presentation we consider the following classes of vectors of an operator.

12.2.4 DEFINITION For an operator T we denote the following:

1. $D^a = \{x \in C^\infty \mid \Sigma\ (z^n/n!)\|T^n x\|$ has a positive radius of convergence$\}$ = analytic vectors of T.
2. $D^{qa} = \{x \in C^\infty \mid \Sigma\ \|T^n x\|^{-1/n} = \infty\}$ = quasi-analytic vectors of T.
3. $D^{sa} = \{x \in C^\infty \mid \Sigma\ (z^n/(2n)!)\|T^n x\|$ has a positive radius of convergence$\}$ = semi-analytic vectors of T.
4. $D^{(1/p)a} = \{x \in C^\infty \mid \Sigma\ (z^n/(pn!))\|T^n x\|$ has a positive radius of convergence$\}$ = p-analytic vectors of T.
5. $D_S = \{x \in C^\infty \mid \Sigma\ \|T^n x\|^{-1/2n} = \infty\}$ = Stieltjes vectors of T.
6. $D_S^p = \{x \in C^\infty \mid \Sigma\ \|T^n x\|^{-1/np} = \infty\}$ = p-Stieltjes vectors of T.

The following relations hold between the above classes.

12.2.5 THEOREM We have the relations

$$\begin{array}{ccc} D^a & \subset & D^{qa} \\ \cap & & \cap \\ D^{sa} & \subset & D_S \\ \cap & & \cap \\ D^{(1/p)a} & \subset & D_S^p \end{array}$$

and $p_1 \le p_2$ implies $D^{(1/p_1)a} \subset D^{(1/p_2)a}$, which implies $D_S^{p_1} \subset D_S^{p_2}$.

Proof. All the assertions are obvious except possibly the last. For this we remark that it is a simple consequence of the following lemma.

12.2.6 LEMMA If $\Sigma\ a_n$, $a_n \ge 0$, is a divergent series, then for any $\alpha \in (0, 1)$ the series $\Sigma\ a_n^\alpha$ is also divergent.

The proof is very simple and thus we omit it.

Now the proof of the above inclusion follows from the relation

$$\sum \frac{1}{\|T^n x\|^{1/np_2}} = \sum \left(\frac{1}{\|T^n x\|^{1/np_1}}\right)^{p_1/p_2}$$

If T is an operator and x is a vector of C^∞ type in some class defined above, then it appears natural to consider the problem of whether Tx is in the same class. In what follows we give some results in this direction.

12.2.7 *THEOREM* If T is an operator and $x \in C^\infty$, then

1. If $x \in D^a$, then $Tx \in D^a$.
2. If $x \in D^{qa}$, then $Tx \in D^{qa}$.

Proof. The first assertion follows from the relation

$$\sum \frac{z^n}{n!}\|T^{n+1}x\| = \frac{1}{z}\sum \frac{z^{n+1}}{(n+1)!}\|T^{n+1}x\|(n+1)$$

but the second assertion follows from an inequality due to Carleman (1922) and is as follows.

12.2.8 *LEMMA* If $a_n \geq 0$, then

$$\sum_{i=2}^{n} a_i^{1-1/i} \leq \sum_{i=2}^{n} a_i + 2\left(\sum_{i=2}^{n} a_i\right)^{1/2}$$

Proof. Let $r > 1$ and we consider the terms in the sum in two subsets: $a_i^{-1/i} \geq r$ and $a_i^{-1/i} < r$.

Now the sum corresponding to the first set is inferior or equal to

$$rS = r \sum a_i$$

and the second, since for i in the complementary set we have

$$a_i \leq \left(\frac{1}{r}\right)^i \qquad a_i^{1-1/i} \leq \left(\frac{1}{r}\right)^{i-1}$$

and this implies that

$$\sum_{i=2}^{n} a_i^{1-1/i} \leq rS + \sum_{i=2}^{n} \left(\frac{1}{r}\right)^{i-1} \leq rS + \frac{1}{r-1}$$

The assertion follows if we take $r = 1 + 1/S^{1/2}$.

Now the proof of the inclusion follows from this lemma if we remark that it implies that $\sum s_n^{-1/n}$ diverges if and only if

$\Sigma s_{n+1}^{-1/n}$ diverges, and in our case $s_n = \|T^n x\|$.

For certain classes of operators including the hermitian operators we can prove a more general result.

12.2.9 THEOREM If T is an operator satisfying the relation

$$\|Tx\|^2 \le \|T^2 x\| \qquad \|x\| = 1$$

and $x \in D_T \cap D_{T^2}$, then

1. If x is a C^∞ vector for T and p(z) is any polynomial, then p(T)x is a C^∞ vector.
2. If x is analytic or quasi-analytic, then p(T)x is analytic or quasi-analytic, respectively.

For the proof we need the following lemma which has the same proof as in the case of bounded operators of class (N).

12.2.10 LEMMA If T has the property that

$$\|Tx\|^2 \le \|T^2 x\| \qquad \|x\| = 1$$

for any $x \in D_T \cap D_{T^2}$, then for any C^∞ vector x the sequence $\{\|T^n x\|^{1/n}\}$ is monotone.

For the proof of the theorem we remark that we can suppose, without loss of generality, that $p(T)x \ne 0$, and then we have

$$\|T^n p(T)x\|^2 \le \|T^{2n} p(T)x\| \cdot \|p(T)x\|$$

and this implies immediately the assertion of the theorem.

12.2.11 REMARK As we know, every hyponormal operator satisfies the condition of Theorem 12.2.9.

12.3 ANALYTIC AND QUASI-ANALYTIC VECTORS AND ESSENTIALLY SELF-ADJOINT OPERATORS

A fundamental result about the connection between analytic vectors and essentially self-adjoint operators was proved by Nelson (1959). The result was extended to quasi-analytic vectors by Nussbaum (1965) and to Stieltjes vectors by Nussbaum and independently by Masson and McClary (1972).

12.3.1 THEOREM If A is a hermitian operator on a Hilbert space, and if A has a dense set of analytic vectors, then the closure of A is self-adjoint (i.e., A is essentially self-adjoint).

12.3.2 THEOREM If A is a hermitian operator on a Hilbert space, and if A has a dense set of quasi-analytic vectors, then A is essentially self-adjoint.

12.3.3 THEOREM Suppose A is a hermitian operator having the following properties:

1. A is semibounded.
2. A has a dense set of Stieltjes vectors.

Then A is essentially self-adjoint.

Proof of Theorem 12.3.1. As we know, A has an extension by a theorem by Naimark, say A_0, in a possible larger space $K \supset H$. In this case the operator A_0 generates a group of unitary operators $\{U_t\}$, where $U_t = e^{itA_0}$. From the preceding results to prove that A is self-adjoint, it will be sufficient to show that the operators A + i and A - i have dense ranges.

We prove now that A + i has a dense range; similarly we can prove for A - i.

Suppose, on the contrary, that A + i has no dense range and thus we can find a nonzero element of H, say y, orthogonal to the range of A + i. Thus for all $x \in D_T$ we have that

$$\langle (A + i)x, y\rangle = 0$$

Suppose now that x is a C^∞ vector for A and thus we obtain

$$\langle A^n x, y\rangle = (-i)^n \langle x, y\rangle \qquad n = 1, 2, 3, \ldots$$

We consider now the function defined on the real axis by the relation

$$f(t) = \langle U_t x, y\rangle$$

and since if x is an analytic vector, f(t) is an analytic function, we have that

$$f^{(n)}(0) = i^n\langle A^n x, y\rangle = \langle x, y\rangle$$

which gives that

$$f(t) = e^t\langle x, y\rangle$$

Now f is bounded on the real axis and thus $\langle x, y\rangle$ must be zero. Since the set of x with this property is dense, we have that A is self-adjoint.

Proof of Theorem 12.3.2. As above we can define the function $f(t) = \langle U_t x, y\rangle$, and if x is a quasi-analytic vector of A, it follows that f is a quasi-analytic function. As above the derivative of f and $g(t) = e^t\langle x, y\rangle$ are the same at $t = 0$ and from quasi-analyticity we obtain that $f(t) = e^t\langle x, y\rangle$. However this implies that $\langle x, y\rangle = 0$. Thus the conclusion of the theorem follows as above.

Following an idea of Chernoff (1972), we obtain results about semi-analytic and Stieltjes vectors from Theorems 12.3.1 and 12.3.2.

Suppose now that A is a semibounded hermitian operator and we may assume that $A > a$ for some $a > 0$. Indeed, A is self-adjoint if and only if $A + b$ is self-adjoint for all real b, and we show that

$$\Sigma\ \|(A + a)^n x\|^{-1/2n} = \infty$$

if and only if

$$\Sigma\ \|A^n x\|^{-1/2n} = \infty$$

To prove this, it is sufficient to show that the following assertion holds:

$$\Sigma\ \|A^n x\|^{-1/2n} = \infty$$

implies

$$\Sigma\ \|(A + a)^n x\|^{-1/2n} = \infty$$

Since we have

$$\begin{aligned}\|(A + a)^n x\| &= \|\Sigma \binom{n}{k} a^{n-k} A^k x\| \le \Sigma \binom{n}{k} |a| \|A^k x\| \\ &\le \Sigma \binom{n}{k} |a|^{n-k} \|A^n x\|^{k/n} = (\|A^n x\|^{1/n} + |a|)^n \\ &\le \|A^n x\|(1 + |a| \|A^n x\|^{-1/n})^n \le \|A^n x\| \left(\frac{1 + |a|}{\|Ax\|^n}\right)^n\end{aligned}$$

since, as we know, for any operator of class (N) the sequence $\{\|T^n x\|^{1/n}\}$ is monotone.

The result proved can be formulated as follows.

12.3.4 THEOREM If T is an operator with the property that for all real or complex numbers s, T + s is of class (N), then T and T + s have the same set of Stieltjes vectors.

Thus if A is a hermitian and semibounded operator, we can suppose, without loss of generality, for our purpose, that $A > I$.

Suppose that A is a hermitian operator with a dense set of semi-analytic vectors. Let A_0 be the restriction of A to this set. Since $A > I$, then there exists the Friederichs extension, which is denoted by A_1, with the property $A \ge I$ and, of course, is self-adjoint. From the construction of the extension it follows that the set of semi-analytic vectors is dense in the set $D^{1/2}_{A_1}$.

The following lemma gives a useful property of a matrix operator constructed with A_1.

12.3.5 LEMMA Let $A_1 \ge I$ be a self-adjoint operator on a Hilbert space $H = H \oplus H$. Let $K = D^{1/2}_{A_1} \oplus H$, which is a Hilbert space with the inner product defined as

$$\left\langle \begin{bmatrix} u \\ v \end{bmatrix}, \begin{bmatrix} u' \\ v' \end{bmatrix} \right\rangle = \langle A_1 u, u' \rangle + \langle v, v' \rangle$$

and let B_1 be the operator defined on K by the matrix

$$B_1 = \begin{bmatrix} 0 & I \\ -A_1 & 0 \end{bmatrix}$$

with the domain $D^{1/2}_{A_1} \oplus D^{1/2}_{A_1}$. Then B_1 is skew-symmetric, i.e.,

$$\left\langle B_1 \begin{bmatrix} u \\ v \end{bmatrix}, \begin{bmatrix} u' \\ v' \end{bmatrix} \right\rangle = -\left\langle \begin{bmatrix} u \\ v \end{bmatrix}, B_1 \begin{bmatrix} u' \\ v' \end{bmatrix} \right\rangle$$

Proof. Since the computation is simple, we omit it.

Also it is not difficult to see that the following assertion holds.

12.3.6 LEMMA The operator B_1 has the property that $I + B_1$ is an injective and surjective operator.

The operator B_1 generates a unitary group of operators on K defined by the formula

$$e^{tB_1} = \begin{bmatrix} \cos tA_1^{1/2} & -A_1^{1/2}(\sin tA_1^{1/2}) \\ -A_1^{1/2} \sin tA_1^{1/2} & \cos tA_1^{1/2} \end{bmatrix}$$

and now if A_1 is the Friederichs extension of the operator A, we define the operator B_0 on K by the matrix

$$B_0 = \begin{bmatrix} 0 & I \\ -A_0 & 0 \end{bmatrix}$$

and the domain is

$$D_{A_0} \oplus D_{A_1}^{1/2}$$

12.3.7 LEMMA The closure of B_0 is the operator with the matrix

$$\bar{B}_0 = \begin{bmatrix} 0 & I \\ -\bar{A}_0 & 0 \end{bmatrix}$$

with the domain

$$D_{\bar{B}_0} = D_{\bar{A}_0} \oplus D_{A_1}^{1/2}$$

Proof. Let us consider the operator

$$C = \begin{bmatrix} 0 & I \\ -\bar{A}_0 & 0 \end{bmatrix}$$

and first we prove that this is a closed operator on the space K. Indeed, if

$$\begin{bmatrix} u_n \\ v_n \end{bmatrix} \to \begin{bmatrix} u \\ v \end{bmatrix}$$

and

$$C\begin{bmatrix} u_n \\ v_n \end{bmatrix} \to \begin{bmatrix} w \\ z \end{bmatrix}$$

then from the form of C we have that

$$\begin{bmatrix} v_n \\ -\bar{A}_0 u_n \end{bmatrix} \to \begin{bmatrix} w \\ z \end{bmatrix}$$

and this gives that

$$v = w$$

Since $u_n \to u$ and $\bar{A}_0 u_n \to -z$, we obtain (since $\bar{A}_0$ is a closed operator) that $u \in D_{\bar{A}_0}$ and $\bar{A}_0 u = -z$. From these it is clear that $\begin{bmatrix} u \\ v \end{bmatrix} \in D_C$ and also that $C\begin{bmatrix} u \\ v \end{bmatrix} = \begin{bmatrix} w \\ z \end{bmatrix}$.

Now, C is closed and extends $\bar{B}_0$. The inclusion in the opposite direction is proved as follows. Let

$$\begin{bmatrix} u \\ v \end{bmatrix} \in D_C$$

This means that $u \in D_{\bar{A}_0}$ and $v \in D_{A_1}^{1/2}$. We can find a sequence of elements in D_{A_0}, say $\{u_n\}$, such that $u_n \to u$ and $A_0 u_n \to \bar{A}_0 u$. But $\bar{A}_0 \subset A_1$, and thus $A_1^{1/2} u_n \to A_1^{1/2} u$. From this we have

$$B\begin{bmatrix} u_n \\ v \end{bmatrix} = \begin{bmatrix} v \\ -A_0 u_n \end{bmatrix} \to \begin{bmatrix} v \\ -\bar{A}_0 u \end{bmatrix} = C\begin{bmatrix} u \\ v \end{bmatrix}$$

and the assertion is proved.

From this result it follows that if we show that $\bar{B}_0 = B_1$, where the operators B_0 and B_1 are the above operators, it gives the relation $\bar{A}_0 = A_1$, and thus $\bar{A} = A_1$ is a self-adjoint operator.

To prove the assertion in the case of semi-analytic vectors, we need the following result.

12.3.8 LEMMA If x and y are semi-analytic vectors for A_0, then $\begin{bmatrix} x \\ y \end{bmatrix}$ is an analytic vector for B_0.

Proof. First we note that

$$B_0^{2k} = (-1)^k \begin{bmatrix} A_0^k & 0 \\ 0 & A_0^k \end{bmatrix}$$

and

$$B_0^{2k+1} = (-1)^k \begin{bmatrix} 0 & A_0^k \\ -A_0^{k+1} & 0 \end{bmatrix}$$

and $\begin{bmatrix} x \\ y \end{bmatrix}$ is a C^∞ vector for the operator B_0.

Since we have

$$B_0^{2k}\begin{bmatrix} x \\ y \end{bmatrix} = (-1)^k \begin{bmatrix} A_0^k x \\ A_0 y \end{bmatrix}$$

this implies that

$$\left\| B_0^{2k}\begin{bmatrix} x \\ y \end{bmatrix} \right\| \le \|A_0^{k+1}x\| + \|A_0^k y\|$$

Similarly we have the relation

$$\left\| B_0^{2k+1}\begin{bmatrix} x \\ y \end{bmatrix} \right\| \le \|A_0^{k+1}x\| + \|A_0^{k+1}y\|$$

These estimates show that $\begin{bmatrix} x \\ y \end{bmatrix}$ is an analytic vector.

The proof of the theorem about semi-analytic vectors which we give below follows easily from the results just obtained.

First we give a theorem.

12.3.9 THEOREM Let A be an operator on a Hilbert space with the following properties:

1. A is semibounded.
2. A has a dense set of semi-analytic vectors.

Then A is essentially self-adjoint.

Proof. From the above result we know that $D^{sa} \oplus D^{sa}$ is a set of analytic vectors for the operator B_0, and it is also dense in K since D^{sa} is dense in H by the hypothesis and D^{sa} is dense in $D_{A_1}^{1/2}$ by the construction. In this case by Theorem 12.3.1, the operator iB_0 has a self-adjoint closure and thus $\bar{B}_0 = B_1$ and this implies that $\bar{A}_0 = A_1$.

The following lemma gives us the opportunity to treat the case of Stieltjes vectors.

12.3.10 LEMMA If x is a Stieltjes vector for A_0, then $\begin{bmatrix} x \\ 0 \end{bmatrix}$ and $\begin{bmatrix} 0 \\ y \end{bmatrix}$ are both quasi-analytic vectors for B_0.

Proof. Since A_0x is also a Stieltjes vector for A_0, it follows that

$$\Sigma \ \|A_0^{k+1}\mathbf{x}\|^{-1/2k} = \Sigma \ \|A_0^kA_0x\|^{-1/2k} = \infty$$

and this implies, by the estimates just obtained, that

$$\Sigma \ \left\|B_0^n\begin{bmatrix} x \\ 0 \end{bmatrix}\right\|^{-1/n} \qquad \Sigma \ \left\|B_0^{2k}\begin{bmatrix} x \\ 0 \end{bmatrix}\right\|^{-1/2k} = \infty$$

Similarly, we can show that $\begin{bmatrix} 0 \\ x \end{bmatrix}$ is a quasi-analytic vector for B_0.

From the form of the domains it follows that the set of quasi-analytic vectors for B_0 is dense, and as above, we obtain the conclusion of Theorem 12.3.3.

12.3.11 REMARK In Chernoff (1972) it is asserted that a similar result is false for the case of 4-analytic vectors.

12.4 QUASI-ANALYTIC VECTORS AND SEMIGROUPS OF OPERATORS

Let H be a Hilbert space and T be an operator.

12.4.1 DEFINITION The operator T is called *dissipative* if

$$\langle Tx, x\rangle + \langle x, Tx\rangle \leq 0$$

for all $x \in D_T$.

For all results about semigroups of operators needed in what follows we refer to the book of Hille and Phillips (1957) and the papers of Phillips (1959) and Lumer and Phillips (1961).

The first result about semigroups and quasi-analytic vectors was the following theorem of Hasegawa (1971).

12.4.2 THEOREM If T is a closed dissipative operator with a dense domain, then T generates a strongly continuous semigroup when T has a dense set of quasi-analytic vectors.

Since the proof of Hasegawa uses directly the properties of quasi-analytic functions, we give these here.

For the proof we use the following result about generators of semigroups of contractions which are strongly continuous.

12.4.3 THEOREM A dissipative operator T with dense domain generates a strongly continuous contraction semigroup if and only if $R(kI - T) = H$ for some $k > 0$.

For a proof, see Phillips (1959) and Lumer and Phillips (1961).

Proof of Theorem 12.4.2. Suppose that the assertion of the theorem is false. In this case by Theorem 12.4.3 there exists k_1 and k_2 such that the ranges of $k_1 - T$ and $k_2 - T$ are properly contained in H.

We can take $k_1 = 1$ and $k_2 = 2$ and note that

$$R(I - T)^{-} \cap R(2I - T)^{-} = \{0\}$$

and that D_1 and D_2 are two sets containing D_A properly. If $\{T_t\}$ and $\{S_t\}$ are the semigroups whose generators are A_1 and A_2, then for any quasi-analytic vector x of A we consider the function

$$f(t) = \begin{cases} \langle T_t x - S_t x, y\rangle & t \geq 0 \\ 0 & t < 0 \end{cases}$$

and from

$$|f^{(n)}(t)| \leq |\langle (T_t A^n x - S_t A^n x), y\rangle| \leq 2\|y\| \cdot \|A^n x\|$$

we see that f is a quasi-analytic function and since it is zero, for $t < 0$ it is identically zero. This implies that for any quasi-analytic vector x of T,

$$T_t A^i x = S_t A^i x$$

and this implies, of course, that $T_t = S_t$ for all t. This gives further that $A_1 = A_2$, which is a contradiction and the assertion is proved.

This theorem was extended to Banach space operators by Chernoff (1972). His result is the following.

12.4.4 THEOREM Let T be a closed operator on a Banach space X and T has an extention $T^{\sim}$ which is the generator of a C_0 semigroup. If T has a dense set of quasi-analytic vectors, then $T = T^{\sim}$.

For the proof we need the following lemma about sequences of positive numbers.

12.4.5 LEMMA If $\{s_n\}$ is a sequence of nonnegative numbers and $\Sigma\, s_n^{-1/n} = \infty$, $0 \leq k_n \leq as_n + b^n$ with $a,b > 0$. Then $\Sigma\, k_n^{-1/n} = \infty$.

Proof. We can suppose, without loss of generality, that a = 1 and define $M = \{n \mid s_n \leq b^n\}$.

Case 1. M is infinite. In this case we have

$$k_n \leq b^n + b^n = 2b^n$$

and the assertion is obvious.

Case 2. M is finite. In this case for an infinite n we have that

$$s_n \geq b^n$$

and this implies that for such n,

$$k_n \leq 2s_n$$

and this implies the assertion of the lemma.

Proof of Theorem 12.4.4. Let $U_t = e^{itT^{\sim}}$ be the semigroup generated by $T^{\sim}$. We can find the constants M and β such that

$$\|U_t\| \leq Me^{\beta t}$$

for all $t > 0$. Let $s > \beta$ and in this case the operator $s - T^{\sim}$ is surjective and bounded below, and since it is the extension of the operator $s - T$, we obtain that $s - T$ is bounded below. Since T is closed, $R(s - T)$ is closed. The assertion of the theorem is proved if we show that $s - T$ has a dense range.

Let $x^* \in X^*$ which annihilates the range of $s - T$. If x is a quasi-analytic vector for T, we define the function

$$f(t) = x^*(U_t x)$$

and from the property of x^* we obtain that

$$f^{(n)}(0^+) = x^*(T^n x) = s^n x^*(x)$$

and the function

$$g(t) = \begin{cases} f(t) - e^{st}x^*(s) & t \geq 0 \\ 0 & t < 0 \end{cases}$$

which is a C^∞ function. We show now that it is a quasi-analytic function. If we set

$$k_n = \sup_{0 \leq t \leq T} |g^{(n)}(t)| \leq a(\|T^n x\| + s^n)$$

the above lemma implies that $\Sigma\, k_n^{-1/n} = \infty$. Thus g is in a quasi-analytic class for any interval and since it is zero for $t < 0$, $g(t)$ is identically zero. This implies that

$$x^*(U_t x) = e^{st} x^*(x)$$

for all $x \in D^{qa}$. This implies that

$$U_t^* = e^{st} I^*$$

and if x^* is not zero, from this we obtain that

$$\|U_t^*\| = e^{st}$$

which is a contradiction and the assertion is proved.

12.5 ANALYTIC AND QUASI-ANALYTIC ELEMENTS IN COMMUTATIVE BANACH ALGEBRAS

If A denotes the set of all continuous complex-valued functions on [0, 1], then it is clear that this is a commutative Banach algebra with unit. If we consider the operator $T = d/dx$, then its domain D_T is equal to the set of all C^∞ functions and for any $f,g \in D_T$ we have that

$$\frac{d}{dx}(fg) = \left(\frac{d}{dx\ f}\right) g + f \frac{d}{dx} g$$

12.5.1 DEFINITION An operator defined on a commutative Banach algebra is called a *derivation* if D_d is an algebra and d satisfies thc relation

$$d(xy) = x\ dy + y\ dx$$

for all $x,y \in D_d$.

An element x of a commutative Banach algebra is said to be a C^∞ *element* if $x \in D_d n$.

Let K be a fixed integer and consider the set C^k of all elements in $D_d i$, $i \le k$.

12.5.2 LEMMA The set C^k is a Banach algebra under the norm

$$\|x\|_k = \sum_{i=0}^{k} \frac{1}{i!} \|d^i x\|$$

As in the case of operators on Hilbert or Banach spaces, we can define the notion of analytic, quasi-analytic, semi-analytic vectors, etc. For example, the definition of analytic elements is as follows.

12.5.3 DEFINITION An element x is called *analytic* if

1. $x \in D_{d^i}$.
2. $\Sigma_{n=0}^{\infty} (z^n/n!)\|d^n x\|$ has a positive radius of convergence.

It is worth noting the following property of analytic vectors.

12.5.4 LEMMA The set of analytic vectors forms an algebra.

Suppose now that we have a commutative Banach algebra A and a finite or infinite set of derivations with the property that their domains contain a dense subset of A. Let $D_{d_i^n}$ be this set.

12.5.5 DEFINITION Let $d_1, \ldots, d_n$ be derivations in the above family. An element $x \in D_{d_i^n}$ is called *quasi-analytic* for the family if there exists the constants L_k such that

$$\|d_1^{m_1} \cdots d_n^{m_n} x\| \le C^{k+1} L_k^k$$

where $k = m_1 + \cdots + m_n$, $k = 0, 1, 2, \ldots,$ and

$$\int_0^{\infty} \log \left(\sum_{k=0}^{\infty} \frac{t}{L_k} \right)^k \frac{dt}{1 + t^2}$$

is equal to ∞ (equivalently $\Sigma\, 1/L_k = \infty$).

From this definition it is clear that the set of analytic vectors corresponds to the case $L_k = k$ and $L_0 = 1$.

It is of interest to know a theory which has some similarity with the theory of complex functions exists in this abstract setting.

Chapter 13

SCHWARZ NORMS

Suppose that f is an analytic function in the open disk

$$\{z \mid |z| < 1\} = U$$

and bounded, i.e.,

$$\|f\|_\infty = \sup \{|f(z)| \mid z \in U\} < \infty$$

If f has the following additional property

$$f(0) = 0 \qquad \|f\|_\infty < \infty$$

then the famous Schwarz lemma asserts that

1. $|f(z)| \leq |z| \qquad z \in U$
2. $|f'(0)| \leq 1$

and the equality appears in 1 for one $z \in U\{0\}$, then $f(z) = az$ and also if the equality appears in 2, f has the same form.

For the proof we refer to any book on the theory of complex variables. This lemma has an interesting application in the theory of operators, namely, the following assertion holds: if f is analytic and $f(0) = 0$, $\|f\|_\infty < 1$, then for any operator $T \in L(H)$, $\|T\| < 1$, we have $\|f(T)\| < \|T\|$. Clearly, if we have an equality for some T, then F is of the form $f(z) = az$.

The above considerations lead us to consider a class of norms on the Banach algebra L(H) of all bounded linear operators on H, which have similar properties. More precisely, following Williams (1968) we define the notion of a Schwarz norm as follows: a norm

$\|\cdot\|^*$ on L(H) is called a *Schwarz norm* if it is equivalent to the norm $\|\cdot\|$ and the Schwarz lemma holds for this norm, i.e., for any f analytic with f(0) = 0 and $\|f\|_\infty < 1$, then for any $T \in L(H)$ with $\|T\|^* < 1$, $\|f(T)\|^* < 1$.

In what follows we give some results about this class of norms and we indicate a way to construct new Schwarz norms which are not between the norms constructed by Williams.

13.1 SCHWARZ NORMS

As we know one of the most interesting norms on L(H) is the norm induced by the numerical radius of an operator. Indeed, if $T \in L(H)$, then the numerical radius is defined as

$$T \mapsto w(T) = \sup \{|z| \mid z \in W(T)\}$$

where W(T) is the numerical range of T, i.e., the set

$$W(T) = \{\langle Tx, x\rangle \mid \|x\| = 1\}$$

Berger and Stampfli (1967) have proved that this norm is a Schwarz norm. In what follows we give a more general result. First we give a result about an interesting connection between the norms $\|\cdot\|$ and w(T).

13.1.1 LEMMA If a is a complex number such that $|a| < 1$ and $f_a(z) = (z - a)/(1 - a^*z)$, then the following assertions are equivalent:

1. $w(f_a(T) \leq 1$ for all T
2. $\|f_a(T)\| \leq 1$ for all T

Proof. Since for any operator $w(T) \leq \|T\|$, one of implications, namely, 2 implies 1, is obvious. We prove that 1 implies 2. From the hypothesis we obtain that

$$|\langle (T - a)x, (1 - a^*T)x\rangle| \leq \|(1 - a^*T)x\|^2$$

for all x and all a, $|a| < 1$. Also this relation implies that

$$t(\|x\|^2 + \|Tx\|^2) + (t^2 + 1)|\langle Tx, x\rangle| \le \|x\|^2 + 2t|\langle Tx, x\rangle| + t^2\|Tx\|^2 \qquad 0 < t < 1$$

and thus

$$0 \le (t - 1)[t(\|Tx\|^2 - |\langle Tx, x\rangle|) - (\|x\|^2 - |\langle Tx, x\rangle|)]$$

From this it is clear that $w(T) = \|T\|$, and the lemma is proved.

The following lemma gives some information about the norms $\|\cdot\|$ and $w(\cdot)$.

13.1.2 LEMMA If $T \in L(H)$, then the following assertions hold:

1. $\|T\| \le 1$ if and only if $\text{Re } (1 + zT)(1 - zT)^{-1} \ge 0$ for all $|z| < 1$.
2. $w(T) \le 1$ if and only if $\text{Re } (1 - zT)^{-1} \ge 0$ for all $|z| < 1$.

Proof. First we remark the fact that if T has the property $\|T\| \le 1$, then for any z, $|z| < 1$, the operator $1 - zT$ is invertible and thus for $y = (1 - zT)x$, $x \in H$, we have

$$\langle (1 + zT)(1 - zT)y, y\rangle = \langle (1 + zT)x, (1 - zT)x\rangle$$

and thus

$$\text{Re } \langle (1 + zT)(1 - zT)^{-1}y, y\rangle = \|x\|^2 - |z|^2\|Tx\|^2$$

and thus the necessity is proved. From this relation it is obvious that the condition is also necessary.

The second assertion can be proved as follows: First we remark that the following equivalence holds (trivial):

$$w(T) \le 1 \text{ implies } |\langle Tx, x\rangle| \le 1 \qquad \|x\| = 1$$

Since the spectral radius $r_T < 1$, we obtain that for any $|z| < 1$ the operator $1 - zT$ is invertible. From the above equivalence we conclude that $W(T) \le 1$ if and only if $\text{Re } \langle (1 - zT)x, x\rangle > 0$, which is obviously equivalent to $w(T) \le 1$ if and only if $\text{Re } (1 - zT) \ge 0$.

Consider now for each z the operator $B = (1 - zT)$ and thus

$$\text{Re } \langle Bx, x\rangle > 0 \text{ implies Re } \langle Bx, BB^{-1}x\rangle = \text{Re } \langle B^{-1}y, y\rangle$$

which shows B^{-1} has the real part a positive operator.

The converse assertion also follows immediately from the above relations.

From the form of the operators used for the characterization of the operators T for which $\|T\| \leq 1$ or $w(T) \leq 1$ we see that they are of the form

$$1 + c \sum_{n=1}^{\infty} z^n T^n$$

and the conditions refer to such operators.

The following definition introduces the class of operators which plays a fundamental role in the construction of Schwarz norms.

13.1.3 DEFINITION The S_c class of operators is the set of all operators $T \in L(H)$ for which the following properties hold:

1. $\sigma(T) \subset U$
2. $\mathrm{Re}\,(1 + c \sum z^n T^n) \geq 0$

In this definition c is a positive number.

13.1.4 REMARK From the definition and Lemma 13.1.2 it follows that

1. $\|T\| \leq 1$ if and only if $T \in S_2$.
2. $w(T) \leq 1$ if and only if $T \in S_1$.

The following theorem gives information about functional calculus with operators in the class S_c.

13.1.5 THEOREM If $T \in L(H)$ and $T \in S_c$, then for any rational function with no poles in the closed unit disk and with the property $f(0) = 0$, $\|f\|_\infty \leq 1$, we have $f(T) \in S_c$.

Proof. The function

$$z \mapsto \left\langle \left(1 + c \sum_{n=1}^{\infty} z^n T^n\right) x,\ x \right\rangle$$

is with the real part positive, and thus by the Herglotz theorem there exists a positive measure μ_x such that

$$\|x\|^2 + c \sum_{n=1}^{\infty} z^n \langle T^n x, x\rangle = \int_0^{2\pi} \frac{1 + ze^{it}}{1 - ze^{it}} \, d\mu_x(t)$$

for all $|z| < 1$. This gives the following relation:

$$c\langle T^n x, x\rangle = 2 \int_0^{2\pi} e^{int} \, d\mu_x(t)$$

for $n = 1, 2, 3, \ldots$.

From these relations we obtain immediately that for any polynomial $p(z) = \sum a_i z^i$ and any $x \in H$,

$$\langle p(T)x, x\rangle = 2 \int_0^{2\pi} p(e^{it}) \, d\mu_x(t)$$

and if we take $p^n(z)$, we obtain that

$$\langle p^n(T)x, x\rangle = 2 \int_0^{2\pi} p^n(e^{it}) \, d\mu_x(t)$$

This implies that if $\|p\|_\infty \le 1$, $\|p^n(T)\|$ is a bounded operator and for z, $|z| < 1$, we obtain

$$\left\langle 1 + c \sum_{n=1}^{\infty} z^n p^n(T)x, x \right\rangle = \|x\|^2 + 2 \sum_{n=1}^{\infty} z^n \int_0^{2\pi} p^n(e^{it}) \, d\mu_x(t)$$

$$= \int_0^{2\pi} \frac{1 + zp(e^{it})}{1 - zp(e^{it})} \, d\mu_x(t)$$

From this relation we obtain that $f(T) \in S_c$ when p is a polynomial. Now if f is any function which is rational and with no poles in the closed unit disk, then it can be approximated uniformly by polynomials and this gives the assertion of the theorem.

13.1.6 COROLLARY If $\{a_i\}_{i=1}^{\infty}$ is a sequence of complex numbers and

$$f(z) = \sum_{i=1}^{\infty} a_i z^i \qquad \sum_{i=1}^{\infty} |a_i| < \infty$$

then for any $T \in S_c$, $f(T) \in S_c$.

To obtain Schwarz norms from these classes of operators, we need more information about these classes. The most important is that S_c is a convex set for any $c > 1$.

13.1.7 THEOREM For the classes S_c $(c > 0)$ of operators, the following properties hold:

1. $S_c = S_c^* = \{T^* \mid T \in S_c\}$.
2. $S_{c_1} \subset S_{c_2}$ if $c_2 < c_1$.
3. S_c is a convex set if $c \geq 1$.
4. For $c > 1$, $T \in S_c$ if and only if for all $x \in H$,

$$(c - 1)\|Tx\|^2 + |2 - c|\,|\langle Tx, x\rangle| \leq \|x\|^2$$

Proof. First we remark that the definition of the class S_c can be reformulated as follows: $T \in S_c$ if and only if the spectrum $\sigma(T)$ is in the unit disk and for any $|z| < 1$,

$$\operatorname{Re}\langle(1 - zT)^{-1}x, x\rangle \geq (1 - c^{-1})\|x\|^2$$

From this, assertions 1 and 2 are obvious. Suppose now that $c \geq 1$ and in this case we remark that $T \in S_c$ if and only if the spectrum of T is in the unit disk and

$$\operatorname{Re}\langle x, (1 - zT)x\rangle > \frac{(c - 1)}{c}\|(1 - zT)x\|^2$$

for all $|z| < 1$ and any $x \in H$. From this assertion 4 follows.

To prove the convexity of S_c for $c \geq 1$, we use property 4 and if T_1 and T_2 are two operators in S_c, from 4, using the parallelogram law

$$\|T_1x + T_2x\|^2 \leq 2(\|T_1x\|^2 + \|T_2x\|^2)$$

we have that

$$(c - 1)\|(T_1 + T_2)x\|^2 + |2 - c|\,|\langle(T_1 + T_2)x, x\rangle|$$

$$\leq 2(c - 1)\{\|T_1x\|^2 + \|T_2x\|^2\} + |2 - c|\,|\langle T_1x, x\rangle| + |2 - c|\,|\langle T_2x, x\rangle|$$

$$\leq 2\|x\|^2 + 2\|x\|^2 = 4\|x\|^2$$

and thus $(1/2)(T_1 + T_2) \in S_c$.

We prove now that the classes S_c are nonempty and are strictly decreasing. For this we consider the following example used also by Nagy and Fois (1966): For any $s > 0$ we consider the operator sA, where A is an operator on a two-dimensional space with the matrix

$$\begin{bmatrix} 0 & 0 \\ 1 & 0 \end{bmatrix}$$

and we remark that sA is in S_c if and only if

$$0 < \mathrm{Re}\,(1 + c \,\Sigma\, z^n s^n A^n) = \mathrm{Re}\,(1 + cszA)$$

It is easy to see that for $|z| < 1$ the spectrum of the operator $\mathrm{Re}\,(1 + cszA)$ is the set

$$\{1 + \frac{1}{2}\,|csz|,\ 1 - \frac{1}{2}\,|csz|\}$$

and thus sA is in S_c if and only if $cs < 2$. From this we conclude that $(2/c)A$ is in S_c and since for $c_1 > c_2$, $(2/c_1)A$ is not in S_{c_2}, the assertion about inclusion is proved.

This example can be used to show that for $c \in (0, 1)$ the set S_c is not convex. Indeed, if S_c is convex, then by property 1 we have that

$$\frac{1}{2}\left(\frac{2}{c}\,A + \frac{2}{c}\,A^*\right) \in S_c$$

and since this is equal to

$$\frac{2}{c}\,\mathrm{Re}\,A$$

which has the spectrum $\{-1/c, 1/c\}$. Thus if $c < 1$, the set S_c is not convex.

The following lemma summarizes the properties of the set S_c.

13.1.8 *LEMMA* The set S_c for $c < 1$ has the following properties:

1. S_c is bounded and weakly closed.
2. S_c is a circled convex set and is a neighborhood of zero.

The properties in this lemma permit us to define for each $c > 1$ a norm on the $L(H)$.

13.1.9 DEFINITION For any $c \geq 1$ the function on $L(H)$

$$\|T\|_c = \inf \{s \mid T \in sS_c\}$$

is a norm equivalent to the norm $\|\cdot\|$.

The fact that $\|T\|_c$ is a norm equivalent to the norm $\|\cdot\|$ follows from the properties of the class S_c proved above.

Also we note the following properties of the norm $\|T\|_c$ which follows directly from Theorem 13.1.7:

1. $\|T\|_c = \|T^*\|_c$.
2. If $c_1 < c_2$, then $\|T\|_{c_1} \leq \|T\|_{c_2}$.
3. For $c \in [1, 2)$, $\|I\|_c = 1$.

In a paper Williams (1968) expresses the opinion that the norms $\|\cdot\|_c$ introduced above, which are obvious Schwarz norms, do not include all Schwarz norms on $L(H)$. We give now a very simple method for constructing new Schwarz norms on $L(H)$ which are not of type $\|\cdot\|_c$ but are very similar.

However we give further a new method of constructing Schwarz norms using a class of operators generalizing the class S_c.

13.1.10 THEOREM For any $c \geq 1$ we define on $L(H)$ the function

$$\|T\|_c^* = \|T\|_c + \|\hat{T}\|_c$$

where $\hat{T}$ denotes the image of T in the Calkin algebra of H. Then

$$T \mapsto \|T\|_c^*$$

is a Schwarz norm on $L(H)$ and is not in the class considered by Williams.

Proof. First we remark that we can construct a more general Schwarz norm on $L(H)$ by taking the sum of two different Schwarz norms applied to T and to the image of T in the Calkin algebra. Also since $\|T\|_c$ is a norm and $\|\hat{T}\|_c$ is a seminorm, it follows that the sum is a Schwarz norm.

13.2 A NEW CLASS OF SCHWARZ NORMS

Suppose that Q is a positive hermitian operator with the property that

$$0 < mI \leq Q \leq MI$$

and thus we can construct the operator $Q^{1/2}$ which is also positive and invertible.

The following class of operators is a generalization of the class S_c to which it reduces when $Q = cI$.

13.2.1 DEFINITION If Q is a hermitian operator, $0 < mI < Q < MI$, then the class S_Q is the set of all operators $T \in L(H)$ with the following properties:

1. $\sigma(T)$ is in the unit disk.
2. For all $|z| < 1$, $\mathrm{Re}\ (1 + \Sigma\ Q^{1/2}T^nQ^{1/2}z^n) \geq 0$.

We can prove some results about this class as for the class S_c.

13.2.2 THEOREM If f is a rational function with no poles in the closed unit disk and $\|f\|_\infty < 1$, $f(0) = 0$, then for any $T \in S_Q$, $f(T) \in S_Q$.

The proof is essentially the same as for S_c operators.

13.2.3 THEOREM The operator $T \in L(H)$ is in S_Q if and only if

1. $\sigma(T)$ is in the unit disk.
2. $\mathrm{Re} <(Q^{1/2}(1 - zT)^{-1}Q^{1/2}x, x> - <Qx, x> + \|x\|^2 \geq 0$.

Proof. The condition

$$\mathrm{Re}\ (1 + \Sigma\ Q^{1/2}T^nQ^{1/2}z^n) \geq 0$$

is equivalent to the following one,

$$\mathrm{Re} <(1 + \Sigma\ Q^{1/2}T^nQ^{1/2}z^n)x, x> = \mathrm{Re}\ [<Q^{1/2}(1 - zT)^{-1}Q^{1/2} - Q + I)x, x>] \geq 0$$

which is our assertion.

From this characterization we obtain the following result.

13.2.4 THEOREM If $Q \geq I$, then $T \in S_Q$ if and only if

1. $\sigma(T)$ is in the unit disk.
2. $\mathrm{Re} \langle Q^{1/2}(1 - zT)Q^{1/2}x, x\rangle > \|Q^{1/2}x\|^2 - \|x\|^2 = \langle (Q - I)x, x\rangle$.

Proof. This follows directly from the above theorem.

The following theorem gives information about the class S_Q which is similar to that given in Theorem 13.1.7 for the class S_c.

13.2.5 THEOREM If Q is a positive hermitian operator, then the following assertions hold:

1. $S_Q = S_Q^* = \{T^* \mid T \in S_Q\}$.
2. If $Q_1 < Q_2$, then $S_{Q_2} \subset S_{Q_1}$.
3. For $Q \geq I$, S_Q is a convex bounded, circled, and weakly compact set in $L(H)$ (it is also a neighborhood of zero).

Proof. The assertions follow as in the case of the class S_c from the above results about the class S_Q and we omit the details.

From the properties of S_Q in Theorem 13.2.5 we obtain the following useful theorem.

13.2.6 THEOREM For any bounded hermitian operator $Q \geq I$, the function

$$T \mapsto \|T\|_Q = \inf \{s \mid T \in sS_Q\}$$

is a Schwarz norm on $L(H)$.

From this class of Schwarz norms we can obtain, using the Calkin algebra, another class of Schwarz norms.

13.2.7 THEOREM Let Q_1 and Q_2 be two bounded hermitian operators and $Q_i \geq I$, $i = 1, 2$. In this case the function on $L(H)$

$$T \mapsto \|T\|_{Q_1} + \|\hat{T}\|_{Q_2}$$

where $\hat{T}$ denotes the image of T in the Calkin algebra of H, is a Schwarz norm on $L(H)$.

13.2.8 REMARK The above construction of Schwarz norms can be given in the case of B^*-algebras. For the construction of Schwarz norms we can use the representations of the B*-algebras in the algebra of L(H) for some H.

13.3 SCHWARZ NORMS ON BANACH SPACES

It is quite natural the problem about the existence of Schwarz norms on the algebra L(X) of all bounded operators on a Banach space X.

For this we recall that a function [,] on $X \times X$ is called a *semi-inner product* if the following conditions are satisfied:

1. $[x_1 + x_2, y] = [x_1, y] + [x_2, y]$
2. $[ax, by] = ab^*[x, y]$
3. $|[x, y]| \leq \|x\| \cdot \|y\|$
4. $[x, x] > 0$ for $x \neq 0$

for all $x_1, x_2, x, y \in X$ and a, b complex numbers.

Also we recall that an operator $T \in L(X)$ is called *hermitian* if $\|e^{itT}\| = 1$ for all real numbers t, or equivalently, if

$$W(T) = \{[Tx, x] \mid \|x\| = 1\}$$

is a subset of real numbers.

The operator $T \in L(X)$ is called *positive* if T is hermitian and the spectrum is in R^+. Now the definition of the class S_Q can be as follows.

13.3.1 DEFINITION An operator $T \in S_Q$ if and only if

1. The spectrum is in the unit disk.
2. For any $x \in X$ and $|z| < 1$,

$$\text{Re } [(1 + \Sigma\, Q^{1/2}T^nQ^{1/2}z^n)x, x] \geq 0$$

where Q is a hermitian positive operator such that $Q^{1/2}$ is also a hermitian operator.

The following results give indications about the possible existence of Schwarz norms.

13.3.2 EXAMPLE There exists a Banach space X and an operator T such that Re $[Tx, x] > 0$ does not imply Re $[T^{-1}x, x] \geq 0$.

For this proof we consider the Banach space ℓ_2^p of all pairs $x = (x_1, x_2)$ with the norm

$$x \mapsto \{|x_1|^p + |x_2|^p\}^{1/p} \qquad 1 < p < \infty$$

In this case it is easy to see that the semi-inner product compatible with the norm (i.e., $[x, x] = \|x\|^2$) is given by (see Chapter 6)

$$[x, y] = x_1|y_1|^{p-1} + x_2|y_2|^{p-2}$$

where $x = (x_1, x_2)$ and $y = (y_1, y_2)$.

Consider now an operator on this space with the matrix

$$\begin{bmatrix} a & 0 \\ c & b \end{bmatrix}$$

where a, b, c are complex numbers. The problem is to find conditions for a, b, c such that Re $[Tx, x] \geq 0$. An easy computation shows that these are

1. Re $a \geq 0$ Re $b \geq 0$
2. $|c| \leq (p \text{ Re } a)^{1/p}(q \text{ Re } b)^{1/q}$ $(1/p + 1/q = 1)$

and the conditions such that Re $[T^{-1}x, x] \geq 0$ are the following:

$$\left|\frac{c}{ab}\right| \geq (p \text{ Re } a^{-1})^{1/p}(\text{Re } q\, b^{-1})^{1/q}$$

and thus if Re $[Tx, x] > 0$, then Re $[T^{-1}x, x] > 0$ if and only if

$$|c| \leq |a|^{1-2/p}|b|^{1-2/q}(\text{Re } p\, a)^{1/p}(\text{Re } q\, b)^{1/q}$$

and this gives easily that Re $[Tx, x] \geq 0$ does not imply that Re $[T^{-1}x, x] \geq 0$.

13.3.3 REMARK In the case of Hilbert space operators the condition Re $T \geq 0$ implies the condition Re $T^{-1} \geq 0$.

We give now an example of a Banach space with the property that the induced norm on L(X) is not a Schwarz norm.

13.3.4 EXAMPLE If $X = \ell_2^1$, then the induced norm on L(X) is not a Schwarz norm.

We consider the operator T with the matrix

$$\begin{bmatrix} a & 0 \\ c & b \end{bmatrix}$$

and a simple computation shows that

$$\|T\| = \max \{|a| + |c|, |b|\}$$

We take now $0 < a < 1$ and in this case the operator with the matrix

$$\begin{bmatrix} a & 0 \\ 1 - a & 1 \end{bmatrix}$$

is a contraction operator. We consider now the function $f_\alpha(z)$ as defined in Lemma 13.1.1. The computation of the norm of the operator $f_\alpha(T)$ shows that this is equal to

$$\|f_\alpha(T)\| = a|\alpha + a| + (1 - a)\left|\frac{1 + \alpha + a(1 + \alpha^*)}{(1 + \alpha^* a)(1 + \alpha)}\right|$$

and thus for $\|f_\alpha(T)\| \le 1$, when α is a real number, we obtain

$$a|\alpha + a| + (1 - a)(1 + a) \le |1 + \alpha a|$$

which is not true if $\alpha = -(1/2)(a + 1)$.

In connection with the results presented above we mention the following characterization of Hilbert spaces obtained by Foias (1957).

13.3.5 THEOREM If X is a complex Banach space and for any contraction T, f(T) is also a contraction for all $|f| \le 1$, then X is a Hilbert space.

Proof. Let $x_0 \in X$ be arbitrary and $x_0^* \in X^*$ such that $\|x_0\|\|x_0^*\| \le 1$ and define the operator on X by the relation

$$Tx = x_0^*(x)x_0$$

It is clear that T is a contraction. From the hypothesis it follows that for any f_α, $f_\alpha(T)$ is also a contraction. This gives the relation

$$\|(T + \alpha)(I + \alpha^*T)^{-1}x\| < \|x\|$$

which is equivalent to the relation

$$\|(T + \alpha)x\| \le \|(I + \alpha^*T)x\|$$

From the form of the operator T is follows that

$$\|x_0^*(x)x_0 + x\| \le \|x + \alpha^*x^*(x)x_0\|$$

Now, if $x,y \in X$ and $\|x\| \ge \|y\| > 0$, we obtain from the Hahn-Banach-Sobczyc-Bohnenblust theorem that there exists $x_0^* \in X^*$ such that

$$\|x_0^*\| = \|x\|^{-1} \qquad x_0^*(x) = 1$$

We take $x_0 = y$ and remark that the operator T constructed with these elements gives us

$$\|y + \alpha x\| \le \|x + \alpha^*y\| \qquad |\alpha| < 1$$

and from the continuity argument it follows that this relation holds also for $|\alpha| = 1$. Now if $\|x\| = \|y\|$, changing the role of x with y and α with α^*, we obtain

$$\|x + \alpha^*y\| \ge \|y + \alpha x\|$$

Thus we have the equality

$$\|x + \alpha^*y\| = \|y + \alpha x\|$$

Now if $|\alpha| > 1$, then for $\beta = 1/\alpha$ we have, by the above result

$$\|x + \alpha^*y\| = |\alpha|\|\beta x + y\| = |\alpha|\|x + \beta^*y\| = \|\alpha x + y\|$$

and thus the relation is true for any α.

Now for $\alpha = p/q$, p and q real numbers, we obtain that

$$\|px + qy\| = |q|\|\frac{p}{q}y + x\| = |q|\|y + \frac{p}{q}x\| = \|qy + px\|$$

and thus for any x and y, $\|x\| = \|y\|$, and any p,q real numbers we obtain that

$$\|px + qy\| = \|qx + py\|$$

and by a famous result of F. A. Ficken this relation is characteristic for a norm to be an inner product norm, i.e., there exists an inner product $\langle\ ,\ \rangle$ on X such that for all $x \in X$,

$$\|x\|^2 = \langle x, x\rangle$$

Chapter 14

MAXIMUM THEOREMS FOR OPERATOR-VALUED HOLOMORPHIC FUNCTIONS

As is known, a great part of the theory of functions of a complex variable extends to vector-valued functions, but there exist cases in which the extension to this general setting is no longer valid. Such an example is given by Hille and Phillips (1957) in connection with the maximum principle. In recent years, in connection with some problems in operator theory, interesting results were obtained in connection with some numerical functions associated with an operator-valued function. The aim of this chapter is to give an account of the results obtained.

We first give some preliminaries about holomorphic and subharmonic functions.

14.1 HOLOMORPHIC FUNCTIONS

We suppose that the reader is familiar with the classical complex function theory. For this see, for example, Titchmarsh (1932).

Let X be a complex Banach space and D be a domain in the complex plane and $F: D \to X$.

14.1.1 DEFINITION The function f is called *holomorphic* if the function

$$x^*(f): D \to C$$

defined by $x^*(f)(z) = x^*(f(z))$ is a holomorphic function.

As a consequence of the principle of uniform boundedness, we have that f: $D \to X$ is holomorphic if and only if f is strongly continuous and strongly differentiable in D uniformly with z in any compact part of D.

Also, the following Cauchy formula holds: If f is holomorphic in D and C is a simple closed rectifiable curve in D, the interior of which is contained in D and z_0 such that arg $(z - z_0)$ increases by 2π when z describes C, then

$$f^{(n)}(z_0) = \frac{n!}{2\pi i} \int_C \frac{f(z)}{(z - z_0)^{n+1}} \, dz$$

for all n = 0, 1, 2, 3, ...

Also if we have a power series

$$\sum_{n=0}^{\infty} a_n (z - z_0)^n \qquad a_n \in X$$

then the series is absolutely convergent if $|z - z_0| < \rho$ and divergent for $|z - z_0| > \rho$ and if $\{z \mid |z - z_0| < \rho\}$ represents a holomorphic function. Here

$$\rho^{-1} = \lim \sup \{\|a_n\|^{1/n}\}$$

Now if f is a holomorphic function and

$$C = \{z \mid |z - z_0| = r\}$$

then we have

$$f(z_0) = \frac{1}{2\pi i} \int_C \frac{f(z)}{z - z_0} \, dz = \frac{1}{2\pi i} \int_0^{2\pi} f(z_0 + re^{it}) \, dt$$

which gives immediately that

$$\|f(z_0)\| \leq \frac{1}{2} \int_0^{2\pi} f(z_0 + re^{it}) \, dt$$

From this formula it follows that when f: $D \to X$ is a holomorphic function, then

$$z \mapsto \|f(z)\|$$

can have no maximum in D except the case when $\|f(z)\|$ is a constant function. The following example given in Hille and Phillips (1957, p. 100) shows that this case can appear.

Indeed, let $X = \ell_2^1$ of all pairs $z = (z_1, z_2)$ with the norm

$$\|z\| = \max \{|z_1|, |z_2|\}$$

and the function f defined as follows:

$$f(z) = a_0 + a_1 z$$

where $a_0 = (1, 0)$ and $a_1 = (0, 1)$, has the property that for all z, $|z| \le 1$,

$$\|f(z)\| = \|(1, z)\| = 1$$

14.2 SUBHARMONIC FUNCTIONS

Let D be a domain in the complex plane and f be a real function defined on D.

14.2.1 DEFINITION The function f is called *subharmonic* if the following conditions are satisfied:

1. $-\infty \le f(z) < \infty$.
2. f is upper semicontinuous.[†]
3. For any $z_0 \in D$ and such that $\{z \mid |z - z_0| < r\} \subset D$

$$f(z_0) \le \frac{1}{2\pi} \int_0^{2\pi} f(z_0 + re^{it})\, dt$$

[†]A real-valued function defined on D is called *upper semicontinuous* in the point $z_0 \in D$ if

$$\limsup_{z \to z_0} f(z) = f(z_0)$$

f is called *upper semicontinuous* in D if at each point it is upper semicontinuous.

4. For any z_0 as in 3,

$$\int_0^{2\pi} f(z_0 + re^{it})\, dt > -\infty$$

From the definition and the properties of convex functions we have the following result.

14.2.2 THEOREM If f defined on a domain D is subharmonic and g: R → R is any monotonically increasing convex function,† then $g_0 f$ is a subharmonic function.

Proof. First we remark that we define $g_0 f$ at the points z where $f(z) = -\infty$ as $\lim g(f(f(z_n)))$, where $z_n \to z$ and $f(z_n) > -\infty$. It is clear that $g_0 f$ is upper semicontinuous since g is an increasing function. Now, if $\{z \mid |z - z_0| < r\}$ is in D, then

$$g(f(z_0)) \le g\left(\frac{1}{2\pi}\int_0^{2\pi} f(z_0 + re^{it})\, dt\right)$$

$$\le \frac{1}{2\pi}\int_0^{2\pi} g(f(z_0 + re^{it}))\, dt$$

since g is a convex function. The other properties of subharmonic functions are obvious.

14.2.3 COROLLARY If f is a subharmonic function, then

1. e^f is a subharmonic function.
2. If $f \ge 0$, then f^p is a subharmonic function.

(Here $p \ge 1$ and $p < \infty$.)

The following assertion refers to the case of holomorphic functions.

14.2.4 COROLLARY If f: D → C is a holomorphic function, then

1. $|f(z)|^p = e^{p \ln|f(z)|}$

---

†The function h:R → R is *convex* if for all x,y ∈ R and t ∈ [0, 1] the following inequality holds:

$$h(tx + (1 - t)y) \le th(x) + (1 - t)h(y)$$

2. $\ln^+ |f(z)| = \max\{0, \ln |f(z)|\}$

are subharmonic functions.

Proof. The assertion follows from Corollary 14.2.3 since $\ln |f(z)|$ is a subharmonic function. Indeed, if $\ln |f(z)| = -\infty$ the assertion is trivial; in the contrary case since $\ln |f(z)| = \mathrm{Re} \ln f(z)$ which is an harmonic function in a neighborhood of z.

The following theorem is useful in obtaining subharmonic functions from operator-valued functions.

14.2.5 THEOREM If $\{f_n\}$ is a sequence of subharmonic functions and is monotone, then

$$f(z) = \lim f_n(z)$$

exists and is subharmonic.

Proof. From the monotonicity the existence follows.

We prove now that f is upper semicontinuous. Indeed, let

$$M = \sup_{x \in K} f(x)$$

where K is a compact, then we can find a sequence $\{x_n\}$ such that $M = \lim f(x_n)$ and we can suppose, without loss of generality, that $x_n \to x_0$. In this case we have

$$f(x_0) = \lim \sup f(x_0) \geq \lim f(x_n) = M \geq f(x_0)$$

and thus f is upper semicontinuous.

Suppose now that

$$\{z \mid |z - z_0| < r\}$$

is in D and thus we have

$$f(z_0) = \lim f_n(z_0) \leq \lim \frac{1}{2\pi} \int_0^{2\pi} f_n(z_0 + re^{it})\, dt$$

$$\leq \frac{1}{2\pi} \int_0^{2\pi} \lim f_n(z_0 + re^{it})\, dt = \frac{1}{2\pi} \int_0^{2\pi} f(z_0 + re^{it})\, dt$$

and thus f is a subharmonic function since the rest of the properties of subharmonic functions are trivially satisfied.

The following theorem can be used to give a new proof of Theorem 14.2.5.

14.2.6 THEOREM If $\{f_n\}$ is a sequence of subharmonic functions which are locally uniformly upper bounded in D, then

1. $\bar{f}(z) = \overline{\lim_{z^* \to z}} \sup_n f_n(z^*)$

2. $\underline{f}(z) = \overline{\lim_{z^* \to z}} \overline{\lim_n} f_n(z^*)$

are subharmonic functions.

Proof. We prove the assertion for $\bar{f}$ only; for $\underline{f}$ the proof is similar.

First we remark that $\bar{f}(z) < \infty$ and by Beppo Levi's theorem we obtain successively

$$\bar{f}(z) = \overline{\lim_{z^* \to z}} \sup_n f_n(z^*) \le \lim_{z^* \to z} \frac{1}{2\pi} \int_0^{2\pi} \sup_n f_n(z^* + re^{it})\,dt$$

$$\le \frac{1}{2\pi} \overline{\lim_{z^* \to z}} \sup_n \int_0^{2\pi} f_n(z^* + re^{it})\,dt$$

$$= \frac{1}{2\pi} \int_0^{2\pi} \bar{f}(z + re^{it})\,dt$$

and since all the conditions for subharmonicity are obviously satisfied, $\bar{f}$ is a subharmonic function.

The following theorem gives a very important property of subharmonic functions.

14.2.7 THEOREM If f is a subharmonic function in D, then f can have no maximum in D.

Proof. Suppose that there exists a point z_0 in the interior of D such that

$$f(z_0) = \sup_{z \in D} f(z)$$

and we define the set

$$M_{z_0} = \{z \mid f(z) = f(z_0)\}$$

which is clearly nonempty. Also this set is obviously a closed set and since D is open, it follows that $D \setminus M_{z_0}$ is open. We show now that in fact we have the equality $D = M_{z_0}$. Indeed, in the contrary case let C be a connected component of the set $D \setminus M_{z_0}$, and since D is a domain, it follows that $\partial C \cap M_{z_0} \neq \emptyset$. Let a_0 be an arbitrary point in $C \cap M_{z_0}$ and consider $r > 0$ so that the circle with the center at a_0 and radius t intersects C. Now we find t_1 and t_2 in $[0, 2\pi]$ such that

$$t \mapsto a_0 + re^{it}$$

applies the interval $[t_1, t_2]$ in a set in C. If z is in C, we have that

$$f(z) < f(a_0) = f(z_0)$$

and thus we can find $\varepsilon > 0$ such that for all $t \in [t_1, t_2]$ the following inequality holds

$$f(a_0 + re^{it}) < f(a_0) - \varepsilon$$

and thus

$$\frac{1}{2\pi}\int_0^{2\pi} f(z_0 + re^{it})\, dt$$

$$= \frac{1}{2\pi}\int_0^{t_1} f(a_0 + re^{it})\, dt + \frac{1}{2\pi}\int_{t_1}^{t_2} f(a_0 + re^{it})\, dt$$

$$+ \frac{1}{2\pi}\int_{t_2}^{2\pi} f(a_0 + re^{it})\, dt \leq f(a_0)t_1 + \frac{1}{2\pi}(f(a_0) - \varepsilon)(t_2 - t_1)$$

$$+ \frac{1}{2\pi}(f(z_0)(2\pi - t_2) = f(a_0) - \varepsilon\,\frac{t_2 - t_1}{2\pi} < f(a_0)$$

But this is a contradiction and the theorem is proved.

Suppose now that f is a subharmonic function defined on D and $z_0 \in D$; for $r > 0$ we define the number

$$J(r, z_0, f) = \frac{1}{2\pi} \int_0^{2\pi} f(z_0 + re^{it})\, dt$$

which is named, sometimes, the *mean* of the subharmonic function f.

The following lemma gives some information about the function

$$r \mapsto J(r, z_0, f)$$

14.2.8 LEMMA The function

$$r \mapsto J(r, z_0, f)$$

is monotonically increasing function of r in [0, 1] and a convex function with respect to ln r.

Since the proof is not difficult, we omit it. See also Rado (1937).

The following lemma gives a local property of subharmonic functions.

14.2.9 THEOREM If f is a subharmonic function, then f is local integrable.

Proof. Clearly f is integrable over any subdomain of D. We show that the integral is not $-\infty$. Indeed, in the contrary case we can find $z_0 \in D$ such that

$$\iint_{\{z \mid |z-z_0|<r_0\}} f(z)\, dx\, dy = -\infty$$

and r_0 is such that $\{z \mid |z - z_0| \leq 3r_0\}$ is in D. Since f is a subharmonic function in D, we have

$$f(z_0) \leq \frac{1}{2\pi} \int_0^{2\pi} f(z_0 + re^{it})\, dt$$

and since

$$\int_0^{r_0} f(z_0) r\, dr = \frac{1}{2} f(z_0) r_0^2 \le \frac{1}{2\pi} \int_0^{r_0} r\, dr \int_0^{2\pi} f(z_0 + re^{it})\, dt$$

$$= \frac{1}{2\pi} \iint_{\{z \mid |z-z_0| \le r_0\}} f(z)\, dx\, dy = -\infty$$

it follows that $f(z_0) = -\infty$. Let $z^* \in \{z \mid |z - z_0| < r_0\}$ and since

$$\{z \mid |z - z_0| < r_0\} \subset \{z \mid |z - z_0| < 3r_0\}$$

it follows that the integral of f on $\{z \mid |z - z^*| < r_0\}$ is also $-\infty$ and this gives that $f(z^*) = -\infty$. Thus $f(z) = -\infty$ in D and this is a contradiction. The assertion of the theorem is proved.

The following theorem gives another interesting property of subharmonic functions.

14.2.10 THEOREM If f is a subharmonic function in D and g is any upper semicontinuous function in D and

$$D \setminus \{z \mid f(z) = g(z)\}$$

has a Lebesgue measure zero, then for all $f \in D$ for which $g(z) < \infty$ $f(z) \le g(z)$.

Proof. Suppose that the assertion is false. We find a point $z_0 \in D$ such that $g(z_0) < f(z_0)$. Since g is upper semicontinuous at z_0, there exists $d > 0$ such that for $|z - z_0| < d$

$$g(z) < f(z_0)$$

and since f is subharmonic, we obtain that

$$g(z) < f(z_0) < \frac{1}{2\pi} \int_0^{2\pi} f(z_0 + re^{it})\, dt$$

If $z = z_0 + re^{it}$, then

$$\int_0^{2\pi}\int_0^{d} rg(z)\, dr\, dt = \iint_{|z-z_0| \le d} g(z)\, dx\, dy < \iint_{|z-z_0| \le d} f(z)\, dx\, dy$$

which is a contradiction.

To obtain necessary and sufficient conditions for a function to be subharmonic, we need some structure theorems.

The following two theorems will be used to obtain some useful criteria for subharmonicity of a function.

14.2.11 THEOREM A function defined on a domain D is subharmonic if and only if there exists a decreasing sequence of subharmonic functions $\{f_n\}$, $f_n \in C^\infty(G_n)$, $G_n \subset G_{n+1}$, and $f_n \to f$ (pointwise).

Proof. That the condition is sufficient follows from Theorem 14.2.5. We show now that it is also necessary. If $f(z) = -\infty$, then the sequence

$$f_n(z) = n$$

satisfies the theorem. Now if $f(z) \not\equiv \infty$, then, as we know, it is locally integrable and if we consider the function $w \in C^\infty(R)$

$$w: [0, 1] \to R \qquad w(t) = 0 \text{ for } t \geq 1$$

and

$$2\int_0^1 w(t)\, dt = 1$$

then the sequence

$$f_n(z) = \iint f\left(z + \frac{z'}{n}\right) w(|z'|)\, dx'\, dy'$$

satisfies the theorem.

Indeed, since f_n may be written in the form,

$$f_n(z) = \iint w(z')w(n|z - z'|)n^2\, dx'\, dy'$$

$$= \int_0^1 rw(r) \left[\int_0^{2\pi} f\left(z + \frac{re^{it}}{n}\right) dt\right] dr = 2\int_0^1 J\left(\frac{r}{n}, z, f\right) rw(r)\, dr$$

and this gives the fact that each f_n is a subharmonic function.

Also from this representation it follows that each f_n is of C^∞ class on G_n, where

$$G_n = \{z \mid \text{dist}(z, \partial D) > \frac{1}{n}\}$$

From the definition of the sequence $\{G_n\}$ it follows that

1. $G_n \subset G_{n+1}$
2. $\bigcup_n G_n = D$

Let $z \in D$ and $r > 0$ such that the circle with the center at z and radius r, $D(z, r) \subset D$. Let N such that $D(z, r)$ is also in G_n for $n > N$. Since the function $J(r/n, z, f)$ is monotone with respect to n, it follows that $\{f_n\}$ is a monotone sequence.

We show now that the sequence is pointwise convergent to f. Let $\varepsilon > 0$ and $d_\varepsilon > 0$ such that for $|z - z| < d_\varepsilon$,

$$f(z') < f(z) + \varepsilon$$

holds for all $z' \in D(z, d_\varepsilon)$. From the above relations we have

$$f_n(z) \leq f(z) + \varepsilon$$

and since

$$f(z) \leq f_n(z)$$

the assertion follows.

For the following theorem we use the notion of distribution in the sense of Schwartz (1952).

14.2.12 THEOREM If F is a function defined on D, then f is subharmonic if and only if

$$4 \frac{\partial^2 f(z)}{\partial z \partial z^*} = \Delta f(z) \geq 0$$

Proof. First we recall that the assertion

$$\Delta f(z) \geq 0$$

means that for any indefinite derivable function with compact support g,

$$\iint f \, \Delta \, g \, dx \, dy > 0$$

Of course, $g > 0$ and we suppose that f is not identically $-\infty$. Since f is not identically $-\infty$, it is locally integrable and the integral

$$<f, g> = \iint fg \, dx \, dy$$

exists and defines a distribution. Since f is a subharmonic function we have that

$$(\mu_r - \delta) \otimes f \geq 0$$

where

$$<\mu_r, h> = \frac{1}{2\pi} \int_0^{2\pi} h(re^{it}) \, dt$$

and since for $r \to 0$

$$\frac{\mu_r - \delta}{r^2} \to \frac{1}{4} \Delta\delta$$

it follows that for each g indefinite derivable and with compact support g

$$\left\langle \frac{\mu_r - \delta}{r^2}, g \right\rangle = \frac{1}{r^2}\left[\frac{1}{2\pi} \int_0^{2\pi} g(re^{it}) \, dt - g(0)\right] = \frac{1}{4} \Delta g(0)$$

But the convolution is continuous and thus

$$0 < \frac{\mu_r - \delta}{r^2} \otimes f \text{ implies } \frac{1}{4} \Delta\delta \otimes f = \frac{1}{4} \Delta f$$

and the necessity in the theorem is proved.

We show now that the condition is sufficient. Suppose that f is a distribution with the property that for any g indefinite derivable with compact support

$$\iint f \, \Delta \, g \, dx \, dy \geq 0$$

We use the known fact that there exists a measure dm such that

$$\Delta f = dm$$

and let $E(z) = (1/2) \ln |z|$ be the fundamental solution of this equation, i.e., $\Delta f = dm$.

If D' is a subdomain in D and dm' the restriction of dm to D' by the famous Riesz's decomposition,

$$f = E \otimes dm' + f_{D'}$$

where $f_{D'}$ is a subharmonic function in D'. We consider now the function

$$z' \mapsto \iint_{D'} \left|\ln |z - z'|\right| dx\, dy$$

and we remark that it represents a continuous function and the measure dm' defines a linear functional on the space of continuous functions on D'. Also it is obvious that this functional is continuous. Thus,

$$\iint_{D'} \left(\iint_{D'} \left|\ln |z - z'|\right| dx\, dy\right) dm'$$

is finite and Fubini's theorem gives that

$$\frac{1}{2\pi} \iint_{D'} \ln |z - z'|\, dm' = E \otimes dm$$

exists and is finite. The function

$$z \mapsto E \otimes dm$$

is a summable function and from Riesz's representation quoted above it follows that f is a locally integrable function and

$$f(z) = \frac{1}{2\pi} \iint_{D'} \ln |z - z'|\, dm' + f_{D'}(z) = h(z) \qquad \text{(a.e.)}$$

The assertion of the theorem is true if we show that the right member is a subharmonic function.

For this we consider the functions

$$f_n = \ln (|z - z'| + 1/n)$$

which is monotone and converges obviously to $\ln |z - z'|$. By the Levi theorem it follows that

$$h(z) = \lim g_n(z) + f_{D'}(z)$$

where

$$g_n(z) = \frac{1}{2\pi} \iint_{D'} \ln \left(|z - z'| + \frac{1}{n}\right) dm'(z')$$

Now the functions g_n are continuous and since

$$\ln (|z| + \varepsilon) = \sup_{|a|\le\varepsilon} \ln |z + a|$$

it follows that g_n is a subharmonic function. Since $f_{D'}$ is subharmonic, the assertion of the theorem follows from the theorem about decreasing sequence of subharmonic functions.

The following theorem gives a very useful criterion for subharmonicity.

14.2.13 THEOREM [Rádo (1937)] The function f defined on D is with ln f(z) subharmonic if and only if

$$z \mapsto |e^{az}| f(z)$$

is subharmonic for all complex numbers a.

Proof. The condition of the theorem is necessary since the function

$$z \mapsto \ln f(z) + \mathrm{Re}\,(az)$$

is subharmonic for all complex numbers a and thus $f(z)e^{az}$ is also a subharmonic function.

We show now that the condition is sufficient. First we remark that the function

$$z \mapsto \ln f(z)$$

is upper semicontinuous at every point $z \in D$ for which $f(z) \neq 0$.

Consider for this the functions

$$f_n(z) = \iint f\left(z + \frac{z'}{n}\right)w(|z'|)\, dx'\, dy' + \frac{1}{n}$$

where $w(\cdot)$ is the indefinite derivable function considered in the proof of Theorem 14.2.11. We mention that $w(\cdot)$ can be defined, for example, as follows:

$$w(r) = \begin{cases} Ce^{-1/(1-r^2)} & 0 \le r \le 1 \\ 0 & r \ge 1 \end{cases}$$

In this case we obtain

$$f_n(z)|e^{az}| = \iint f\left(z + \frac{z'}{n}\right)e^{|a(z+z'/n)|}w(|z'|)e^{-\mathrm{Re}(az'/n)}\, dx'\, dy' + \left(\frac{1}{n}\right)|e^{az}|$$

which are subharmonic functions in G_n (G_n is as in Theorem 14.2.11) for all n. Since

$$\Delta(f_n|e^{az}|) = |e^{az}|\left[\Delta f_n + 2b\frac{\partial f_n}{\partial x} - 2c\frac{\partial f_n}{\partial y} + (b^2 + c^2)f_n\right] \ge 0$$

where $a = b + ic$ and b, c real numbers. Now $f_n \ge 1/n$ and taking $b \to 0$ and $c \to 0$, we obtain

$$f_n\,\Delta\, f_n - \left(\frac{\partial f_n}{\partial x}\right)^2 - \left(\frac{\partial f_n}{\partial y}\right)^2 \ge 0$$

and since

$$\ln f_n = \frac{1}{f_n^2}\left[f_n\,\Delta\, f_n - \left(\frac{\partial f_n}{\partial x}\right)^2 - \left(\frac{\partial f_n}{\partial y}\right)^2\right]$$

it follows that $\Delta \ln f_n \ge 0$, i.e., f_n is a subharmonic function.

Since $\{f_n\}$ is a monotone sequence, the assertion of the theorem follows.

We mention here some very simple properties of subharmonic functions; the assertions follow from the definition and the Jensen inequality.

14.2.14 LEMMA 1. The set of subharmonic functions is closed for addition.

2. If f is subharmonic, then af is subharmonic for all $a > 0$.

3. If f and g are subharmonic functions, then $\sup\{f, g\}$ is also subharmonic.

14.3 MAXIMUM THEOREMS FOR THE NORM

In Section 14.1 we have proved that for any holomorphic function $f: D \to X$, D is a domain in the complex plane and X is a complex Banach space. Then for any $z_0 \in D$

$$\|f(z_0)\| \leq \frac{1}{2\pi} \int_0^{2\pi} \|f(z_0 + re^{it})\| \, dt$$

where $r > 0$ is such that $\{z \mid |z - z_0| \leq r\}$ is a subset of D.

This property can be formulated as follows.

14.3.1 THEOREM For any holomorphic function defined on a domain D with values in a Banach space, the function

$$z \mapsto \|f(z)\|$$

is subharmonic.

Also we have the following additional result.

14.3.2 THEOREM If $f: D \to X$ is a holomorphic function, then the function

$$z \mapsto \ln \|f(z)\|$$

is a subharmonic function.

Proof. For this we remark that the function

$$z \mapsto e^{az} f(z)$$

is holomorphic and by Theorem 14.3.1 the function

$$z \mapsto |e^{az}| \|f(z)\|$$

is a subharmonic function. Rado's theorem implies the assertion of the theorem.

14.3.3 COROLLARY If $f: D \to X$ is a holomorphic function, then for any positive increasing convex function g, the function

$$z \mapsto g(\|f(z)\|)$$

is subharmonic.

As a consequence of the fact that $\|f(z)\|$ is a subharmonic function, we have the following "maximum principle."

14.3.4 COROLLARY If $f: D \to X$ is a holomorphic function, then the function $\|f(z)\|$ can have no maximum in D unless $\|f(z)\|$ is constant in D.

The example quoted in Section 14.1 shows that there exists holomorphic functions f such that $\|f(z)\|$ may have a minimum other than zero in D. We are interested in the existence of the maximum principle as in the classical function theory.

For this we need the following definition.

14.3.5 DEFINITION Let X be a complex Banach space and D be a domain in the complex plane. We say that the *strong maximum principle holds* for X if for any $f: D \to X$ which is holomorphic and $\|f(z)\|$ is constant, then necessarily f is a constant function.

The example quoted above shows that for the space $X = \ell_2^1$ the strong maximum principle does not hold.

Using the notion of "complex extreme point" introduced by Thorp and Whitley (1967) we can characterize all complex Banach spaces for which the strong maximum principle holds.

14.3.6 DEFINITION Let X be a complex Banach space and x be a point of a convex set K. The point x is called a *complex extreme point* of K if $x + ay \in K$ for some $y \in X$ and all $|a| \le 1$, then $y = 0$.

It is clear that the following assertion holds.

14.3.7 LEMMA If x is a real extreme point of K, then x is a complex extreme point of K.

In what follows we give examples of Banach spaces for which a point is a complex extreme point of the unit ball of X and is not a real extreme point of the unit ball.

Let us consider a measure space $\{S, B, m\}$ and we suppose that $m(S) = 1$, that is, m is a probability measure. As is well known, there are classical Banach spaces associated with the measure space. These are the L^p spaces for $1 < p < \infty$.

The following theorem gives basic information about the complex extreme points for the spaces L^p for $1 \leq p < \infty$.

14.3.8 THEOREM If $\{S, B, m\}$ is a measure space and L^p is the space of all functions (classes) such that

$$f \mapsto \|f\|_p = \left[\int |f(s)|^p \, dm\right]^{1/p} < \infty$$

then every point of norm one is a complex extreme point of the unit ball.

Proof. We have two cases.

Case 1. $1 < p < \infty$. In this case we show that every point of norm one is a real extreme point of the unit ball. Indeed, if f is of norm 1 and is not an extreme point of the unit ball, then we can find a line segment joining two points of norm 1 and f is on this segment. Let $f = tf_1 + (1 - t)f_2$ with $\|f_1\| = \|f_2\| = 1$. Since $\|f\| = 1$, from the Minkowski inequality it follows that $f = f_1 = f_2$.

Case 2. $p = 1$. For this case we need some auxiliary results which are also of independent interest.

14.3.9 LEMMA If f and g are in L^1, then

$$\|f + g\| = \|f\| + \|g\|$$

if and only if there exists a positive function h on $S(f) \cap S(g)$ such that $f = hg$.

(We denote by $S(k)$, k a measurable function, the set $\{s \mid k(s) \neq 0\}$.)

Proof. Let h be a strictly positive function on $S(f) \cap S(g)$ and $f = hg$ a.e. on $S(f) \cap S(g)$. In this case we have

$$\|f + g\| = \int_{S(f)\setminus S(f)\cap S(g)} |f|\ dm + \int_{S(g)\setminus S(f)\cap S(g)} |g|\ dm$$

$$+ \int_{S(f)\cap S(g)} (1 + h)|g|\ dm = \|f\| + \|g\|$$

Conversely, if $\|f + g\| = \|f\| + \|g\|$, we have

$$\int (|f + g| - |f| - |g|)\ dm = 0$$

which gives that

$$|f + g|(s) = |f(s)| + |g(s)| \qquad \text{a.e. } (dm)$$

and from this it is clear that there exists a positive function h on $S(f) \cap S(g)$ such that $f = hg$.

14.3.10 LEMMA If f and g are in L^1 and

1. $\|f\| = 1$
2. $\|f + zg\| \leq 1 \qquad$ for all $|z| \leq 1$

then $\|f + zg\| = 1$ for all $|z| \leq 1$ and also $s(g) \subset s(f) \cup P$, where $dm(P) = 0$.

Proof. First we remark that

$$f = \frac{f + zg}{2} + \frac{f - zg}{2}$$

which gives, taking the norm,

$$1 \leq \left\|\frac{f + zg}{2}\right\| + \left\|\frac{f - zg}{2}\right\| \leq 1$$

which gives that for all $|z| \leq 1$, $\|f + zg\| = 1$. Now

$$1 = \|f\| = \int_{S(f)} |f|\ dm = \frac{1}{2}\int_{S(f)} |f - g|\ dm$$

$$+ \frac{1}{2}\int_{S(f)} |f - g|\ dm \leq 1$$

and since

$$\int_S (|f + g| + |f - g|)\, dm = \int_{S(f)} (|f + g| + |f - g|) + 2\int_{S\setminus S(f)} |g|\, dm$$

we obtain that

$$\frac{1}{2}\int_{S(f)} (|f + g| + |f - g|)\, dm = \frac{1}{2}\int_{S(f)} (|f + g| + |f - g|)\, dm + \int_{S\setminus S(f)} |g|\, dm$$

From this we have

$$\int_{S\setminus S(f)} |g|\, dm = 0$$

which proves the last assertion of the lemma.

Now we are in a position to prove Theorem 14.3.8 in Case 2. Suppose that f and g are two elements in L^1 with the properties:

1. $\|f\| = 1$
2. $\|f + zg\| \leq 1 \qquad$ for all $|z| \leq 1$

In this case by Lemma 14.3.10 $\|f + zg\| = 1$ for all $|z| \leq 1$. Further, we obtain

$$2 = \|f + zg\| + \|f - zg\| = \|f + zg + f - zg\| = 2\|f\|$$

and by Lemma 14.3.9 for each z, $|z| \leq 1$, we can find a positive function h_z on the set $S(f + zg) \cap S(f - zg)$ such that

$$f + zg = h_z(f - zg) \qquad \text{a.e. on } S(f + zg) \cap S(f - zg)$$

From this we obtain that

$$zg = (h_z - 1)\frac{f}{h_z + 1}$$

a.e. on the same set. Suppose now that $s \in S(f)$ and $|z|$ is so small that $s \in S(f + zg)$. If the above relation about zg holds at the point s, we obtain

$$2f(s)^*G(s) = [h_z(s) - 1]\frac{|f(s)|^2}{h_z(s) + 1}$$

and since the right-hand side of this equality is real, we must have $g(s) = 0$. Thus g vanishes a.e. on $S(f)$. Lemma 14.3.10 implies that g vanishes a.e. on S.

14.3.11 REMARK As can be seen, any point of norm 1 is not a real extreme point in the case of L^1. Thus the above theorem furnishes an example of a Banach space whose unit ball has no real extreme points, but every point of norm 1 is a complex extreme point.

The following theorem gives a characterization of those Banach spaces for which the strong maximum principle holds.

14.3.12 THEOREM In a Banach space the strong maximum principle holds if and only if each point of norm 1 is a complex extreme point of the unit ball of the Banach space.

This theorem will be proven as a consequence of some lemmas which also have their own interest.

14.3.13 LEMMA If X is a complex Banach space and $x \in X$, $\|x\| = 1$, is not a complex extreme point, then there exists a nonconstant holomorphic function on $\{z \mid |z| \leq 1\}$ such that $\|f(z)\| = 1$ for all $z \in \{z \mid |z| < 1\}$.

Proof. Since x is not a complex extreme point, we have that for some $y \in X$ and all z, $|z| \leq 1$,

$$\|x + zy\| \leq 1$$

and exactly as in Lemma 14.3.10 we can show that for all z, $|z| \leq 1$, $\|x + zy\| = 1$. Now we can define the function on the complex plane by the relation

$$f(z) = x + zy$$

which satisfies the lemma.

14.3.14 LEMMA If for any integer n,

$$P_n(z) = \sum_{p=0}^{n} \frac{(nz)^p}{p!} = e^{nz} - \sum_{p=n+1}^{\infty} \frac{(nz)^p}{p!}$$

and z satisfies the following inequalities:

1. $|z| \leq 1$
2. $|ze^{1-z}| \leq 1$

then $P_n(z) \neq 0$ for all n.

Proof. Since for any z we have

$$1 - e^{-nz}P_n(z) = e^{-nz} \sum_{p=n+1}^{\infty} \frac{(nz)^p}{p!} = e^{-n}(ze^{1-z})^n \sum_{p=n+1}^{\infty} \frac{n^p z^{p-n}}{p!}$$

and for z satisfying 1 and 2, this gives

$$|1 - e^{-nz}P_n(z)| < e^{-n} \sum_{p=n+1}^{\infty} \frac{n^p}{p!} = 1 - e^{-n}P_n(1) < 1$$

since $P_n(1)$ is a positive number. Clearly this implies the assertion of the lemma.

14.3.15 COROLLARY There exists a number $c \in (0, 1)$ such that if $|z| < c$, $P_n(z) \neq 0$ for all n.

Proof. This follows from the lemma since

$$0 \in \{z \mid |z| \leq 1\} \cap \{z \mid |ze^{1-z}| \leq 1\}$$

14.3.16 LEMMA If c is a positive number as in Corollary 14.3.15, then for any integer n there exist n complex numbers $z_1, z_2, \ldots, z_n$ such that

1. $|z_i| < 1/c$ $\quad i = 1, 2, \ldots, n$
2. $\Sigma_{i=1}^{n} z_i = n$
3. $\Sigma_{i=1}^{n} z_i^p = 0$ $\quad p = 2, 3, \ldots, n$

Proof. The roots of the polynomial

$$Q_n(z) = z^n - nz^{n-1} + \frac{n^2}{2!} z^{n-2} + \cdots + \frac{(-1)^n}{n!} n^n$$

satisfy conditions 2 and 3. Condition 1 follows from Corollary 14.3.15.

For any $r > 0$ let $D_r = \{z \mid |z| < r\}$. The following lemma gives some information about the image of D_c through holomorphic mappings.

14.3.17 LEMMA If $f: D_1 \to X$ is a holomorphic mapping such that $f'(0) \neq 0$, then there exists a positive constant c as in Corollary 14.3.15 such that the image of D_c through the mapping

$$z \mapsto f(0) + zf'(0)$$

belongs to the closure of the convex envelope of $f(D_1)$.

Proof. We consider the holomorphic mapping defined by

$$g(z) = f(z) - f(0)$$

and we remark that

1. $g'(0) = f'(0)$.
2. Convex envelope of $f(D_1) = f(0)$ + convex envelope of $g(D_1)$.

Now if g has the form

$$g(z) = a_1 z + a_2 z^2 + \cdots$$

we can choose, for any integer n, the complex numbers $z_1, z_2, \ldots$ as in Lemma 14.3.16, and we obtain

$$\frac{1}{n} \sum_{i=1}^{n} g(z_i z) = a_1 z + h_n(z)$$

where

$$nh_n(z) = a_{n+1}\left(\sum_{i=1}^{n} z_i^{n+1}\right) z^{n+1} + a_{n+2}\left(\sum_{i=1}^{n} z_i^{n+2}\right) z^{n+2} + \cdots$$

which gives

$$\|h_n(z)\| \le \sum_{p=n+1}^{\infty} \|a_p\| \left|\frac{z}{e}\right|^p = R_n\left(\frac{|z|}{c}\right)$$

where

$$R_n(|z|) = \|a_{n+1}\| |z|^{n+1} + \|a_{n+2}\| |z|^{n+2} + \cdots$$

Clearly, for $|z| < c$,

$$\lim \|h_n(z)\| = 0$$

which gives that

$$\lim \frac{1}{n} \sum_{i=1}^{n} g(z_i z) = a_1 z$$

and since

$$\frac{1}{n} \sum_{i=1}^{n} g(z_i z)$$

is in the convex envelope of $g(D_1)$, it follows that $a_1 z$ is in the closure if the convex envelope of $g(D_1)$ and the assertion is proved.

14.3.18 LEMMA If $f: D_1 \to X$ is a nonconstant holomorphic function then there exists an element $a_1 \in X$, $a_1 \neq 0$ such that the image of D_c through the mapping

$$z \mapsto f(0) + a_1 z$$

belongs to the closure of the convex envelope of $f(D_1)$.

Proof. We consider g defined as in the proof of Lemma 14.3.17 and put

$$g(z) = a_n z^n + a_{n+1} z^{n+1} + \cdots$$

If $\{\varepsilon_i\}_{i=1}^n$ are the n roots of the unity, then for any $z \in D_1$

$$\frac{1}{n}\left(\sum_{i=1}^{n} g(\varepsilon_i z^{1/n})\right) = a_n z + a_{n+1}(\varepsilon_1 + \cdots + \varepsilon_n) z^{(n+1)/n} + \cdots$$

and the function

$$G(z) = \frac{1}{n}\left(\sum_{i=1}^{n} g(\varepsilon_i z^{1/n})\right)$$

is holomorphic. Since $G'(0) = a_n \neq 0$, by Lemma 14.3.17, $G(D_1)$ belongs to the convex envelope of $g(D_1)$. The assertion of the lemma is implied by Lemma 14.3.17.

14.3.19 COROLLARY If K is a closed convex set in X and $f: D_1 \to X$ is a holomorphic mapping such that f(0) is a complex extreme point, then f is a constant function.

Proof. Suppose that f is nonconstant. Then for suitable $a_1 \in X$ and $c > 0$,

$$f(0) + a_1 D_c \subset K$$

which implies that f(0) is not a complex extreme point. This contradiction proves the assertion.

Proof of Theorem 14.3.12. The assertion of the theorem follows from Lemma 14.3.13 and Corollary 14.3.19.

14.3.20 REMARK In the case of a Hilbert space a more simple proof was found by Agmon given in Vesentini (1970). We show now that the arguments of Agmon can be used for the case of L^p spaces $(1 < p < \infty)$ or more generally for uniformly convex spaces. For simplicity we restrict our considerations to L^p spaces.

Let $f: D_1 \to L^p$ be a holomorphic function such that $\|f(z)\|$ is constant on D_1. We consider the semi-inner product on L^p [,], and for any point $z_0 \in D_1$ we define the holomorphic function

$$z \mapsto [f(z), f(z_0)]$$

which is obviously holomorphic. By the property of the semi-inner product we have that

$$|[f(z), f(z_0)]| \le \|f(z)\| \cdot \|f(z_0)\| = \|f(z_0)\|^2$$

and since the value of the function at z_0 is $\|f(z_0)\|$, this holomorphic function reaches its maximum in D_1. Thus it is a constant function, i.e.,

$$[f(z), f(z_0)] = \|f(z_0)\|^2$$

We use now a very simple characterization of "strictly convex spaces."

14.3.21 *LEMMA* A Banach space is strictly convex if and only if

$$[x, y] = \|x\| \cdot \|y\|$$

then $x = ay$ for some $a > 0$.

Since the proof is simple, we omit it. Also we recall that any L^p space or, more generally, any uniformly convex space is strictly convex. From this, in our case, if $f(z_0)$ is not zero, we find a function $c(z)$ such that

$$f(z) = c(z)f(z_0)$$

Also the complex-valued function

$$z \mapsto c(z)$$

is holomorphic and $|c(z)| = 1$. From the relation between $f(z)$ and $f(z_0)$ it follows that $c(z) = 1$.

We remark that the case $f(z_0) = 0$ is trivial.

14.4 MAXIMUM THEOREMS FOR THE SPECTRAL RADIUS AND FOR THE ESSENTIAL SPECTRAL RADIUS

In what follows we suppose that X is a complex Banach algebra with unit.

Let $f\colon D \to X$ be a holomorphic function and for each $z \in D$ we can consider the function

$$r_{f(z)}\colon D \to R^+$$

where $r_{f(z)}$ is the spectral radius of the element $f(z)$.

As a consequence of the Beurling-Gelfand formula, we have

$$r_{f(z)} = \lim \|f^n(z)\|^{1/n}$$

The following result is basic and has many applications in the theory of Banach algebras.

14.4.1 THEOREM For any f: D → X, holomorphic, the function

$$z \mapsto \log r_{f(z)}$$

is a subharmonic function.

Proof. Indeed, for any $z \in D$,

$$r_{f(z)} = \lim \|f^{2^n}(z)\|^{1/2^n}$$

and thus for every complex number a,

$$|e^{az}|r_{f(z)} = \lim \|[e^{az}f(z)]^{2^n}\|^{1/2^n}$$

From this it is clear that $\log(|e^{az}|r_{f(z)})$ is a limit of a pointwise decreasing sequence of subharmonic functions. Thus by Rado's theorem $\log r_{f(z)}$ is a subharmonic function.

14.4.2 COROLLARY For any f: D → X, holomorphic, the function

$$z \mapsto r_{f(z)}$$

is a subharmonic function.

To obtain new results as a consequence of Theorem 14.4.1, we need a function associated with an upper semicontinuous function.

Let u be an upper semicontinuous function on a neighborhood of the annulus

$$A = \{z \mid r_1 \leq |z| \leq r_2\}$$

with the property that $0 < u(z) < \infty$ and $\log u(z)$ is subharmonic on A. For every $r \in [r_1, r_2]$ the restriction of u to the circumference with center 0 and radius r reaches a maximum

$$M(r) = \max u(re^{it}) \qquad 0 \leq t \leq 2\pi$$

The following lemma gives an important property of the function M(r).

14.4.3 LEMMA The function log M(r) is an increasing and convex function of log r, $r \in [r_1, r_2]$.

Proof. First we remark that since $\log |z|$ is a subharmonic function, for every positive number t, $\log |z|^t u(z)$ is a subharmonic function on A. Thus, for $r \in [r_1, r_2]$ we have

$$r^t M(r) \leq \max \{r_1^t M(r_1), r_2^t M(r_2)\}$$

Now if $\inf \{M(r_1), M(r_2)\} > 0$, we can choose t such that

$$t = \log \frac{(M(r_2)/M(r_1))}{\log r_1 - \log r_2}$$

and thus

$$r^t M(r) \leq r_1^t M(r_1)$$

which implies that

$$t \log r + \log m(r) \leq t \log r_1 + \log M(r_1)$$

But $\log r = s \log r_1 + (1 - s) \log r_2$ for some $s \in [0, 1]$, and thus $\log M(r) \leq \log M(r_1) + t(1 - s) \log (r_1/r_2) < s \log M(r_1) + (1 - s) \log M(r_2)$.

Let $z_0 \in D$ and r_0 such that

$$\{z_0 + re^{it}\} \subset D$$

for all $0 \leq r \leq r_0$ and $0 \leq t \leq 2\pi$. If we put

$$m(r_f, z_0, r) = \max r_f(z_0 + re^{it})$$

then by the above lemma $\log M(r_f, z_0, r)$ is a convex increasing function of $\log r$ if $r \in (0, r_0)$. Let us consider the convex set

$$C = \{(x, y) \mid x = \log r \text{ for some } 0 < r < r_0, y > \log M(r_f, z_0, r)\}$$

and we draw the line

$$y = px + q$$

with $p > 0$. Clearly this line is disjoint from C. Thus we have

$$\log M(r_f, z_0, r) \geq p \log r + q$$

which is equivalent with

$$M(r_f, z_0, r) \geq e^q + r^p$$

This relation proves the following assertion.

14.4.4 THEOREM If f: D → X is a holomorphic function and r_f does not vanish identically on D, then the zeros of $r_{f(z)}$ in D have finite order.

14.4.5 COROLLARY If f is as in the above theorem, then

$$\{z \mid r_{f(z)} = 0\}$$

contains no interior points.

For a function f: D → X we can define another function by the formula

$$\theta(f)(z) = \max \{\mathrm{Re}\, z,\ z \in \sigma(f(z))\}$$

We can prove the following property of the function considered.

14.4.6 THEOREM If f: D → X is holomorphic, then the function $\theta(f(z))$ is a subharmonic function on D.

Proof. From the spectral mapping theorem, for any z ∈ D,

$$\sigma(e^{f(z)}) = e^{\sigma(f(z))}$$

and this gives that

$$r_e f(z) = \max \{|e^z| \mid z \in \sigma(f(z))\} = \max \{e^{\mathrm{Re}\, z} \mid z \in \sigma(f(z))\}$$

$$= e^{(\max \mathrm{Re}\, z, z\ \sigma(f(z))} = e^{\theta(f(z))}$$

But

$$z \mapsto \log r_e f(z)$$

is a subharmonic function and this gives the assertion of the theorem.

14.4.7 COROLLARY If F is a linear form on the complex numbers with real values, then for any holomorphic function $f: D \to X$ the function

$$G(f(z)) = \max \{F(a) \mid a \in \sigma(f(z))\}$$

is a subharmonic function.

Proof. We find two real numbers r and s such that

$$F(z) = s \operatorname{Re} z + r \operatorname{Im} z$$

and in this case we obtain

$$\begin{aligned} G(f(z)) &= \max \{\operatorname{Re} (za) \mid a \in \sigma(f(z))\} \\ &= \max \{\operatorname{Re} w \mid w \in a\sigma(f(z))\} \\ &= \max \{\operatorname{Re} \bar{z} \mid \bar{z} \in \sigma(af(z))\} \end{aligned}$$

and by Theorem 14.4.6 the assertion follows.

From the properties of the functions associated with a holomorphic mapping defined on D with values in a Banach algebra with unit we can prove some maximum theorems for the spectral radius as well as for the max Re z with z in the spectrum.

14.4.8 THEOREM If $f: D \to X$ is a holomorphic mapping and $r_{f(z)}$ is a constant in D, say c, then the intersection of $\sigma(f(z))$ with the circumference of center 0 and radius c is independent of $z \in D$.

Proof. We may assume, without loss of generality, that $c > 0$.

Now if S is the circumference with the center at 0 and radius c, we have that for all $z \in D$,

$$\emptyset \neq S \cap \sigma(f(z))$$

Let $\bar{z}_0 \in \sigma(f(z_0)) \cap S$ for a $z_0 \in D$. In this case for any $z \in D$

$$\sigma(f(z_0) + \bar{z}_0) = \sigma(f(z_0)) + \bar{z}_0$$

and since for all $z \in D$, $\sigma(f(z))$ is contained in the disk with center 0 and radius c, it follows that $\sigma(f(z) + \bar{z}_0)$ is contained in the disk with center $\bar{z}_0$ and radius $|\bar{z}_0|$.

Thus we have

$$r_{f(z)+\bar{z}_0} \leq |\bar{z}_0| + c = 2c \qquad \text{for all } z \in D$$

Clearly

$$r_{f(z)+\bar{z}} = 2|\bar{z}_0| = 2c$$

and this happens if and only if, for all $z \in D$, $2\bar{z}_0 \in \sigma(f(z) + \bar{z}_0)$ which is equivalent with the fact that $\bar{z}_0 \in \sigma(f(z))$ for all $z \in D$.

14.4.9 COROLLARY Suppose that X is a Banach algebra of all bounded linear operators on a Banach space L. If for each $z \in D$, $f(z)$ is a contraction and if a complex number $\bar{z}_0$ of modulus 1 is in $\sigma(f(z_0))$ for some $z_0 \in D$, then it belongs to $\sigma(f(z))$ for all $z \in D$.

14.4.10 COROLLARY If f is as in the above corollary and $1 + f(z_0)$ is invertible for some $z_0 \in D$, then $1 + f(z)$ is invertible for all $z \in D$.

14.4.11 REMARK Corollary 14.4.9 can be slightly generalized using another norm and another function, the so-called numerical norms and the operators are numerical contractions.

14.4.12 THEOREM If $f: D \to X$ is holomorphic and $\theta(f)(z)$ is equal to a constant, say a, on D, then the intersection of $\sigma(f(z))$ with the line

$$L = \{z \mid \text{Re } z = a\}$$

is independent of $z \in D$.

Proof. Let t be a positive number, and since by the spectral mapping theorem

$$\sigma(tf(z)) = t\sigma(f(z))$$

by Theorem 14.4.8, the intersection of $\sigma(f(z))$ with the circumference C_{at} with the center 0 and radius e^{at} is independent of $z \in D$. Since the function $z \mapsto e^z$ maps the line L into the circumference C_{at}, if a point $\bar{z} = a + ib$, a and b real numbers, belongs to $\sigma(f(z_0))$ for some z_0, then $t\bar{z} \in \sigma(tf(z_0))$. Thus, for any $z_1 \in D$ there is an integer $n(t)$ such that $t\bar{z} + 2\pi n(t)/i \in \sigma(tf(z_1))$, i.e.,

$$\bar{z} + \frac{2\pi n(t)i}{t} \in \sigma(f(z_1))$$

Now, if $\bar{z}$ is not in $\sigma(f(z_1))$, then $n(t) = 0$ for all $t > 0$ and

$$\frac{|2n(t)i|}{t} \geq \frac{2}{t} \to \infty$$

as $t \to 0$. This contradiction proves the theorem.

As in the case of Corollary 14.4.7 with the same notations we have the following.

14.4.13 COROLLARY If F has its maximum a at a point of D, then the intersection of $\sigma(f(z))$ with the line

$$\{z \mid F(z) = a\}$$

is independent of $z \in D$.

We consider now holomorphic mappings defined on the complex plane with values in a Banach algebra X.

We need the following result about convex functions.

14.4.14 LEMMA If f is a convex function on R^+ and nondecreasing and bounded, then f is constant.

Proof. Suppose that f is nonconstant. Of course, we may suppose that for $f(0) = 0$. If f is not constant, let $c = \sup f(s)$. In this case, for any $t_1 > 0$ such that $f(t_1) > 0$ there exists $t_2 > 0$ such that

$$\frac{f(t_2)}{t_2} < \frac{f(t_1)}{t_1}$$

(any straight line passing through 0 and $(t_1, f(t_1))$ intersects the line $f = c$ at a point $(ct_1/f(t_1), c)$ and any $t_2 > ct_1/f(t_1)$ satisfies our requirement).

Since f is not constant, the point t_1 can be chosen such that $0 < f(t_1) < c$. Since f is a convex function, for any $s \in [0, 1]$ we have

$$f(st_2) = f((1 - s)0 + st_2) \leq (1 - s)f(0) + sf(t_2) = sf(t_2)$$

Now for $s = t_1/t_2$ we obtain

$$f(t_1) \leq f(t_2) \frac{t_1}{t_2}$$

and this contradiction proves the assertion.

14.4.15 LEMMA If f is a subharmonic function on C and bounded, then f is constant.

Proof. Let z_0 be an arbitrary complex number and consider the means

$$J(z_0, r) = J(f, z_0, r) = \frac{1}{2\pi} \int_0^{2\pi} f(z_0 + re^{it})\, dt$$

which is a convex nondecreasing function of log r. Since f is bounded, J is bounded. Thus J is a constant function and let J_0 be the constant. In this case $f(z_0) \leq J_0$. But f is upper semicontinuous, and thus for every $\varepsilon > 0$ we find that the set

$$A_\varepsilon = \{z \mid f(z) < f(z_0) + \varepsilon\}$$

is open. Let $r > 0$ such that the set

$$\{z_0 + re^{it} \mid \{0 \leq t \leq 2\pi\}$$

is in A_ε. Since we have the following inequalities,

$$f(z_0 + re^{it}) \leq f(z_0) + \varepsilon$$

we have that

$$J_0 \leq f(z_0) + \varepsilon$$

and since $\varepsilon > 0$ was arbitrary, it follows that $f(z_0) = J_0$. Thus f is a harmonic function which is bounded. But in this case f is constant.

As a consequence, we obtain the following corollary.

14.4.16 COROLLARY If $f: C \to X$ is holomorphic and $r_{f(z)}$ is bounded on C, then r_f is constant.

14.4.17 COROLLARY If $f: C \to X$ is holomorphic and X is a semisimple commutative Banach algebra, then if

$$r_{f(z)} = 0(|z|^N)$$

for some $N > 0$, then f is of the form

$$f(z) = a_0 + a_1 z + \cdots + a_N z^N$$

with $a_i \in X$.

Proof. From the condition of the theorem it follows that for some constant $c > 0$,

$$r_{f(z)} \le c|z|^N$$

and if m is any multiplicative functional on X, $m(f(z))$ is an entire function satisfying the relation

$$|m(f(z))| \le r_{f(z)} < c|z|^N$$

and by a well-known theorem in the theory of functions, $m(f(z))$ is a polynomial of degree $\le N$. Since f has the power expansion

$$f(z) = a_0 + a_1 z + \cdots + a_n z^n + \cdots$$

it follows that for any multiplicative functional m and any $n > N$,

$$m(a_n) = 0$$

and since X is semisimple, it follows that $a_i = 0$ for $i > N$.

For the next result we need the following result of Gleason.

14.4.18 LEMMA (A. M. Gleason) [Browder, (1969)] If g is an entire function and

$$\text{Re } g(z) = 0(|z|^N)$$

then g is of the form $g(z) = b_0 + b_1 z + \cdots + b_n z^N$.

Proof. We may suppose, without loss of generality, that $g(0) = 0$ and let $g(z) = \Sigma\, a_i z^i$. In this case,

$$\text{Re } g(z) = \Sigma\, r^n(\text{Re } a_n \cos nt - \text{Im } a_n \sin nt)$$

which implies that

$$\frac{1}{2\pi}\int_0^{2\pi} \text{Re } g(re^{it}) \cos kt\, dt = r^k \text{ Re } a_k$$

$$\frac{1}{2\pi}\int_0^{2\pi} \text{Re } g(re^{it}) \sin kt\, dt = r^k \text{ Im } a_k$$

for all $k = 1, 2, 3, \ldots$. From these it is clear that

$$\pm\text{Re } a_k \leq 2\,\frac{M}{r^{k-N}} \qquad \pm\text{Im } a_k \leq 2\,\frac{M}{r^{k-N}}$$

since $g(re^{it}) \leq Mr^N$. From these relations we obtain for $r \to \infty$ that $a_i = 0$ for all $i > N$.

14.4.19 THEOREM If $f: D \to X$ is a holomorphic mapping and the function θ defined as in Theorem 14.4.6 has a property that

$$\theta(f(z)) = O(|z|^N)$$

and X is semisimple commutative Banach algebra, then f is a polynomial.

Proof. Indeed, from the hypothesis of the theorem it follows that there exists $c > 0$ such that

$$\theta(f(z)) \leq c|z|^N$$

and since by the definition of θ, we conclude that for any multiplicative linear functional m,

$$\text{Re } m(f(z)) \leq c|z|^N$$

and by the Gleason result, f is a polynomial of degree at most N. From the fact that X is a semisimple Banach algebra we have the assertion of the theorem.

As an application of the results about the spectral radius and function θ, we can obtain some maximum theorems for the spectrum as well as for the real part.

14.4.20 DEFINITION If f is a function defined on a domain D with values sets which are all subsets of a set T, then we say that f attains the maximum in $z_0 \in D$ if for all $z \in D$,

$$f(z) \subset f(z_0)$$

We need the following lemma, which also has independent interest, since it contains some information about the behavior of the boundary of $\sigma(f(z))$.

14.4.21 LEMMA If $f: D \to X$ is a holomorphic mapping and there is a point $\bar{z}_0$ such that $\bar{z}_0 \notin \sigma(f(z))$ for all $z \in D$, and if the distance between $\bar{z}_0$ and $\sigma(f(z))$ attains its minimum value c for some $z_0 \in D$, then the intersection of $\sigma(f(z))$ with the circumference of center $\bar{z}_0$ and radius c is independent of $z \in D$.

Proof. Let us consider the following mapping

$$G: C \setminus \bar{z}_0 \to C$$

defined by

$$G(z) = \frac{1}{z - \bar{z}_0}$$

which is clearly holomorphic. Since $\bar{z}_0 \notin \sigma(f(z))$, the function $(f(z) - \bar{z}_0)^{-1}$ has the values in X, and by the spectral mapping theorem we have

$$\sigma((f(z) - \bar{z}_0)^{-1} = G(\sigma(f(z)))$$

for all $z \in D$. We obtain further that

$$\begin{aligned}
r_{((f(z)-\bar{z}_0)^{-1})} &= \max\{|s| \mid s \in \sigma((f(z)) - \bar{z}_0)^{-1})\} \\
&= \max\{|s| \mid s \in G(\sigma(f(z)))\} \\
&= \max\left\{\frac{1}{|s - \bar{z}_0|} \mid s \in G(\sigma(f(z))\right\} \\
&= \min_{s\in\sigma(f(z))} |s - \bar{z}_0|^{-1} = \frac{1}{d(\bar{z}_0, \sigma(f(z)))}
\end{aligned}$$

where d(a, b) denotes the distance between a and b.

Since

$$d(\bar{z}_0, \sigma(f(z)) \geq d(\bar{z}_0, \sigma(f(z_0)) = c$$

we obtain

$$\begin{aligned}
r_{(f(z)-\bar{z}_0)^{-1})} &= \frac{1}{d(\bar{z}_0, \sigma(f(z)))} \leq \frac{1}{d(\bar{z}_0, \sigma(f(z_0)))} \\
&= r_{((f(z_0)-\bar{z}_0)^{-1})} = \frac{1}{c}
\end{aligned}$$

Thus, for all $z \in D$ we have

$$r_{(f(z)-\bar{z}_0)^{-1})} = \frac{1}{c}$$

and the results proved above imply that

$$\sigma(f(z)) \cap S_{(0,1/c)}$$

is independent of $z \in D$.

From this result we can obtain, in the case when the maximum is attained, some information about boundaries.

14.4.22 THEOREM If $f: D \to X$ is holomorphic and $z \mapsto \sigma(f(z))$ assumes its maximum at a point $z_0 \in D$, then the boundary of $\sigma(f(z_0))$ belongs to the boundary of $\sigma(f(z))$ for all $z \in D$.

Proof. Let $s \in \partial\sigma(f(z_0))$ and we prove that $s \in \partial\sigma(f(z))$ for all $z \in D$. Suppose on the contrary, that there exists a point $z_1 \in D$

such that $s \notin \partial\sigma(f(z_1)))$. As we know, $\sigma(f(z_1))$ is a closed and bounded set, and thus for some $\varepsilon > 0$ the disk with center at s and radius ε is disjoint from $\sigma(f(z_1)))$, i.e.,

$$\sigma(f(z_1)) \cap D_{(s,\varepsilon)} = \emptyset$$

Since $s \in \partial\sigma(f(z_0))$, there exists s_0 such that

$$s_0 \in \{C \setminus \sigma(f(z_0))\} \cap D_{(s,\varepsilon/3)}$$

and thus for all $z \in D$,

$$d(s_0, \sigma(f(z))) \geq d(s_0, \sigma(f(z_0))) \geq |s - s_0| < \frac{\varepsilon}{3}$$

and

$$d(s_0, \sigma(f(z_1))) > \frac{2}{3}\varepsilon$$

which contradicts Lemma 14.4.21.

14.4.23 DEFINITION We say that for a set-valued function f defined on a domain D, the *maximum principle holds* if for some $z_0 \in D$,

$$f(z) \subset f(z_0)$$

for all $z \in D$ implies that f(z) is constant.

The following examples show that in the case of a holomorphic function with values in a Banach algebra the set-valued function

$$z \mapsto \sigma(f(z))$$

admits no maximum principle.

14.4.24 EXAMPLE Let us consider the Banach algebra of all bounded sequences with coordinatewise multiplication and the well-known norm

$$x = (x_i) \mapsto \|x\| = \sup |x_i|$$

Clearly we have a commutative Banach algebra with unit e = (1) and an element is invertible if and only if

$$\inf |x_i| > 0$$

for $x = (x_i)$. From this it is clear that the spectrum of an element x is exactly the closure of the set $\{x_i\}$.

Let a be an element in this algebra defined as follows:

$$a = (a_0, 0, a_2, 0, a_4, 0, \ldots, 0, a_{2n}, 0, \ldots)$$

such that

$$\sigma(a) = \left\{ z \mid \tfrac{1}{3} \le |z| \le 1 \right\}$$

Let b be another element defined as follows:

$$b = (0, b_1, 0, b_3, \ldots, 0, b_{2n+1}, 0, \ldots)$$

with

$$\sigma(b) = \left\{ z \mid |z| \le \tfrac{2}{3} \right\}$$

We consider now the holomorphic function on C, the set of complex numbers, defined as follows:

$$f(z) = a + zb$$

If $D = \{z \mid |z| \le 1\}$, then a computation shows the following relations:

$$\sigma(f(z)) \subset D \qquad \text{for all } z,\ |z| \le \tfrac{2}{3}$$

$$\sigma(f(z)) = D \qquad \text{for all } z,\ \tfrac{1}{3} \le |z| \le \tfrac{2}{3}$$

and thus for z, $|z| < 1/2$, $\sigma(f(z)) \ne D$.

14.4.25 EXAMPLE Let ℓ^p, $1 \le p < \infty$, be the Banach space of all sequences $x = \{s_i\}_{-\infty}^{\infty}$ with the norm

$$x \mapsto \|x\| = \{\Sigma\ |x_i|^p\}^{1/p}$$

If e_i denotes the element

$$e_i = \{\ldots, 0, \underset{i}{1}, 0, \ldots\}$$

then we define an operator T on ℓ^p by the relations

$$Te_i = \begin{cases} 0 & i = 0 \\ e_{i-1} & i \neq 0 \end{cases}$$

and it is easy to see that T is bounded. More precisely, $\|T\| = 1$.

From this it is clear that the spectrum of T is contained in

$$D = \{z \mid |z| \leq 1\}$$

and we show now that in fact is equal to this set. Let z_0 be any point with the property $|z_0| < 1$. Consider now the element

$$x_{z_0} = \sum_{n=0}^{\infty} z_0^n e_n$$

and since

$$Tx_{z_0} = z_0 x_{z_0}$$

it follows that $z_0 \in \sigma(T)$. Now since $\sigma(T)$ is closed, the assertion follows. Consider now another operator S defined on ℓ^p by the relations

$$Se_i = \begin{cases} e_{-1} & i = 0 \\ 0 & i \neq 0 \end{cases}$$

which is also bounded. If we consider $L(\ell^p)$ the Banach algebra of all bounded linear operators on ℓ^p, then we can define a holomorphic mapping by

$$f(z) = T + zS$$

A simple computation shows that

$$\|f(z)\| = \max \{|z|, 1\}$$

and also for any integer n

$$\|f^n(z)\| = \max \{|z|, 1\}$$

From these relations we obtain that

$$r_{f(z)} = 1$$

for all z. From Theorem 14.4.8 it follows that for all z, $\sigma(f(z))$ contains the unit circumference

$$S = \{z \mid |z| = 1\}$$

We show now that for all $z \neq 0$, $\sigma(f(z)) = S$. Let $z \neq 0$ and G_z be the shift operator defined on ℓ^p by

$$G_z e_i = \begin{cases} \dfrac{1}{ze_0} & i = -1 \\ e_{i+1} & i \neq -1 \end{cases}$$

and we can easily see that

$$G_z f(z) = f(z) G_z = I$$

and this gives that f(z) is invertible. From this we obtain that the spectrum of G_z is the image of $\sigma(f(z))$ by the holomorphic mapping defined on $C \setminus \{0\}$,

$$z \mapsto \frac{1}{z}$$

and since this mapping leaves S invariant and $r_{f(z)} = 1$,

$$\sigma(f(z)) = S$$

and the assertion is proved in the case $z \neq 0$. Also, since $\sigma(f(0))$ is $\{z \mid |z| \leq 1\}$, the mapping

$$z \mapsto \sigma(f(z))$$

has no maximum principle.

As a positive result we mention the following consequence of Theorem 14.4.12.

14.4.26 THEOREM If $f: D \to X$ is a holomorphic mapping on D with values in a Banach algebra X and

$$z \mapsto \sigma(f(z))$$

assumes its maximum at a point $z_0 \in D$ and if $\sigma(f(z_0))$ has no interior points, then $z \mapsto \sigma(f(z))$ is constant on D.

Now we give some results about the essential spectral radius. We use the same method, i.e., the fact that the spectral radius is a limit of a special sequence, which was introduced in Istrăţescu (1969).

As a main tool in studying the essential radius, we use a formula proved by Lebow and Schechter (1971). Let us consider the Banach space X and for any bounded set A of X we define the number q(A) as the greatest lower bound of the set of numbers $r > 0$ such that A can be covered by a finite set of open spheres of radius r. As in the case of Kuratowski's measure of noncompactness $\alpha(A)$, the following properties can be proved for $q(\cdot)$:

1. $0 \leq q(A) \leq \text{diam } A$.
2. $q(A) = 0$ if and only if A is relatively compact.
3. $q(A \cup B) = \max \{q(A), q(B)\}$.
4. $q(A + B) \leq q(A) + q(B)$.

We can define a seminorm on the Banach space L(X) by the formula

$$\|T\|_q = q(TS_{(0,1)})$$

where $S_{(0,1)}$ is the closed unit sphere of X. Clearly $\|T\|_q = 0$ if and only if T is a compact operator. Also it is easy to see that we have the relation

$$\|PQ\|_q \leq \|P\|_q \cdot \|Q\|_q$$

Another useful seminorm on L(X) is considered in the following.

14.4.27 DEFINITION For any operator $T \in L(X)$ the number $\|T\|_m$ is defined as the *greatest lower bound* of all numbers $r > 0$ having the property that there is a subspace X_1 of X with finite codimension and such that

$$\|Tx\| \leq r\|x\|$$

for all $x \in X_1$.

The following theorem gives information about the numbers $\|T\|_q$ and $\|T\|_m$.

14.4.28 THEOREM The function on L(X)

$$T \mapsto \|T\|_m$$

is a seminorm equivalent to the seminorm

$$T \mapsto \|T\|_q$$

and in fact the following relation holds:

$$\frac{1}{2}\|T\|_q \leq \|T\|_m \leq \|T\|_q$$

Proof. Let $\varepsilon > 0$ be given. We can find the elements $y_1, y_2, \ldots, y_n \in X$ such that

$$\min \|Tx - y_i\| \leq \|T\|_q + \varepsilon$$

for all x, $\|x\| \leq 1$. By the extension theorem there exists $\{y_i'\}$ in X* such that

1. $\|y_i'\| = 1$
2. $y_i'(y_i) = \|y_i\|$

for $1 \leq i \leq n$. Let M be the set of all x annihilated by T^*y_i' for $1 \leq i \leq n$. Clearly this subspace has finite codimension in X. For any $x \in M \cap S_{(0,1)}$ let y_k be one of the elements which is the closest to Tx. Since $x \in M$, we have $y_k'(Tx) = T^*y_k'(x) = 0$. Thus,

$$\|y_k\| = y_k'(y_k) = y_k'(y_k - Tx) \leq \|y_k - Tx\|$$

From the definition of the elements y_k we have

$$\|Tx\| = \|Tx - y_k + y_k\| \leq \|Tx - y_k\| + \|y_k\|$$
$$\leq \|T\|_q + \varepsilon + \|T\|_q + \varepsilon = 2(\|T\|_q + \varepsilon)$$

and since $\varepsilon > 0$ is arbitrary, the right-hand side of the inequality is proved. For the rest, let $\varepsilon > 0$ be arbitrary. Then we can find a subspace M in X having a finite codimension and

$$\|Tx\| \leq (\|T\|_m + \varepsilon)\|x\|$$

for all $x \in M$. Let P be the bounded projection onto M and thus

$I - P$ is a finite operator (i.e., the rank of $I - P$ is finite). In this case for any $x \in X$ we have

$$\|Tx\| = \|(TP + T(I - P))x\| \le \|TPx\| + \|T(I - P)x\|$$

$$\le \|T\| + \varepsilon\| \cdot \|Px\| + \|T\| \cdot \|(I - P)x\|$$

and since

$$\|Px\| = \|(I - P)x - x\| \le \|x\| + \|(I - P)x\|$$

we obtain further that

$$\|Tx\| \le (\|T\|_m + \varepsilon) + (2\|T\| + \varepsilon)\|(I - P)x\|$$

Now the projector P is a compact operator, and thus we can find $x_1, x_2, \ldots, x_n$ in $S_{(0,1)}$ of X such that

$$\min \|(I - P)(x - x_k)\| \le \frac{\varepsilon}{2} (\|T\| + \varepsilon)$$

for all $x \in S_{(0,1)}$. If x is an arbitrary element of $S_{(0,1)}$ and x_k is the element satisfying the above inequality, we obtain

$$\|T(x - x_k)\| \le (\|T\|_m + \varepsilon)\|(x - x_k)\| + (2\|T\| + \varepsilon)\|(I - P)(x - x_k)\|$$

$$\le 2(\|T\|_m + \varepsilon) + \varepsilon$$

and since $\varepsilon > 0$ is arbitrary, we obtain

$$\|T\|_q \le 2\|T\|_m$$

We prove now that

$$T \mapsto \|T\|_m$$

is a seminorm with the property

$$\|ST\|_m \le \|T\|_m \|S\|_m$$

for all $T, S \in L(X)$. Since all the properties of the seminorm are easy to see, we prove the last property only. Let $\varepsilon > 0$ and we can find a subspace M of finite codimension such that

$$\|Tx\| \le (\|T\|_m + \varepsilon)\|x\|$$

for all $x \in M$. Now if N is also a subspace of finite codimension such that

$$\|Sx\| \le (\|S\|_m + \varepsilon)\|x\|$$

for all $x \in N$ we define W to be the set of all $x \in M$ such that $Tx \in N$. Clearly W is a subspace of X having also finite codimension.

For $x \in W$ we have

$$\|STx\| \le (\|S\|_m + \varepsilon)(\|T\|_m + \varepsilon)\|x\|$$

and since $\varepsilon > 0$ is arbitrary the assertion follows.

For any operator $T \in L(X)$ we define the Weyl spectrum as

$$\sigma_w(T) = \cap\, \sigma(T + k)$$

where the intersection is taken over all compact operators in $L(X)$. The essential spectral radius of T is defined as

$$r_{\text{ess } T} = \sup \{|z| \mid z \in \sigma_{\text{ess } T}\}$$

The following basic relation about the seminorms $\|\cdot\|_m$ and the essential spectral radius was proved by Lebow and Schechter (1971, Corollary 6.6) (see also Chapter 11):

$$r_{\text{ess } T} = \lim \|T^n\|_m^{1/n}$$

In what follows we use the sequence

$$k_n = 2^n$$

and in this case we have

$$r_{\text{ess } T} = \lim \|T^{2^n}\|_m^{1/2^n}$$

Exactly as in the case of a norm we can prove the following.

14.4.29 THEOREM If $f\colon D \to L(X)$ is a holomorphic function, then the function $z \mapsto \|f(z)\|_m$ is subharmonic.

Using this and the Rado theorem, as well as the above result about $r_{ess\ T}$, we obtain

14.4.30 THEOREM If f is as in the above theorem, then

$$z \mapsto \log r_{ess\ f(z)}$$

is a subharmonic function.

14.4.31 COROLLARY If f is as above, then

$$z \mapsto r_{ess\ f(z)}$$

is a subharmonic function.

Let f be as in the above theorems and define the function

$$\theta_{ess}(f(z)) = \sup\{\mathrm{Re}\ z \mid z \in \sigma_{ess\ f(z)}\}$$

In this case we can prove, as in the case of the function θ, the following result.

4.4.32 THEOREM If f is as in the above theorems, then the function

$$z \mapsto \theta_{ess}(f(z))$$

is subharmonic.

Now it is clear that we can obtain results about the above functions which are the analogs of the results about the spectral radius and function θ. These are left to the reader.

14.5 MAXIMUM THEOREMS FOR OTHER OPERATOR-VALUED HOLOMORPHIC FUNCTIONS

As we know for any operator $T \in L(X)$, where X is a Hilbert space or a Banach space, we can define some quantities which are useful in the study of operators.

In what follows we give some results about these functions in connection with the property of a function being subharmonic. First we consider the numerical radius.

14.5.1 THEOREM If $f: D \to L(X)$ is a holomorphic function with values in $L(X)$, then the function

$$z \mapsto \log w(f(z))$$

is subharmonic.

Proof. For any $x \in X$, $\|x\| = 1$, we consider the complex-valued holomorphic function

$$f_a(z) = e^{az}[f(z)x, x]$$

and clearly the modulus of this function is subharmonic. Since for all $z \in D$,

$$|e^{az}[f(z)x, x]| \le \|e^{az}f(z)x\| \le \|e^{az}f(z)\|$$

to prove that the numerical radius is a subharmonic function it is sufficient to prove that it is an upper semicontinuous function. Let $z_0 \in D$ be arbitrary and $c > 0$ such that

$$\sup_{\|x\|=1} |e^{az}[f(z)x, x]| \le c$$

Since the function

$$z \mapsto \|e^{az}f(z)\|$$

is continuous, we can find $d > 0$ such that if $|z - z_0| < d$, then

$$\|e^{az}f(z) - e^{az_0}f(z_0)\| \le \varepsilon$$

Thus for any $x \in X$, $\|x\| = 1$, and any $z \in D$, $|z - z_0| < d$ we have

$$|[e^{az}f(z)x, x] - [e^{az_0}f(z_0)x, x]| \le |[(e^{az}f(z) - e^{az_0}f(z_0))x, x]|$$
$$\le \|e^{az}f(z) - e^{az_0}f(z_0)\| < \varepsilon$$

and this gives that

$$|[e^{az}f(z)x, x]| \le |[e^{az}f(z)x, x] - [e^{az_0}f(z_0)x, x]|$$
$$+ |[e^{az_0}f(z_0)x, x]| \le \varepsilon + w(e^{az_0}f(z_0))$$

and for z satisfying the relation $|z - z_0| < d$, we obtain

$$w(e^{az}f(z)) = \tilde{c} < C$$

and this proves our assertion since we can take $a = 0$.

14.5.2 COROLLARY If f is as in Theorem 14.5.1, then

$$z \mapsto w(f(z))$$

is a subharmonic function.

14.5.3 THEOREM If f is as in Theorem 14.5.1 and X is a Hilbert space, then the intersection of the closure of W(f(z)) with the circumference

$$S = \{z \mid |z| = c\}$$

is independent of $z \in D$ when the function

$$z \mapsto w(f(z))$$

is a constant function (and equal to c).

Proof. Since for any $a \in C$,

$$W(f(z)) + a = W(f(z) + a)$$

the equality follows also for the closure. We can suppose, without loss of generality, that $c > 0$. For all $z \in D$,

$$\text{Cl } W(f(z)) \cap S \neq \emptyset$$

Suppose now that there exists two points z_0 and z_1 such that

$$\text{Cl } W(f(z_0)) \cap S \neq \text{Cl } W(f(z_1)) \cap S$$

and we take s_0 in Cl $W(f(z_0))$ and not in Cl $W(f(z_1))$. Then for $a > 0$

$$\text{Cl } W(f(z) + as_0)) = \text{Cl } W(f(z)) + as_0$$

is in the disk with center as_0 and radius c. This disk is also contained in the closure of the disk C with the center 0 and the radius $c + a|s_0| = (a + 1)c$. Since the boundary of this circle is tangent to the boundary of the circle C at the point $s_0 + as_0$, we obtain

$$w(f(z_0) + as_0) \leq (a + 1)c$$

for all $z \in D$. Since $s_0 + as_0 \in \mathrm{Cl}\ W(f(z_0) + as_0)$, we have that in fact in the above inequality,

$$w(f(z_0) + as_0) = (a + 1)c$$

and by Corollary 14.5.2

$$w(f(z) + as_0) = (a + 1)c$$

for all $z \in D$.

Now since s_0 is not in $\mathrm{Cl}\ W(f(z_1))$, the point $s_0 + as_0$ is not in $\mathrm{Cl}\ W(f(z_1) + as_0)$ and this gives

$$w(f(z_1) + as_0) < (a + 1)c$$

which is a contradiction and thus the theorem is proved.

For any operator $T \in L(H)$, H is a Hilbert space, we define the number

$$\theta_w(T) = \sup \{\mathrm{Re}\ z \mid z \in W(T)\}$$

and for the study of this number in the case of holomorphic functions we need the following lemma.

14.5.4 LEMMA If $T \in L(H)$, then

$$\theta_w(T) = \lim \{\|T + t\| - t\}$$

Proof. First we remark that the function

$$t \mapsto \|T + t\| - t = h(t)$$

has the property that $t_1 < t_2$ implies $h(t_1) \geq h(t_2)$. Indeed,

$$\begin{aligned} h(t_2) &= \|T + t_2\| - t_2 = \|(T + t_1) + (t_2 - t_1)\| - t_2 \\ &\leq \|T + t_1\| + (t_2 - t_1) - t_2 = h(t_1) \end{aligned}$$

From this remark it is clear that $\lim h(t)$ exists for $t \to \infty$.

The assertion of the lemma is now exactly Lemma 6.11.14 for the operator iT.

From this lemma we obtain the following important property of the number θ_W.

14.5.5 THEOREM If f: D → L(H) is a holomorphic function, then the function

$$z \mapsto \theta_W(f(z))$$

is subharmonic.

A more general theorem can be obtained from the above lemma and the Rado theorem.

14.5.6 THEOREM If f is as in Theorem 14.5.5, then the function

$$z \mapsto \log \theta_W(f(z))$$

is subharmonic.

14.5.7 REMARK As a consequence of these results, as in the case of the spectral radius, the numerical radius, etc., we can obtain some maximum theorems. These are left to the reader.

14.5.8 PROBLEM Is it possible to obtain similar information in the case when we use the maximal numerical range?

We give now a maximum theorem for the numerical range. For this we need some preliminary results which are given as lemmas.

14.5.9 LEMMA If f: D → L(H), H is a Hilbert space and f is a holomorphic function, then if "ℓ" is the line support of the convex set Cl $W(f(z_0))$ for some $z_0 \in D$ and the function

$$z \mapsto \text{Cl } W(f(z))$$

assumes its maximum at z_0, then

$$W(f(z_0)) \cap \ell \subset W(f(z)) \cap \ell$$

for all $z \in D$.

Proof. Suppose that $W(f(z_0)) \cap \ell \neq \emptyset$. Consider a point s_0 in $W(f(z_0)) \cap \ell$, and thus for any complex number b,

$$W(f(z) + b) = W(f(z)) + b$$

and since

$$W(d(f(z)) = dW(f(z))$$

for suitable b and d, we can suppose, without loss of generality, that

$$\ell = \{z \mid \operatorname{Re} z = 0\}$$
$$\operatorname{Cl} Wf(z_0) = \{z \mid \operatorname{Re} z \geq 0\}$$
$$s_0 = 0$$

Since $s_0 = 0 \in \operatorname{Cl} W(f(z_0))$, we can find an element $x_0 \in H$ of norm one such that

$$\langle f(z_0)x_0, x_0\rangle = 0$$

and if we define the holomorphic function g by

$$g(z) = \langle f(z)x_0, x_0\rangle$$

we have

1. $g(z) \in W(f(z))$ for all $z \in D$
2. $\operatorname{Re} g(z) \geq 0$ for all $z \in D$

In this case

$$|e^{-g(z)}| = e^{-\operatorname{Re} g(z)} \leq 1$$

for all $z \in D$. Since $g(z_0) = 0$, $e^{-g(z_0)} = 0$ and thus $e^{-g(z)} = 1$ for all $z \in D$. This implies that $g(z) = 0$ for all $z \in D$, which is equivalent with

$$\langle f(z)x_0, x_0\rangle = 0$$

for all $z \in D$. This implies that $s_0 \in W(f(z))$ for all $z \in D$.

14.5.10 LEMMA If ℓ, z_0, and f are as in the above lemma, then

$$\operatorname{Cl} W(f(z_0)) \cap \ell = \operatorname{Cl} W(f(z)) \cap \ell$$

for all $z \in D$.

Proof. Let $s_0 \in \mathrm{Cl}\ W(f(z_0)) \cap \ell$ and as above we can suppose ℓ, s_0 as above. But $s_0 = 0 \in \mathrm{Cl}\ w(f(z_0))$, and thus we can find a sequence of unit elements such that

$$\lim \langle f(z_0)x_n, x_n\rangle = 0$$

and we consider now the sequence of holomorphic functions

$$g_n(z) = \langle f(z)x_n, x_n\rangle$$

and since

$$|g_n(z)| \leq w(f(z) \leq w(f(z_0))$$

for all $z \in D$, we can apply Vitali's theorem to find a subsequence converging to a holomorphic function g. We may suppose that g has the above property. Since Re $g_n(z) \geq 0$, we have that Re $g(z) \geq 0$. Now $g(z_0) = \lim g_n(z_0) = 0$ and this implies that $g(z) = 0$ for all $z \in D$. From this we have that $s_0 \in \mathrm{Cl}\ W(f(z))$ for all $z \in D$. From the hypothesis

$$\mathrm{Cl}\ W(f(z)) \subseteq \mathrm{Cl}\ W(f(z_0))$$

for all $z \in D$, we obtain that

$$\mathrm{Cl}\ W(f(z)) \cap \ell \subseteq \mathrm{Cl}\ W(f(z_0)) \cap \ell$$

for all $z \in D$.

Now we are ready to prove a maximum theorem for the numerical radius.

14.5.11 THEOREM If $f: D \to L(H)$ is a holomorphic function and for some z_0.

$$\mathrm{Cl}\ W(f(z)) \subseteq \mathrm{Cl}\ W(f(z_0))$$

for all $z \in D$, then $W(f(z))$ is a constant set.

Proof. Let s_0 be a point of the boundary of the set $W(f(z_0))$, and since $W(f(z_0))$ is a convex set, by the Hausdorff-Toeplitz theorem we can find at least one line support of Cl $W(f(z_0)$ passing through

s_0. In this case Lemma 14.5.10 implies that s_0 is in Cl W(f(z)) for all $z \in D$. Thus we have

$$\partial W(f(z_0)) \subset \partial W(f(z))$$

for all $z \in D$. Since Cl W(f(z)) is a bounded convex set, $C \setminus \mathrm{Cl}\, W(f(z))$ is connected, and thus, by the lemma which follows (Lemma 14.5.12),

$$\mathrm{Cl}\, W(f(z)) = \mathrm{Cl}\, W(f(z_0))$$

for all $z \in D$.

We prove now that every boundary point s_0 of W(f(z)) belongs to $W(f(z_1))$ for some $z_1 \in D$, then $s_0 \in W(f(z))$ for all $z \in D$, and this clearly implies that W(f(z)) is independent of $z \in D$.

From Lemma 14.5.9 we may replace z_0 by any other point z_1 of D. If ℓ is a support line of $W(f(z_1))$ through s_0, then from Lemma 14.5.9 it follows that $s_0 \in W(f(z))$ for all $z \in D$.

Now we give the following topological lemma used above.

14.5.12 LEMMA If S is a topological space and E, F two closed subsets of S and the following properties hold:

1. $E \subset F$
2. $\partial F \subset \partial E$

then F is the union of E and of some connected components of $S \setminus E$.

Proof. Let B be a connected component of $S \setminus E$. We show that if $F \cap B \neq \emptyset$, then $B \subset F$. Consider for this a point a of the boundary of $F \cap B$ in B. If V is a neighborhood of a in A, then $V \cap B$ is not disjoint from both $F \cap B$ and the complement of $F \cap B$ in B. Thus V is not disjoint from both F and $A \cap F$. This shows that a belongs to the boundary of F in A, i.e., the boundary of $F \cap B$ in B is in ∂F. Since $a \in \partial F$, we have a E which is a contradiction. Thus the boundary of $F \cap B$ in B is empty, i.e., $F \cap B$ is open in B.

But F is closed and therefore A and $F \cap B$ are closed in B for the induced topology. Now B is connected and thus

$$F \cap B = B$$

which shows that $B \subset F$ and the lemma is proved.

We close with some questions.

14.5.13 QUESTION 1. Is it possible to obtain maximum theorems for the maximal numerical range?

2. Since other numbers can be associated with a closed convex set, is it possible to obtain maximum theorems for these numbers?

3. Is it possible to extend the arguments in the last result for the case of Banach spaces or for Banach algebra elements?

Chapter 15

UNIFORM ERGODIC THEOREMS FOR SOME CLASSES OF OPERATORS

We shall now discuss the behavior of the averages of the iterates of a linear operator on a Banach space when the operator is in some specific class. Aa is well known, a very important class of operators with deep applications in the theory of Markov processes is the class of quasi-compact operators. Basic results concerning ergodic theorems for the quasi-compact operators were obtained by Yosida and Kakutani (1941). For an excellent account as well as new results we refer to the Dunford-Schwartz treatise (1967). In what follows we show that almost all results known for quasi-compact operators can be extended to a more general class of operators, namely, the classes considered in Chapter 2. These classes were defined using a new and very fruitful notion introduced by Kuratowski (1929), the measure of noncompactness of a bounded set in a complete metric space.

For ergodic theory we refer to Dunford and Schwartz (1967).

15.1 CLASSES OF OPERATORS

We repeat here the definitions of the classes of operators which we shall discuss.

In what follows X is a complex Banach space and L(X) denotes the set of all bounded operators on X.

15.1.1 DEFINITION An element $T \in L(X)$ is called quasi-compact if

1. $\|T^n\| \leq M < \infty$.

2. For some compact operator Q and some $m \geq 1$, $\|T^m - Q\| < 1$.

15.1.2 REMARK In some cases we relax the first condition to the following one: $\{T^n/n\}$ converges to zero in the weak operator topology. It is easy to conclude as an application of the principle of uniform boundedness that this condition implies that there exists $M < \infty$ such that

$$\|T^n\| \leq nM \qquad n = 1, 2, 3, \ldots$$

Also it is obvious that $\|T^n\| \leq nM$ for all positive integers implies that the spectral radius of T satisfies the inequality $r_T \leq 1$.

15.1.3 DEFINITION An operator $T \in L(X)$ is called an α-set contraction if there exists $k \in [0, 1)$ such that for all bounded subsets A of X,

$$\alpha(TA) \leq k\alpha(A)$$

when $\alpha(A) > 0$.

15.1.4 DEFINITION An element $T \in L(X)$ is called a *locally power α-set contraction* if there exists $k \in [0, 1)$ such that for any bounded set A there exists an integer $n = n(A)$ with the property

$$\alpha(T^n A) \leq k\alpha(A)$$

when $\alpha(A) > 0$.

We need also the following result given in Dunford and Schwartz (1967, Lemma VIII.8.1).

15.1.5 LEMMA If $T \in L(X)$ is an operator such that $\{T^n/n\}$ converges to zero in the weak operator topology, then every point in the spectrum and of modulus 1 is a pole of T of order 1.

Proof. First we recall that a point in the spectrum of a bounded operator is called a *pole of the operator* if it is an isolated point of the spectrum, and the resolvent has a pole at that point. The order of the pole is the order of the point as a pole of T.

The proof of the lemma is as follows. As we remarked the spectrum of T is in the set $\{z \mid |z| \leq 1\}$. We may suppose, without loss of generality, that the point is 1. Let $z = 1$ be a pole of T and suppose that the order of the pole is at least 2. In this case we can find an element x_0 in the range of the projection associated with the pole $z = 1$ such that

1. $(I - T)x_0 \neq 0$
2. $(I - T)^2 x_0 = 0$

Now since

$$\frac{1}{n} T^n x_0 = \frac{1}{n} x_0 + (I - T)x_0$$

for $n \to \infty$ we obtain

$$x^*(I - T)x_0 = 0$$

and thus

$$(I - T)x_0 = 0$$

which is a contradiction. Thus the order is 1.

Suppose that T is an operator with the following properties:

1. $\{T^n/n\}$ converges weakly in the weak operator topology to zero.
2. T is a locally power α-set contraction.

In the latter case we can prove the following result.

15.1.6 THEOREM If $T \in L(X)$ has the above two properties, then

$$\left\{\frac{1}{n} \sum_{i=0}^{n-1} T^i\right\}$$

converges to a projection associated with 1, $P_1(T)$ in the uniform topology.

Proof. Let $T_n = 1/n\ \Sigma_{i=0}^{n-1}$ and let σ' be the part of $\sigma(T)$ in the open $\{z \mid |z| < 1\}$. Define also $\sigma_1 = \sigma(T) \setminus \sigma'$. If $P_{\sigma'}$ is the projection associated with the set σ', then we have

$$\lim T_n P_{\sigma'} = 0$$

Now since for each $z = 1$,

$$\frac{1}{n} \sum_{i=0}^{n-1} z^i \to 0$$

we obtain that

$$T_n P_{\sigma_1} \to P_1(T)$$

Since

$$T_n = T_n P_{\sigma'} + T_n P_{\sigma_1}$$

the assertion follows.

We give now an application of this theorem in the case of a class of operators on the space $C(S)$, where S is a compact space. First we give the following definition.

15.1.7 DEFINITION Let $T \in L(C(S))$. The operator T is called *positive* if for all $f \geq 0$, Tf is also positive.

15.1.8 DEFINITION Let $T \in L(L^1(\Omega, B, \mu))$. The operator T is called *positive* if for all $f \in L^1(\Omega, B, \mu)$

$$f \geq 0 \qquad \text{(a.e. } \mu)$$

implies

$$Tf \geq 0 \qquad \text{(a.e. } \mu)$$

The following result is a generalization of Lemma VIII.8.3 in Dunford and Schwartz (1967).

15.1.9 LEMMA Let $T \in L(C(S))$ be a positive operator and satisfying the following properties:

1. $\{T^n/n\}$ converges to zero in the weak operator topology.
2. T is a locally power α-set contraction.

Then there exists an integer N such that if $z_1, z_2, \ldots, z_n$ are the points in $\sigma(T)$ of unit modulus, then $z_i^N = 1$, $i = 1, 2, \ldots, n$.

Proof. From the results in Chapter 2 about locally power α-set contractions, we know that we can apply Theorem 15.1.6.

For this we need some preliminary information about certain sets in S. Let $z \in \sigma(T)$, $|z| = 1$, and $f \in C(S)$ arbitrary but not equal to 0, and in the range of the projection associated with z. Thus $Tf = zf$. If s_0 is a point in S such that $\|f\| = |f(s_0)|$, we may suppose, without loss of generality, that $f(s_0) = 1$. Let δ_{s_0} be the linear functional on C(S) defined by the evaluation at s_0,

$$\delta_{s_0}(g) = g(s_0)$$

for all $g \in C(S)$. Clearly this is a positive, linear, and bounded functional on C(S). By the Riesz-Kakutani representation theorem we can find a measure μ such that

$$\delta_{s_0}(g) = \int_S g(s)\, d\mu(s)$$

Similarly, if we consider the adjoint T^* of T and the linear functional defined on C(S) by the relation

$$g \mapsto (T^{*n}\delta_{s_0})$$

then it is clear that these functionals are the evaluation at s_0 of $T^n g$. By the same Riesz-Kakutani theorem we can find a positive measure μ_n such that

$$(T^n g)(s_0) = \int_S g(s)\, d\mu_n(s)$$

From the fact that the point s_0 is such that $1 = |f(s_0)| = \|f\|$ it follows that since

$$(T^n f)(s_0) = z^n f(s_0) = \int_S f(s)\, d\mu_n(s)$$

we have

$$\mu_n(S) \geq 1$$

We prove now that for each n the set

$$A_n = \{s \mid z^{-n}f(s) \neq 1\}$$

has μ_n-measure zero.

Indeed, $T^n f = z^n f$ implies that

$$0 = f(s_0) - z^{-n}(T^n f)(s_0) = \int_S (1 - z^{-n}f(s))\, d\mu_n(s)$$

and thus

$$0 = \text{Re } (f(s_0) - z^n(T^n f)(s_0) = \int_S (1 - \text{Re } z^{-n}f(s))\, d\mu_n(s)$$

$$= \int_S (1 - \text{Re } (z^{-n}f(s))\, d\mu_n(s)$$

Since $|\text{Re } z^{-n}f(s)| \leq 1$ the integrand is nonnegative and since the measure μ_n is positive, it follows that

$$B_n = \{s \mid \text{Re } z^{-n}f(s) \neq 1\}$$

has the property that $\mu_n(B_n) = 0$. Since B_n contains A_n, it follows that $\mu_n(A_n) = 0$. Define now the sets

$$C_{A_n} = S \setminus A_n$$

and we show that the sequence $\{C_{A_n}\}$ has the important property that the members are not pairwise disjoint. Indeed, if this property does not hold, then if $A' = \cap C_{A_i}$, we have $\mu_i(C_{A_j}) = 0$ for $i \neq j$. This implies that

$$0 \leq \lim \frac{1}{n} \sum_{j=1}^{n} \mu_j(C_{A_k}) = \lim \frac{1}{n} = 0$$

From the sequence of measures $\{\mu_n\}$ we define a new measure μ^* by the relation

$$\mu^*(\ ) = \lim \frac{1}{n} \sum_{j=0}^{n-1} \mu_j = \lim \frac{1}{n} \sum_{j=0}^{n-1} (T^{*j}\delta_{s_0})$$

and we remark that since $\mu_n(S) > 1$, then clearly $\mu^*(S) \geq 1$. From the above property for the sequence C_{A_i} it follows that

$$\mu^*(A') = \Sigma\, \mu^*(C_{A_i}) = 0$$

Now if $A = \bigcap A_i$, then clearly $S = A \cup A'$ and

$$0 = \mu^*(A) + \mu^*(A') = \mu^*(S) \geq 1$$

and this contradiction proves the assertion about the property of the sequence C_{A_n}. Since the sets C_{A_i} cannot be pairwise disjoint, there exist the integers p and q such that a point $s_1 \in S$ is in $C_{A_p} \cap C_{A_q}$. Thus $z^{-p}f(s_1) = 1 = z^{-q}f(s_1)$ which gives that $z^{p-q} = 1$. Since z is a point in $\sigma(T)$ with modulus 1, we know that there exists only a finite number of such points and the integer N with the required property exists.

From the result obtained we can describe the structure of operators which are as in the lemma. This is given in the following theorem, which is an extension of the structure theorem in Dunford and Schwartz (1967, Lemma VIII.8.6).

15.1.10 THEOREM Let T be an operator as in Lemma 15.1.9. In this case (T) can be decomposed into the union of a closed set σ in $\{z \mid |z| < 1\}$ and a finite number of simple poles $e^{2\pi it_k}$, where t_k are rational numbers. The spectral projections associated with $e^{2\pi it_k}$ are finite dimensional and T has the form

$$T^m = \sum_{i=1}^{n} e^{2\pi it_k}P_k + D_m$$

where $D_m = (TP_\sigma)^m$.

Proof. All the assertions were proved except the assertion about the form of T^m which is a consequence of the structure theorem for locally power α-set contractions.

15.1.11 COROLLARY Let T be as in Theorem 15.1.10. If E_p is $\Sigma\, P_k$, then

1. E_PX is finite dimensional and $X = C(S)$.
2. If $C_D = P_\sigma X$, then $C(S)$ is the direct sum of E_P and C_D and the subspaces E_PX and C_D are invariant for T.
3. There exists an integer N such that $T^N x = x$ for all $x \in E_PX$.
4. $T^n x \to 0$ exponentially fast for $x \in C_D$.
5. The subspaces E_PX and C_D are uniquely defined by properties 3 and 4.

Proof. Since $E_P = \Sigma P_k$ is a projection and $E_PC_D = C_DE_P$, $I = E_P + C_D$, the assertions in 1 and 2 follow from the above theorem as well as assertions 3 and 4.

Suppose that $x \in C(S)$ has the property that $T^n x \to 0$. Since $x = x_p + x_d$ with $x_p \in E_PX$ and $x_d \in C_D$, we obtain that

$$T^n x_p \to 0 \qquad T^n x_d \to 0$$

and since $\{T^n x_p\}$ has infinitely many terms equal to x_p, it follows that $x_p = 0$. This gives that

$$C_D = \{x \mid T^n x \to 0\}$$

Similarly, if $x \in C(S)$ and $T^n x = x$ for some n, it follows that $x \in C_P = E_PX$. Thus 5 is proved.

As is well known, an important space in ergodic theory is $L^1(\Omega, B, \mu)$, μ a positive measure. We are interested in extending the properties proved before for $C(S)$ to operators on L^1 which are positive. Following the presentation in Dunford and Schwartz, we know that the conjugate space is representable as a space of continuous functions. Since the structure theorems for locally power α-set contractions are preserved by the conjugate, then we obtain as a consequence of these the following.

15.1.12 THEOREM Suppose $T \in L(L^1(\Omega, B, \mu))$ and satisfies the following properties:

1. $\{T^n/n\}$ converges to zero in the weak operator topology.
2. T is positive.

3. T is a locally power α-set contraction.

Then the operator T has the properties of the operator T from Theorem 15.1.10 and Corollary 15.1.11.

Proof. The adjoint operator T^* in the space L_∞ is positive and is a locally power α-set contraction; also T^{*n}/n converges in the weak operator topology. Since L_∞ is isometrically isomorphic to a space of the form C(S) for some Hausdorff compact space S, we can consider the operator T^* on this space. Since the positiveness is preserved, the assertion of the theorem follows.

15.2 APPLICATIONS TO MARKOV PROCESSES

In the theory of Markov processes one of the most important problems is the asymptotic behavior of conditional probabilities.

An important and efficient method for studying these conditional probabilities is based on operator theory. If $\{P^n(t, A)\}$ are the family of transition probabilities, then we suppose that

1. $P^1(\ , A)$ is a measurable function for any $A \in \mathcal{B}$.
2. $P^1(t, \ )$ is a probability measure on $\mathcal{B}$.
3. $P^{n+m}(t, A) = \int_R P^n(t, dy)P^m(y, A)$.

The number $P^1(t, A)$ is interpreted as giving the probability that after a unit of time, the point t will be in the set A. In this case the point t will be in the set A after the elapse of n unit times, with the probability $P^n(t, A)$.

Associated with the family of transition probabilities $\{P^n(t, A)\}$ are two bounded linear operators. To define these operators we consider two Banach spaces:

1. The set of all complete additive functions defined on $\mathcal{B}$, organized as a linear space in an obvious way. The norm is defined as

$$\|\mu\| = \text{var } \mu = \sup_{\|f\|\le 1} \left| \int f(s) \quad d\mu(s) \right|$$

and this space is denoted as W.

2. The space of all complex-valued measurable functions which are bounded and the norm is sup. This space is denoted G.

Using the family $\{P^n(t, A)\}$, we define the operators T and $\bar{T}$ by

$$T: W \to W \qquad Tx(A) = \int_R P^1(t, A)\, dx(t)$$

and

$$\bar{T}: G \to G \qquad \bar{T}y(t) = \int_R y(s)P^1(t, ds)$$

It is easy to verify the following properties.

15.2.1 LEMMA For operators T and $\bar{T}$ the following properties are true:

1. $\|T^n\| = \|\bar{T}^n\| = 1$, for all $n = 1, 2, 3, \ldots$.
2. $\bar{T}$ is a positive operator.

15.2.2 REMARK If we take as function x the function $P^1(t, A)$, where t is fixed, then Tx is exactly the function $P^2(t, A)$ and for any n, $T^n x$ is exactly the function $P^{n+1}(t, A)$. We consider now the operator $\bar{T}$ and take as function y the function $P^1(t, A)$ with A fixed; then $\bar{T}x$ is $P^2(t, A)$ and also for any n, $\bar{T}^n x$ is $P^{n+1}(t, A)$.

In this way we are led to consider for the study of the family of transition probabilities, the operators T and $\bar{T}$.

Concerning the operators T and $\bar{T}$ an important condition was introduced by Krilov and Bogoliubov [see Kakutani and Yosida (1941)]:

K: The operator T is quasi-compact.

This condition was studied by Kakutani and Yosida (1941) whose results were shown above in a generalized form (i.e., for the class of locally power α-set contractions).

As a consequence of the results proved for locally power α-set contractions, we obtain the following results if we suppose that the operator T satisfies the following:

K*: The operator T is a locally power α-set contraction and $\{T^n/n\}$ converges to zero in the weak operator topology.

15.2.3 THEOREM If T satisfies K*, then

1. $$\left|\frac{1}{n}\sum_{i=1}^{n} P^{i}(t, A) - P^{\infty}(t, A)\right| \le \frac{\mu}{n}$$

   for all $t \in R$, $A \in \mathcal{B}$ and where $P^{\infty}(t, A)$ is a probability measure for each $t \in R$, and M is an appropriate constant.

2. If 1 is the single eigenvalue of T of modulus 1, then for any $t \in R$ and $A \in \mathcal{B}$,

   $$|P^{n}(t, A) - P^{\infty}(t, A)| \le \frac{M}{(1 + \varepsilon)^{n}} \qquad \text{for all } n$$

3. If 1 is an eigenvalue of multiplicity 1, then

   $$P^{\infty}(t, A) = P^{\infty}(A)$$

Proof. All the assertions are clear from the above results except 3. For this we remark that if 1 is an eigenvalue of modulus 1 and multiplicity 1, then

$$P^{\infty}(t, A) = c(t)P^{\infty}(A)$$

and since for the entire space Ω, $P^{\infty}(t, \Omega) = 1$, it follows that c(t) is a constant function. Changing eventually $P^{\infty}(A)$, we can suppose that $c(t) = 1$.

In connection with the condition K it is worth mentioning the condition D introduced by W. Doeblin [see Kakutani and Yosida (1941)], and this is formulated as follows:

D: There exists $\delta > 0$ and $\eta > 0$ such that

$$P^{n}0(t, A) \le 1 - \delta$$

if the measure of A is less than η.

Kakutani and Yosida (1941) have proved that D implies K.

In connection with the condition of Doeblin we can consider the following condition:

D*: There exists an integer $n = n(\eta)$ such that for all A with $A < \eta$, $P^{n}(t, A) \le 1 - \delta$ for some $\delta > 0$ independent of A.

The problem concerns the relation between these conditions, more precisely, whether D* implies K*.

15.2.4 REMARK Our use of operators defined with the measures of noncompactness can also be useful in the following case: Suppose that we have two families of transition probabilities, say $\{P^n(t, A)\}$ and $\{Q^n(t, A)\}$, with the properties:

1. The family $\{sP^1(t, A) + (1 - s)Q(t, A)\}$ is again a family of transition probabilities.
2. The operators T_P and T_Q associated with these families are as follows:
   a. T_P is a contraction mapping and noncompact; the contraction constant for T_P is less than 1.
   b. T_Q is a quasi-compact operator such that for some compact operator C, $\|T_Q - C\| < 1$.

In this case the operator T associated with the family $\{sP^1(t, A) + (1 - s)Q^1(t, A)\}$ is an α-set contraction. Thus we can apply the results obtained above for the Markov process.

Since the associated operator is of the form

$$sT_P + (1 - s)T_Q$$

it is clear that this is not a quasi-compact operator.

Appendix

C_p CLASSES

Let H be a Hilbert space and L(H) be the set of all bounded linear operators on H. The set of all compact operators on H will be denoted by K(H) and it is easy to see that this is a closed subset of L(H).

Let $T \in K(H)$ and in this case the operator T^*T is compact and positive. Then there exists a unique operator which is the square root of T^*T. We denote this square root by $(T^*T)^{1/2}$. From the form of the sequence of operators which converges to $(T^*T)^{1/2}$ it follows that this is also a compact operator. The elements in the spectrum of $(T^*T)^{1/2}$ are countable and we arrange them in a decreasing sequence

$$\mu_1 \geq \mu_2 \geq \mu_3 \geq \cdots \qquad \lim \mu_n = 0$$

In this sequence the eigenvalues are enumerated according to their multiplicity.

Let $p \in [1, \infty)$ and we define the number (finite or not)

$$\|T\|_p = \left(\sum_1^\infty \mu_n^p \right)^{1/p}$$

and for $p = \infty$,

$$\|T\|_\infty = \sup \{\mu_n\} = \|T\|$$

A.1 DEFINITION The class C_p is the set of all $T \in K(H)$ such that

$$\|T\|_p < \infty$$

Since, as is easy to see, the spectrum of $(T^*T)^{1/2}$ and the spectrum of $(TT^*)^{1/2}$ are the same (except possibly zero) we have

A.2 PROPOSITION If $T \in C_p$, then $T^* \in C_p$.

At the same time we observe that $\|T\|_p = \|T^*\|_p$.

Before presenting some structure theorems on this class of operators, we need information concerning the positive operators.

A.3 PROPOSITION Let $T \geq 0$ and s be a positive number. Then

1. If $0 < s \leq 1$, $<T^s x, x> \leq (<Tx, x>)^s \|x\|^{2(1-p)}$.
2. If $1 \leq s < \infty$, $<T^s x, x> \geq (<Tx, x>)^s \|x\|^{2(1-s)}$.

Proof. Let $s \geq 1$ and thus using the Hölder inequality we get

$$<Tx, x> = \int_0^\infty t<(E(dt)x, x> \leq \left|\int_0^\infty t^s<(E(dt)x, x>\right|^{1/s}$$

$$\left|\int_0^\infty <(E(dt)x, x>\right|^{(s-1)/s} = <T^s x, x>^{1/s}\|x\|^{2(s-1)/s}$$

The case when $s \in (0, 1)$ is obtained from this by applying the result to the operator T^s and the number $1/s > 1$.

From this proposition we obtain easily the following result.

A.4 PROPOSITION Let $T \in C_p$. Then the following assertions hold:

1. If $p \in [1, 2]$, then

$$\|T\|_p = \inf \{\Sigma \|Te_i\|^p\}^{1/p}$$

2. If $p \in [2, \infty)$, then

$$\|T\|_p = \sup \{\Sigma \|Te_i\|^p\}^{1/p}$$

where the inf and sup are taken over all orthonormal bases of H.

Proof. Let $\{e_i\}$ be an arbitrary orthonormal basis of H. In this case for each e_i we have

$$<(T^*T)^{p/2}e_i, e_i> \leq <(T^*T)e_i, e_i>^{p/2} = \|Te_i\|^p \qquad p \in [1, 2]$$

and

$$<(T^*T)^{p/2}e_i, e_i> \geq <(T^*T)e_i, e_i>^{p/2} = \|Te_i\|^p \qquad p \in [2, \infty)$$

and summing on i, we get

$$\|T\|_p \leq (\Sigma \|Te_i\|^p)^{1/p}$$

and

$$\|T\|_p \geq (\Sigma \|Te_i\|^p)^{1/p}$$

The assertions follow obviously from these relations taking into account the eigenvectors of T^*T.

From this we obtain the following result.

A.5 PROPOSITION The set C_p is a linear normed space.

The norm on C_p is defined with the number $\|T\|_p$, i.e.,

$$T \to \|T\|_p$$

We note also the following important property of the normed space C_p, which follows from the definition of the norm $\|,\|_p$.

A.6 PROPOSITION The normed space C_p is complete, i.e., C_p is a Banach space.

We note also that the class C_p has a very interesting generalization to the case of Banach space using the so-called approximation numbers of an operator, which are defined as follows: For any bounded linear operator on a Banach space X,

$$a_n(T) = \inf \{\|T - A_n\|\}$$

where the inf is taken over all mappings A_n whose range is at most n-dimensional. It is not difficult to see that for the case of elements in K these numbers are exactly the numbers $\{\mu_n\}$ considered above when the space X is a Hilbert space.

We also mention that interesting and important generalizations of this class were considered by Gochberg and Krein using the analogy with Orlicz space. Other generalizations were also considered using the analogy with other sequence spaces.

REFERENCES

Akhiezer, N. I., and I. M. Glazman (1961, 1963). *Theory of Linear Operators in Hilbert Space*, Vols. I and II, Ungar, New York.

Ambrosetti, A. (1967). Una teorema di esistenza per le equazioni differenziali negli di spazi di Banach, *Rend. Mat. Sem. Padova* 40:349-361.

------ (1969). Proprieta spatrali di certi operatori lineari non compatti, *Rend. Mat. Sem. Padova* 42:189-200.

Ando, T. (1963a). Note on invariant subspaces of a compact normal operator, *Arch. Math.* 14:337-340.

------ (1963b). On hyponormal operators, *Proc. Amer. Math. Soc.* 14:290-291.

------ (1972). Operators with a norm condition, *Acta Math.* (Szeged) 33:169-178.

Arens, R. (1947). Representations of Banach *-algebras, *Duke Math. J.* 14:269-282.

Aronszajn, N., and K. T. Smith (1954). Invariant subspaces of completely continuous operators, *Ann. Math.* 60:345-350.

------ (1961). Quadratic forms on vector spaces, *Proc. Int. Symp. on Linear Space,* Jerusalem.

------ and R. D. Brown (1970). Finite dimensional perturbations of spectral problems and variational approximation methods for eigenvalue problems I. Finite dimensional perturbations. *Studia Math.* 36:1-76.

------ and U. Fixman (1968). Algebraic spectral problems, *Studia Math.* 30:273-338.

Arveson, W., and J. Feldman (1968). A note on invariant subspaces, *Mich. Math. J.* 15:61-64.

Atkinson, F. V. (1951). The normal solvability of equations in normed spaces, *Mat. Sb.* 28(70):3-14.

Banach, S. (1922). Sür les operations dans les ensembles abstraits et leurs applications aux equations integrales, *Fund Math.* 3:133-181.

------ (1932). *Theorie des Operations Lineaires*, Warzawa.

Baxley, J. (1971). Some general conditions implying Weyl's theorem, *Rev. Roum. Math. Pure Appl.* 16:1163-1166.

------ (1972). On the Weyl spectrum of a Hilbert space operator, *Proc. Amer. Math. Soc.* 34:447-452.

Beals, R. (1931). *Topics in Operator Theory*, Chicago Lectures in Mathematics, Univ. Chicago Press, Chicago.

Beck, W. A., and C. R. Putnam (1956). A note on normal operators and their adjoint, *J. Lond. Math. Soc.* 31:213-216.

Berberian, S. K. (1959). Note on a theorem of Fuglede and Putnam, *Proc. Amer. Math. Soc.* 10:403-404.

------ (1961). Introduction to Hilbert space, Oxford Univ. Press, New York.

------ (1962a). A note on operators unitarily equivalent to their adjoints, *J. London Math. Soc.* 37:403-404.

------ (1962b). Approximate proper vectors, *Proc. Amer. Math. Soc.* 13:111-114.

------ (1966). A note on operators whose spectrum is a spectral set, *Acta Math. Sci.* (Szeged) 27:201-203.

------ (1967a). Compact sets in inner product spaces, *Amer. Math. Mon.* 74:702-705.

------ (1969). An extension of Weyl's theorem to a class of not necessarily normal operators, *Mich. Math. J.* 17:529-544.

------ (1970a). Some conditions on an operator implying normality, I, *Math. Ann.* 184:188-192.

------ (1970b). Some conditions on an operator implying normality, II, *Proc. Amer. Math. Soc.* 26:277-281.

------ (1970c). On some conditions implying normality, III, *Proc. Jap. Acad. Sci.* 46:630-632.

------ (1970d). The Weyl spectrum of an operator, *Indiana Math. Univ. J.* 20:529-544.

------ (1971). Conditions on an operator implying Re $\sigma(T) = \sigma(\text{Re } T)$, *Trans. Amer. Math. Soc.* 154:267-272.

------ and G. H. Orland (1967b). On the closure of the numerical range of an operator, *Proc. Amer. Math. Soc.* 18:499-503.

Berg, I. D. (1971). An extension of the Weyl-von Neumann theorem to normal operators, *Trans. Amer. Math. Soc.* 160:365-368.

Berger, C. A. (1965). A strange dilation theorem, *Notices Amer. Math. Soc.* 12:590.

Berger, C. A., and J. G. Stampfli (1967). Mapping theorems for the numerical range, *Amer. Math. J.* 89:1047-1055.

Berkson, E. (1966). Some characterizations of C*-algebras, *Ill. J. Math.* 10:1-8.

Bernau, S. J., and F. Smithies (1963). A note on normal operators, *Proc. Camb. Philos. Soc.* 59:727-729.

Bernstein, A. R., and A. Robinson (1966). Solution of an invariant subspace problem of K. T. Smith and P. Halmos, *Pacif. J. Math.* 16:421-431.

Beurling, A. (1947). On two problems concerning linear transformations in Hilbert spaces, *Acta Math.* 81:239-255.

Birkhoff, G. (1935). Orthogonality in linear spaces, *Duke Math. J.* 1:169-172.

Bohnenblust, H. F., and A. Sobczyc (1938). Extension of functional on complex linear spaces, *Bull. Amer. Math. Soc.* 44:91-93.

Bonsall, F., and M. Crabb (1970). The spectral radius of a hermitian element in a Banach algebra, *Bull. Lond. Math Soc.* 2:178-180.

------ and J. Duncan (1971, 1973). *Numerical Ranges,* I and II, Cambridge Univ. Press, New York.

Bram, J. (1955). Subnormal operators, *Duke Math. J.* 22:75-94.

Breuer, M. (1968). Fredholm theories in von Neumann algebras, I, *Math. Ann.* 178:243-254.

------ (1969). Fredholm theories in von Neumann algebras, II, *Math. Ann.* 180:313-325.

Brodskii, M. S. (1961). Unicellularity criteria for Volterra operators, *D.A.N.* 138:512-514.

------ (1965). On the unicellularity of integration operator and a theorem of Titchmarsh, *Uspehi Mat. Nauk* 20:189-192.

------ and M. S. Livshitz (1960). Spectral analysis of non-selfadjoint operators and intermediate systems, *Uspehi Mat. Nauk; Amer. Math. Soc. Trans.* 13(2):265-346.

Bronson, R., and G. Steiner (1973). A note on conjugate spaces, *Math. Mag.* 46:158.

Browder, A. (1969). *Introduction to Function Algebras,* Benjamin, New York.

------ (1971). On Bernstein's inequality and the norm of hermitian operators, *Amer. Math. Monthly* 78:871-873.

Calkin, J. W. (1941). Two-sided ideals and congruences in the ring of bounded operators, *Ann. Math.* 2(42):839-873.

Campbell, J. E. (1897). On a law of combination of operators bearing in the theory of continuous transformations groups, *Proc. Lond. Math. Soc.* 28:381-390.

Campbell, S. L. (1972a). Operator-valued inner functions analytic on the closed disk, *Pacif. J. Math.* 41:57-62.

------ (1972b). Linear operators for which T*T and TT* commute, *Proc. Amer. Math. Soc.* 34:177-180.

------ (1974). Linear operators for which T*T and TT* commute, II, *Pacif. J. Math.* 53:355-361.

------ (1975). Linear operators for which T*T and T + T* commute, *Pacif. J. Math.* 61:53-57.

------ . Linear operators for which T*T and T + T* commute, *Pacif. J. Math.*

------ and R. Gellar (1976). Linear operators for which T*T and T + T* Commute, *Proc. Amer. Math. Soc.* 60:197-202.

------ and R. Gellar (1977). Linear operators for which T*T and T + T* commute, *Trans. Amer. Math. Soc.* 226:305-319.

Carleman, T. (1922). Les fonctions quasi-analytiques, Gauthier-Villars, Paris.

Chernoff, P. R. (1972). Some remarks on quasi-analytic vectors, *Trans. Amer. Math. Soc.* 167:105-113.

------ (1975). Quasi-analytic vectors and quasi-analytic functions, *Bull. Amer. Math. Soc.* 81:637-644.

Civin, P., and B. Yood (1957). Quasi-reflexive spaces, *Proc. Amer. Math. Soc.* 8:906-911.

Clarkson, J. A. (1936). Uniformly convex spaces, *Trans. Amer. Math. Soc.* 40:396-414.

Coburn, L. A. (1965). Weyl's theorem for nonnormal operators, *Mich. Math. J.* 13:285-288.

Constantin, G. (1969). On some classes of operators, *Rend. Accad. Naz. dei Lincei* 46:241-244.

------ (1970). Some remarks on a class of operators, *Yokohama Math. J.* 18:1-3.

------ (1972). Some properties of locally α-contraction operators, *Boll. Union. Mat. Ital.* 4:323-330.

------ and I. Istrăţescu (1969). On Riesz operators with uniformly bounded iterates, *Math. Vestnik* 6:376-378.

------ and V. I. Istrăţescu. On quasi-analytic vectors for some classes of operators.

Crabb, M. (1970a). Some results on the numerical range of an operator, *J. Lond. Math. Soc.* 2(2):741-745.

------ (1970b). Numerical estimates for the norms of iterated operators, *Glasgow Math. J.* 11:85-87.

------ (1971a). The power inequality in normed spaces, *Proc. Edinburgh Math. Soc.* 17(2):237-240.

Crabb, M. (1971b). The powers of an operator and the numerical radius, *Mich. Math. J.* 18:253-256.

------, J. Duncan, and C. M. McGregor (1972a). Some extremal problems in the theory of numerical ranges, *Acta Math.* 128:123-142.

------ (1972b). Mappings theorems and the numerical radius, *Proc. Lond. Math. Soc.* 3(25):486-502.

Danes, J. (1971). Some fixed point theorems in metric and Banach spaces, *Comment. Univ. Carolina* 12:37-51.

------ (1972). On the Istratescu's measure of noncompactness, *Bull. Math. Soc. Rumanie* 64:403-406.

------ (1976). Two fixed point theorems in topological and metric spaces, *Bull. Austral. Math. Soc.* 14:259-267.

Darbo, G. (1955). Punti unity in transformazioni a codominio non compatto, *Rend. Mat. Sem. Padova* 24:84-92.

Daughtry, J. (1975a). An invariant subspace theorem, *Proc. Amer. Math. Soc.* 49:267-268.

------ (1975b). Hyperinvariant subspaces for approximately idem-potent operators, *Proc. Amer. Math. Soc.* 48:262-263.

Day, M. M. (1958). *Normed Linear Spaces*, Springer Verlag, New York.

Dixmier, J. (1949). Les operateurs permutables a l'operateur integral, *Port. Math.* 8:73-84.

------ (1964). Les C*-algebres et leurs representations, Gauthier-Villars, Paris.

Dieudonne, J. (1960). Quasi-hermitian operators, *Proc. Int. Symp. Linear Spaces,* Jerusalem.

Donoghue, W. F. (1957a). The lattice of invariant subspaces of completely continuous quasi-nilpotent operators, *Pacif. J. Math.* 7:1031-1033.

------ (1957b). On the numerical range of a bounded operator, *Mich. Math. J.* 4:261-263.

------ (1963). On a problem of Nieminen, *Inst. Hautes Etud. Sci. Publ. Math.* 16:31-33.

Douglas, R. G. (1972). *Banach Algebra Techniques in Operator Theory*, Academic Press, New York.

------ and P. Rosenthal (1968). A necessary and sufficient condi-tion that an operator be normal, *J. Math. Annal Appl.* 22:10-11.

Dunford, N. (1952). Spectral operators, *Pacif. J. Math.* 2:559-614.

------ (1958). A survey of the theory of spectral operators, *Bull. Amer. Math. Soc.* 64:217-274.

Dunford, N., and J. Schwartz (1965, 1967). *Linear Operators*, I and II, Interscience, New York.

Durszt, E. (1966). Remark on a paper of Berberian, *Duke Math. J.* 35:795-796.

Eberlein, W. F. (1964). A note on spectral theorem, *Bull. Amer. Math. Soc.* 52:328-331.

Embry, M. (1968). Conditions implying normality, *Pacif. J. Math.* 18:457-460.

------ (1970). The numerical range of an operator, *Pacif. J. Math.* 32:647-650.

Feynman, R. P. (1951). An operator calculus having applications in quantum electrodynamics, *Phys. Rev.* 84:108-128.

Fillmore, P. (1970). *Notes on Operator Theory,* Van Nostrand, New York.

------, J. Stampfli, and J. P. Williams (1972). On the essential numerical range, the essential spectrum, and a problem of Halmos, *Acta Sci. Math.* (Szeged) 33:179-192.

------ and J. P. Williams (1971). On operator ranges, *Adv. Math.* 7:254-281.

Fixman, U. (1959). Problems in spectral operators, *Pacif. J. Math.* 9:1029-1059.

Foias, C. (1957a). Sur certains theoremes de von Neumann concernant les ensembles spectraux, *Acta Math. Sci.* (Szeged) 85:15-20.

------ (1957b). La measure harmonique spectrale et la theorie spectrale des operateurs generaux, *Bull. Sci. Math. France* 85:263-282.

Friederichs, K. (1973). *Spectral Theory of Operators in Hilbert Spaces*, Springer Verlag, New York.

Fuglede, B. (1950). A commutativity theorem for normal operators, *Proc. Nat. Acad. Sci.* 36:35-40.

Fujii, M. (1974). Operators essentially similar to their adjoints, *Math. Jap.* 19:381-386.

------ (1976). Some normality conditions for operators and Berberian subrepresentation, *Math. J.* 21:259-263.

Furuta, T. (1967). On a class of paranormal operators, *Proc. Jap. Acad. Sci.* 45:594-598.

Gelfand, I. M. (1936). Sur une lemme de la theorie des espaces lineaires, *Commun. Soc. Math. Kharkov* 4(13:35-40.

------ (1938). Abstrakte Funktionen und lineare Operatoren, *Rec. Math. N. S.* 4:235-284.

------ (1941a). Uber absolutkonvergente trigonometrische Reihen und Integrale, *Rec. Math. N. S.* 9(50):51-66.

Gelfand, I. M. (1941b). Normierte Ringe, *Rec. Math. N. S.* 9(51): 3-24.

------ and M. Naimark (1943). On the imbedding of normed rings into the rings of operators in Hilbert spaces, *Rec. Math. N. S.* 12(54):197-213.

------, G. E. Shilov, and D. Raikov (1964). *Kommutative Normed Algebras*, Berlin.

Giles, J. R. (1967). Classes of semi-inner product spaces, *Trans. Amer. Math. Soc.* 129:436-447.

------, G. Joseph, and B. Sims (1975). On numerical range of operators on locally convex spaces, *Austral. Math. J.* 20:468-482.

------ and D. O. Koehler (1973). On the numerical ranges of elements of locally m-convex algebras, *Pacif. J. Math.* 49:73-91.

Gilfeather, F. (1969). Asymptotic convergence of operators in Hilbert space, *Proc. Amer. Math. Soc.* 22:69-76.

------ (1970). The structure and asymptotic behavior of polynomially compact operators, *Proc. Amer. Math. Soc.* 25:127-134.

Givens, W. (1952). Field of values of a matrix, *Proc. Amer. Math. Soc.* 3:206-209.

Gleason, A. M. (1967). A characterization of maximal ideals, *J. d'Analyse Math.* 19:171-172.

Glickfeld, B. W. (1970a). A metric characterization of C(X) and its generalization to C*-algebras, *Ill. J. Math.* 10:547-556.

------ (1970b). On an inequality of Banach algebra geometry and semi-inner product space theory, *Ill. J. Math.* 14:76-82.

Godement, R. (1947). Theoremes tauberiene et theorie spectrale, *Ann. Ecole. Norm. Sup.* 3(64):119-138.

Gray, J. (1968). Local analytic extension of the resolvent, *Pacif. J. Math.* 27:305-324.

Hahn, H. (1922). Über Folgen linearen Operationen, *Monats. Math. Phys.* 32:18-26.

Halmos, P. (1950). Normal dilations and extensions of operators, *Summa Brasil.* 2:125-134.

------ (1951). *Introduction to Hilbert Space and the Theory of Spectral Multiplicity*, Chelsea, New York.

------ (1952). Spectra and spectral manifolds, *Ann. Soc. Polon. Math.* 25:42-49.

------ (1956). *Lectures on Ergodic Theory*, The Mathematical Society of Japan, Tokyo.

------ (1958a). *Finite Dimensional Vector Spaces*, Van Nostrand, New York.

Halmos, P. (1961). Shifts on Hilbert spaces, *J. Reine Angew. Math.* 208:102-112.

------ (1963). What does the spectral theorem say? *Amer. Math. Monthly* 70:241-247.

------ (1964). Numerical ranges and normal dilations, *Acta Math. Sci.* (Szeged) 25:1-5.

------ (1965). *A Glimpse into Hilbert Space*, Lectures on Modern Mathematics, Vol. I, Wiley, New York, pp. 1-22.

------ (1967). *A Hilbert Space Problem Book*, Van Nostrand, New York.

------ (1968a). Permutation of sequences and the Schroeder-Bernstein theorem, *Proc. Amer. Math. Soc.* 15:215-223.

------ (1968b). Irreducible operators, *Mich. Math. J.* 15:215-223.

------ (1970). Ten problems in Hilbert space, *Bull. Amer. Math. Soc.* 76:887-933.

------ and S. Kakutani (1958b). Products of symmetries, *Bull. Amer. Math. Soc.* 64:77-78.

Hardy, G. H., J. L. Littlewood, and G. Polya (1934). *Inequalities*, Cambridge Univ. Press, New York.

Harris, L. (1972). Banach algebras with involution and Möbius transformation, *J. Funct. Anal.* 11:1-16.

Hausdorff, F. (1919). Der Wertvorrat einer Bilinearform, *Math. Z.* 3:314-316.

------ (1952). *Set Theory*, Chelsea, New York.

Helson, H. (1964). *Lectures on Invariant Subspaces*, Academic Press, New York.

Herstein, I. N. (1957). On a theorem of Putnam and Wintner, *Proc. Amer. Math. Soc.* 8:525-536.

Heuser, H. (1956). *Uber Operatoren mit Endlichen Defecten*, Inaug. Diss. Tubingen.

------ (1959). Uber Eigenwerte und Eigenlösungen symmetrisierbarer finiter Operatoren, *Arch. Math.* 10:12-20.

------ (1960a). On the spectral theory of symmetric finite operators, *Trans. Amer. Math. Soc.* 94:327-336.

------ (1960b). Zur Eigenwertheorie einer Klasse symmetrischer Operatoren, *Math. Z.* 74:167-180.

------ (1968). Z-symmetrisierbare Operatoren, *Rev. Roum. Math. Pure Appl.* 13:177-189.

------ (1969). Die iteration finiter operatoren auf Raumen mit halbskalarprodukt, *Math. Ann.* 182:213-231.

------ (1975). *Funktionalanalysis*, Teubner, Stuttgart.

Hewitt, E., and K. Stromberg (1965). *Real and Abstract Analysis*, Springer Verlag, New York.

Hille, E., and R. S. Phillips (1957). *Functional Analysis and Semigroups*, Amer. Math. Soc., Providence, R.I.

Hoffman, K. (1962). *Banach Spaces of Analytic Functions*, Prentice-Hall, Englewood Cliffs, N.J.

Istrăţescu, A. I., and V. I. Istrăţescu (1971). On the theory of fixed points for some classes of mappings, II, *Rev. Roum. Math. Pure Appl.* 16:1073-1076.

------ (1974). Some remarks on a class of semigroups of operators, *Rend. Acad. Naz. Lincei* 56.

Istrăţescu, I. I. (1969a). On operators with uniformly bounded iterates, *Mat. Vestnik* 6:373-375.

------ (1969b). Structure theorems for some classes of operators, *Proc. Jap. Acad. Sci.* 45:586-589.

------ (1971). On unimodular contractions on Banach spaces and Hilbert spaces, *Rend. Acad. Naz. Lincei* 50:216-219.

------ (1972). Perturbations of the monotone shift, *Yokohama Math. J.* 20:107-113.

------ and V. I. Istrăţescu (1967a). On some classes of operators I, *Proc. Jap. Acad. Sci.* 43:605-606.

------ (1967b). On some classes of operators II, *Proc. Jap. Acad. Sci.* 43:957-960.

------ (1969a). On quasi-normalizable operators I, *Rend. Acad. Naz. Lincei* 46:345-347.

------ (1969b). On quasi-normalizable operators II, *Rend. Acad. Naz. Lincei* 46:523-525.

------ (1970). On bare and semibare points for some classes of operators, *Port. Math.* 29:205-211.

------ (1971). On characters of singly generated C*-algebras, *Proc. Acad. Jap. Sci.* 47:42-43.

Istrăţescu, V. I. (1964). On nuclear operators, *Rev. Roum. Math.* 9:431-434.

------ (1967a). On some hyponormal operators, *Pacif. J. Math.* 22:413-417.

------ (1967b). Remarks on conditions implying normality, *Rev. Roum. Math. Pure Appl.* 12:1293-1295.

------ (1967c). On some subspaces of operators of class (N), I, *Rend. Acad. Naz. Lincei* 43:305-307.

------- (1968). Weyl's theorem for a class of operators, *Rev. Roum. Math. Pure Appl.* 13:343-346.

Istrăţescu, V. I. (1969a). Remark on Schwarz norms for operators, *Rev. Roum. Math. Pure Appl.* 14:359-360.

------ (1969b). On maximum theorems for operators, *Rev. Roum. Math. Pure Appl.* 14:1025-1029.

------ (1969c). On subharmonicity theorem for operator functions, *Boll. Union. Mat. Ital.* 4(2):365-366.

------ (1970a). On Schwarz norms for operators, *Rend. Acad. Naz. Lincei* 48:116-119.

------ (1970b). On some classes of operators, *Math. Ann.* 188:227-232.

------ (1972a). On a class of normaloid operators, *Math. Zet.* 124: 199-202.

------ (1972b). On some classes of operators, I, *Mathematika Balkanika* 2:49-57.

------ (1972c). On the Weyl spectrum of an operator I, *Rev. Roum. Math. Pure Appl.* 7:1049-1059.

------ (1979). On some classes of operators II, *Mathematika Balkanika*.

------ (1977). Local properties of bounded linear operators on Hilbert and Banach spaces, *Rendiconti di Matematica* (4)10:617-630.

------. On subsymmetrizable operators (in preparation).

------ and C. K. Fong (1979). A characterization of hermitian operators and related classes of operators, *Proc. Amer. Math. Soc.* 76:107-112.

------ and I. I. Istrăţescu (1967). On normaloid operators, *Math. Zeit.* 105:153-156.

------ and G. Constantin (1974). On the Weyl spectrum of an operator II, *Mathematika Balkanika* 4.

------, T. Saito, and T. Yoshino (1966). On a class of operators, *Tohoku Math. J.* 18:431-434.

Jacobson, N. (1935). Rational methods in the theory of Lie algebras, *Ann. Math.* 36:875-881.

James, R. C. (1947). Inner products in normed linear spaces, *Bull. Amer. Math. Soc.* 53:559-566.

------ (1951). A nonreflexive Banach space isometric with its second conjugate space, *Proc. Nat. Acad. Sci.* 53:559-566.

Jamison, B. (1965). Eigenvalues with modulus one, *Proc. Amer. Math. Soc.* 16:375-377.

Johnson, B. E. (1967). A commutative semisimple annihilator Banach algebra which is not dual, *Bull. Amer. Math. Soc.* 73:407-409.

Kakutani, S. (1954). An example concerning uniform boundedness of spectral measures, *Pacif. J. Math.* 4:363-372.

Kakutani, S. A proof of Hahn-Banach theorem via fixed point theory.

------ and K. Yosida (1939). Birkhoff's ergodic theorem and the maximal ergodic theorem, *Proc. Imp. Acad. Tokyo,* 15:165-168.

------ (1941). Operator treatment of Markov processes and the mean ergodic theorem, *Ann. Math.* 2(42):188-228.

Kalisch, G. K. (1962). A functional analysis proof of Titchmarsh's theorem on convolution, *J. Math. Ann. Appl.* 5:176-182.

------ (1964). Direct proof of spectral representation theorems, *J. Math. Ann. Appl.* 8:351-354.

Kaplansky, I. (1953). Products of normal operators, *Duke Math. J.* 20:257-260.

Kellog, O. D. (1922). On the existence and closure of sets of characteristic functions, *Math. Ann.* 86:14-17.

Kleinecke, D. C. (1957). On operator commutators, *Proc. Amer. Math. Soc.* 8:535-536.

------ (1963). Almost finite, compact, and inessential operators, *Proc. Amer. Math. Soc.* 14:863-868.

Koehler, D., and P. Rosenthal (1970). On isometries of normed linear spaces, *Studia Math.* 36:213-216.

Köthe, G. (1936a). Das Tragheitsgesetz der quadratischen Formen im Hilbertschen Raum, *Math. Z.* 41:137-152.

------ (1936b). Die Gleichungstheorie im Hilbertschen Raum, *Math. Z.* 41:153-163.

------ (1966). *Topologische Lineare Räume*, Springer Verlag, Berlin.

Krein, M. G. (1947). The theory of self-adjoint extensions of semi-bounded hermitian operators, and its applications, *Mat. Sb.* 20:431-495 (in Russian).

------ and I. C. Gochberg (1966a). Theory of Volterra operators and its applications, Moscow, *Amer. Math. Trans.* 24.

------ (1966b). Introduction to the theory of nonself-adjoint operators, Moscow, *Amer. Math. Soc.* 18.

Krasnoselskii, M. A. (1964). *Positive Solutions of Operator Equations,* Nordhoff, Groningen.

Landau, E. (1907). *Über einen Konvergenzsatz,* Göttinger Nachr, pp. 25-27.

Lavrentiev, M. A. (1936). Sür les fonctions d'une variable complexe representable par des sries de polynomes, Actualites Sci. Indust., Herman, Paris.

Lax, P. D. (1954). Symmetrizable linear operators, *Commun. Pure Appl.* 7:633-647.

Lebow, A. (1963). On von Neumann's theory of spectral sets, *J. Math. Ann. Appl.* 7:64-90.

------ (1968). On power bounded operator which is not polynomially bounded, *Mich. Math. J.* 15:397-399.

------ and M. Finkelstein (1969). A note on Nth roots of operators, *Proc. Amer. Math. Soc.*

------ and M. Schechter (1971). Semigroups of operators and measures of noncompactness, *J. Funct. Anal.* 7(1-26).

Lomonosov, V. I. (1973). On invariant subspaces of families of operators commuting with a completely continuous operator, *Funct. Anal. Appl.* 7:55-56 (in Russian).

Loomis, L. (1953). *An Introduction to Abstract Harmonic Analysis,* Van Nostrand, New York.

Lovaglia, A. R. (1955). Locally uniformly convex Banach spaces, *Trans. Amer. Math. Soc.* 78:225-238.

Luenberger, D. G. (1965). Invertible solutions to the operator equation TA - AT = C, *Proc. Amer. Math. Soc.* 16:1226-1229.

Lumer, G. (1961a). Semi-inner product spaces, *Trans. Amer. Math. Soc.* 100:29-43.

------ (1964). Spectral operators, hermitian operators, and bounded groups, *Acta Math. Sci.* (Szeged) 25:75-85.

------ (1972). Complex methods and the estimates of operator norms and spectra from real numerical ranges, *J. Funct. Anal.* 10:482-495.

------ (1978). Evolution equations and semigroups.

Mandelbrojt, S. (1952). Series adherentes. Regularization des suites. Applications, Gauthier-Villars, Paris.

Masson, D., and W. McClary (1972). Classes of C^{∞} vectors and essential self-adjointness, *J. Funct. Anal.* 10:19-32.

Meng, C. H. (1957). A condition that a normal operator has a closed numerical range, *Proc. Amer. Math. Soc.* 8:57-58.

------ (1963). On the numerical range of an operator, *Proc. Amer. Math. Soc.* 14:167-171.

Michael, E. (1952). Locally multiplicatively convex topological algebras, *Memoir Amer. Math. Soc.* 11.

Moore, R. T. (1969). Adjoint numerical ranges and spectra of operators on locally convex spaces, *Bull. Amer. Math. Soc.* 75:85-90.

------ (1971). Hermitian functional on B*-algebras and duality characterization of C*-algebras, *Trans. Amer. Math. Soc.* 162:253-266.

Moyls, B., and M. Marcus (1955). Field convexity of a square matrix, *Proc. Amer. Math. Soc.* 6:981-983.

Murphy, T., and T. T. West (1972). Inner products in Banach spaces and supports of Radon measures, *Proc. Irish Acad.* 69.

Nagy, B. Sz. (1952). Prolongements des transformations de l'espace de Hilbert qui sortent de cet espace, Appendix to *Lecons d' Analyse fonctionneles* by Fr. Riesz and B. Sz. Nagy. Budapest.

------ (1953). Sur les contractions de l'espace de Hilbert, *Acta Math. Sci.* (Szeged) 15:87-92.

------ (1957). Sur les contractions de l'espace de Hilbert II, *Acta Math. Sci.* 18:1-14.

------ and C. Foias (1966). On certain classes of power bounded operators in Hilbert space, *Acta Math. Sci.* (Szeged) 27:17-25.

------ and ------ (1970). *Harmonic Analysis of Operators on Hilbert Spaces*, North-Holland, Amsterdam.

Naimark, M. (1940). Self-adjoint extension of the second kind of symmetric operators, *Izvestija Acad. Nauk Mat.* 4:53-104.

------ (1943). On spectral functions of a symmetric operator, *Izvestija Acad. Nauk Mat.* 7:285-296.

------ (1960). *Normed Rings*, Nordhoff, Groningen.

Nelson, E. (1959). Analytic vectors, *Ann. Math.* 70:572-615.

------ (1961). The distinguished boundary of the unit ball of operators, *Proc. Amer. Math. Soc.* 12:994-995.

------ and W. F. Stinespring (1959). Representation of elliptic operators in an enveloping algebra, *Amer. J. Math.* 81:547-550.

Neumann, J. von (1929). Zur Algebra der Funktionaloperatoren und theorie der normalen operatoren, *Math. Ann.* 102:370-427.

------ (1935). *Characterisierung des Spectrums eines Integral Operators*, Actualites Sci. Ind., 229 nr. 13, Paris.

------ (1950). *Functional Operators*, Princeton University Press, Princeton.

------ (1951). Eine spectraltheorie für allgemeine Operatoren eines unitären Raumes, *Math. Nachr.* 4:258-281.

------ (1962). *Collected Works, IV*, Pergamon Press, New York.

Newburg, J. D. (1951). On the variation of spectra, *Duke Math. J.* 18:165-176.

Newmann, D. J. (1960). The nonexistence of projections from L^1 to H^1, *Proc. Amer. Math. Soc.* 11:692.

Nieto, J. L. (1968a). On the essential spectrum of symmetrizable operators, *Math. Ann.* 178:148-153.

Nieto, J. L. (1968b). On Fredholm operators and the essential spectrum of singular integral operators, *Math. Ann.* 163:18-49.

------ (1974). Sur une theoreme d'interpolation de J. Lions et J. Peetre, *Canad. Math. Bull.* 14:373-376.

Nikolskii, N. K. (1965). *Invariant Subspaces of Certain Completely Continuous Operators,* Vestnik Leingrad. Univ., pp. 48-77.

Nishitani, I., and Y. Watatani (1976). Some theorems on paranormal operators, *Math. J.* 21:123-126.

Nussbaum, A. E. (1965). Quasi-analytic vectors, *Archiv. Math.* 6:179-191.

------ (1969). A note on quasi-analytic vectors, *Studia Math.* 33:306-309.

Onicescu, O. (1969). *Calcolo delle Probabilita ed Applicazioni,* Veschi Editore, Roma.

------ and S. Guiasu (1971). *Mechanique Statistique,* Udine, Springer Verlag, New York.

Orland, G. H. (1964). On a class of operators, *Proc. Amer. Math. Soc.* 15:75-80.

Palmer, T. W. (1965). *Unbounded Normal Operators on Banach spaces,* Ph.D. Thesis, Harvard University, Cambridge, Mass.

------ (1968a). Unbounded normal operators on Banach spaces, *Trans. Amer. Math. Soc.* 133:385-414.

------ (1968b). Characterizations of C*-algebras, *Bull. Amer. Math. Soc.* 74:538-540.

------ (1970). Characterizations of C*-algebras, II, *Trans. Amer. Math. Soc.* 148:577-588.

Patel, S. M. (1973). Operators with left inverses similar to those of their adjoints, *Proc. Amer. Math. Soc.* 41:127-131.

------ (1974). Operators with powers essentially similar to those of their adjoints, *Proc. Amer. Math. Soc.* 42:243-247.

------ (1975a). On similarities of operators, *Yokohama Math. J.* 23:85-90.

------ (1975b). On classes of nonhyponormal operators, *Math. Nachr.* 73:147-150.

------ (1976). A note on spectraloid operators.

------ and B. C. Gupta (1975c). Operators satisfying certain growth conditions, *Proc. Amer. Math. Soc.* 53:341-346.

Pearcy, C. (Ed.). Topics in operator theory, *Amer. Math. Surveys,* Nr. 13.

Phelps, R. (1960). Uniqueness of the Hahn-Banach extensions and unique best approximation, *Trans. Amer. Math. Soc.* 95:238-255.

Phillips, R. S. (1959). Dissipative operators and hyperbolic systems of partial differential equations, *Trans. Amer. Math. Soc.* 90:193-254.

Putnam, C. R. (1967). *Commutation Properties of Hilbert Space Operators,* Ergeb. der Math. Nr. 36, Springer Verlag, New York.

------ (1970). An inequality for the area of hyponormal operators, *Math. Zeit.* 116:323-330.

------ (1971). The spectra of subnormal operators, *Proc. Amer. Math. Soc.* 28:473-477.

Rado, T. (1937). *Subharmonic Functions*, Ergeb. Math., Springer Verlag, Berlin.

Ragavendran, R. (1969). Toeplitz-Hausdorff theorem on numerical ranges, *Proc. Amer. Math. Soc.*20:284-285.

Ramanujan, P. B. (1970). On operators of class (N,k), *Proc. Camb. Phil. Soc.* 67:141-142.

Reid, W. T. (1951). Symmetrizable completely continuous linear transformations in Hilbert spaces, *Duke Math. J.* 18:41-56.

Rickart, C. (1960). *General Theory of Banach Algebras,* Van Nostrand, New York.

Riesz, F. (1909). Sur les operations fonctionneles linearires, *C. R. Paris* 149:974-977.

------ (1913). *Les Systemes d'Equations Lineaires a une Infinite d'Inconnues,* Paris.

------ and B. Sz. Nagy (1952). *Lecons d'Analyse Fonctionelle,* Budapest.

Ringrose, J. R. (1962). Superdiagonal forms for compact linear operators, *Proc. London Math. Soc.* 3(12):41-56.

------ (1971). Compact Non-Self-Adjoint Operators, *Van Nostrand Reinhold Mat. Stud.* 35.

Rosenblum, M. (1958). On a theorem of Fuglede and Putnam, *J. London Math. Soc.* 33:376-377.

Rosenthal, P. (1968a). A note on unicellular operators, *Proc. Amer. Math. Soc.* 19:505-506.

------ (1968b). Completely reducible operators, *Proc. Amer. Math. Soc.* 19:826-828.

------ (1968). Some recent results on invariant subspaces.

Rota, G. C. (1959). Note on the invariant subspaces for linear operators, *Rend. Circ. Mat. Palermo* 8:1-3.

------ (1960). On models for linear operators, *Commun. Pure Appl. Math.* 13:469-472.

Rubin, H., and M. H. Stone (1953). Postulates for generalizations of Hilbert spaces, *Proc. Amer. Math. Soc.* 4:360-366.

Rudin, W. (1962). *Fourier Analysis on Groups,* Interscience, New York.

------ (1972). *Principles of Mathematical Analysis*, McGraw-Hill, New York.

Russo, B., and H. Dye (1966). A note on unitary operators in C*-algebras, *Duke Math. J.* 33:413-416.

Sadovskii, V. (1967). A fixed point principle, *Funct. Anal. Appl.* 1:74-76.

------ (1972). Limit-compact operators and condensing operators, *Uspehi Mat. Nauk* 27; *Russian Surveys* 27:85-155.

Saito, T. (1966). Some remarks on Ando's theorem, *Tohoku Math. J.* 18.

------ (1968). Hyponormal operators.

------ (1971). Hyponormal operators and related topics, Lecture Notes, Tulane Univ.

Schatten, R. (1960). *Norms Ideals of Completely Continuous Operators,* Springer Verlag, New York.

Schechter, M. (1965). Invariance of the essential spectrum, *Bull. Amer. Math. Soc.* 365-367.

------ (1966). On the essential spectrum of an arbitrary operator, I, *J. Math. Anal. Appl.* 13:205-215.

------ (1967). Basic theory of Fredholm operators, *Ann. Scuola Norm Sup. Pisa* 21:361-380.

------ (1970). Operators obeying Weyl's theorem, *Scripta Math.* 29:67-75.

------ (1971). *Principles of Functional Analysis,* Academic Press, New York.

------ (1975). Proper boundary points of the spectrum, *Osaka Math. J.* 12:41-44.

------ (1976). Remarks on a previous paper, *J. Math. Anal. Appl.* 54:138-141.

Schreiber, M. (1963a). Numerical range and spectral sets, *Mich. Math. J.* 10:283-288.

------ (1963b). Remark on a paper of Kalisch, *J. Math. Anal. Appl.* 7:62-63.

Schur, I. (1921). Über lineare Transformationen in der Theorie der unendlichen Reihen, *J. Rheine Angew. Math.* 151:79-

Schwartz, J. (1962a). On spectral operators in Hilbert spaces with compact imaginary part, *Commun. Pure Appl. Math.* 15:95-97.

-------- (1962b). Subdiagonalization of operators in Hilbert spaces with compact imaginary part, *Commun. Pure Appl. Math.* 15:159-172.

Sheth, I. H. (1960). On normaloid operators, *Pacif. J. Math.* 28:675.

------ (1966). On hyponormal operators, *Proc. Amer. Math. Soc.* 17:998-1001.

------ (1970). Some results on hyponormal operators, *Rev. Roum. Math. Pure Appl.* 15:395-398.

------ (1974). On a conjecture of Istrăţescu, *J. Indian Math. Soc.* 38:337-338.

------ and N. C. Shah (1975). Some results on quasi-hyponormal operators, *J. Indian Math. Soc.* 39:285-291.

Shilov, G. (1961). *Mathematical Analysis*, Moscow (in Russian).

Shirocov, F. (1956). Proof of a Conjecture of Kaplansky, *Uspehi Mat. Nauk* 11:167-168.

Sikonia, W. (1971). The von Neumann converse of Weyl's theorem, *Indiana Univ. J.* 21:121-124.

Singer, I. (1973). *Bases in Banach Spaces*, Springer Verlag, New York.

------ (1974). *Cea mai buna Approximare in Spatii Normate*, Bucharest (in Romanian).

Stampfli, J. (1962). Hyponormal operators, *Pacif. J. Math.* 12:1453-1458.

------ (1965). Hyponormal operators and spectral density, *Trans. Amer. Math. Soc.* 117:469-476.

------ (1966). Analytic extension and spectral localization, *J. Math. Mech.* 16:287-296.

------ (1969a). A local spectral theory of operators, I, *J. Funct. Anal.* 4:1-10.

------ (1969b). A local spectral theory of operators II, *Bull. Amer. Math. Soc.* 84:803-806.

------ and J. Williams (1968). Growth conditions and the numerical range in a Banach algebra, *Tohoku Math J.* 20:417-424.

------ and B. L. Wadhwa (1976). An asymmetric Putnam-Fuglede theorem for dominant operators, *Indiana Univ. J.* 25:359-365.

------, J. P. Williams, and P. Fillmore (1972). On the essential numerical range, the essential spectrum and a problem of Halmos, *Acta Math. Sci.* (Szeged) 33:179-192.

Stinespring, W. F. (1955). Positive definite functions on C*-algebras, *Proc. Amer. Math. Soc.* 6:211-216.

Stone, M. H. (1932a). Linear transformations in Hilbert space and their applications to analysis, *Amer. Math. Soc. Colloq.*

------ (1932b). On one parameter unitary groups in Hilbert spaces, *Ann. Math.* 33:643-648.

------ (1951). On unbounded operators in Hilbert space, *J. Indian Math. Soc.* 15:155-192.

Sukhomlinov, G. (1937). Analytic functionals, *Bull. Moscow Gosud. Univ.* 2.

Suzuki, N. (1969a). The structure of spectral operators with compact imaginary part, *Proc. Amer. Math. Soc.* 22:82-84.

------ (1969b). On the irreducibility of weighted shift, *Proc. Amer. Math. Soc.* 22:579-581.

------ (1971). Reduction theory of operators in Hilbert spaces, *Indiana Univ. Math. J.* 20:953-958.

Szegö, G. (1924). Uber eine Eigenschaft der Exponentialreihe, *Sitz. Berich. Berlin Gess.* 23:50-64.

Taylor, A. E. (1958). *Introduction to Functional Analysis.* John Wiley, New York.

------ (1966). Theorems on ascent, descent, nullity, and defect of linear operators, *Math. Ann.* 163:18-49.

Taylor, M. E. (1968). Functions of several self-adjoint operators, *Proc. Amer. Math. Soc.* 19:91-98.

Thorp, E., and R. Whitley (1968). The strong maximum theorem for analytic functions into Banach spaces, *Proc. Amer. Math. Soc.* 18:640-646.

Titchmarsh, E. C. (1932). *The Theory of Functions*, Clarendon Press, Oxford.

Toeplitz, O. (1918). Das algebraischer analogon zu einem Satze von Fejer, *Math. Zeit.* 2:187-197.

Tulcea, C. T. Ionescu, and A. Ionescu Tulcea (1969). Topics in the Theory of Lifting, Ergeb. Math. Nr. 48, Springer Verlag, New York.

Valentine, F. A. (1964). *Convex Sets* , McGraw-Hill, New York.

Vesentini, E. (1968). On the subharmonicity of the spectral radius, *Boll. Union. Mat. Ital.*

------ (1970a). Maximum theorems for spectra, Essay on Topology and Related Topics, *Memoires dedie a Georges de Rham*, Springer Verlag, New York.

------ (1970b). Maximum theorems for vector-valued holomorphic functions, *Rend. Sem. Mat. e Fisico di Milano* 40:3-34.

------ (1972). On Banach algebras satisfying a spectral maximum principle, *Ann. Scuola Normale Sup. di Pisa*, 26:933-943.

Vesentini, E. (1976). Holomorphic functions on Siegel domains, *Boll. Union. Mat. Ital.* 5(13):225-241.

Vidav, I. (1955). Uber eine Vermutung von Kaplansky, *Math. Z.* 62:330.

------ (1956). Eine metrische Kennzeichnung der selbstadjungierte Operatoren, *Math Z.* 66:121-128.

Vogt, W. G., M. M. Eissen, and G. R. Buis (1969). Contraction groups and equivalent norms, *Nagoya Math. J.* 34:149-151.

Wadhwa, B. L. (1973a). Operators satisfying a sequential growth condition, *Acta Math. Sci.* (Szeged) 35:181-186.

------ (1973b). A hyponormal operator whose spectrum is not a spectral set, *Proc. Amer. Math. Soc.* 38:83-85.

------ (1974). M-hyponormal operators, *Duke Math. J.* 41:655-660.

------ and J. Stampfli. On dominant operators.

Warner, K. K. (1969). A note on a theorem of Weyl, *Proc. Amer. Math. Soc.* 23:469-471.

Wawre, H. (1943). L'iteration directe des operateurs hermitiens et deux theories qui en dependent, *Commun. Math. Helv.* 15:299-317.

Wermer, J. (1952). The existence of invariant subspaces, *Duke Math J.* 19:615-622.

------ (1953). Report on subnormal operators, *Operator Theory*, NAS-NRC.

Weyl, H. (1909). Uber beschrankte quadratische Formen, deren Differenz vollstetig ist, *Rend. Circ. Math. Palermo* 27:373-392.

Whitley, R. (1975). A note on hyponormal operators, *Proc. Amer. Math. Soc.* 49:399-400.

Widom, H. (1964). On the spectra of a Toeplitz operator, *Pacif. J. Math.* 14:365-375.

Wielandt, H. (1944). Das Iterationvefahren bei nich-selbstadjungierten linearen Eigenwertaufgaben, *Math. Z.* 50:93-143.

Wiener, N. (1922). Limit in terms of continuous transformations, *Bull. Soc. Mat. France* 150:124-134.

------ (1933). *The Fourier Integral and Certain of Its Applications*, Cambridge University Press, New York.

Wilansky, A. (1964). *Functional Analysis*, Blaisdell, New York.

Williams, J. P. (1967). Spectra of products and numerical range, *J. Math. Ann. Appl.* 17:214-217.

------ (1968). Schwarz norms for operators, *Pacif. J. Math.* 24:181-189.

------ (1969). Operators similar to their adjoints, *Proc. Amer. Math. Soc.* 20:121-123.

------ (1970). Finite operators, *Proc. Amer. Math. Soc.* 21:129-136.

Williams, J. P., and T. Crimmins ((1967). On the numerical radius of a linear operator, *Amer. Math. Monthly* 74:833-836.

Wintner, A. (1929). Zür Theorie der beschränkten Billinearformen, *Math. Z.* 30:228-282.

Yood, B. (1954). Difference algebras of linear transformations on a Banach space, *Pacif. J. Math.* 4:615-636.

Yorke, J. A. (1969). Permutations and two sequences with the same cluster set, *Proc. Amer. Math. Soc.* 20:606.

Yoshino, T. (1965). On the spectrum of a hyponormal operator, *Tohoku Math. J.* 17:305-307.

------ (1967). Spectral resolution of a hyponormal operator with the spectrum on a curve, *Tohoku Math. J.* 19.

Zaanen, A. C. (1950). Normalizable transformations in Hilbert spaces and systems of linear integral equations, *Acta Math.* 83:197-248.

------ (1951). Characterization of a certain class of linear transformations in an arbitrary Banach space, *Indag. Math.* 13:87-93.

------ (1953). *Linear Analysis*, North-Holland, Amsterdam.

Zabreiko, P. (1969). On a theorem for subadditive functional, *Functional Anal. Appl.* 3:86-89.

Zelasko, W. (1968). A characterization of multiplicative functionals in complex Banach algebras, *Studia Math.* 30:83-85.

Zygmund, A. (1959). *Trigonometric Series*, Cambridge Univ. Press, New York.

SYMBOL INDEX

C^n, 20

c, 20

c_0, 20

ℓ^p, 20

m, 20

C^α, 21

$L^p(X, \mu)$, 21

$\rightharpoonup$, 37

glim, 53

$\beta(L^*, L)$, 57

$\beta(L, L^*)$, 57

$\sigma(L^*, L)$, 57

$\sigma(L, L^*)$, 57

$\xrightarrow{w}$, 57

$\xrightarrow{s}$, 57

F(T), 67

$(\ell^p)^* = \ell^q$, 76

$(\ell^1)^* = m = \ell^\infty$, 77

$c_0(L)$, 78

$\ell^p(L)$, 78

$\ell^1(L)$, 79

$\alpha(\ ,\ )$, 84

$\chi(\ ,\ )$, 84

$K_c(L_1, L_2)$, 87

$K_{c,c}(L_1, L_2)$, 87

α-contractions, 102

r(x, y), 112

α_H, 119

χ_0, 119

$C_0(S)$, 143

A^{-1}, 145

$\rho(a)$, 147

$\sigma(a)$, 147

r_a, 149

M_A, 154

L^∞, 180

W(T), 200

w(T), 204

T^+, 216

$W_{ess}(T)$, 217

$W_{max}(T)$, 220

C(x), 227

e^T, 233

J(L), 241

H(A), 247

$W(A, \|\cdot\|_\alpha, T)$, 263

$J(A)$, 264

G_1^{loc}, 294

γ_T, 306

rest-P, 324

red-P, 324

B_I, 353

(ccc), 356

(+) condition, 361

m_n, 393

$\pi_{0f}(T)$, 413

$\pi_{0i}(T)$, 413

$\pi_{\infty}(T)$, 413

$\sigma_{\mathcal{F}}(T)$, 413

$\omega(T)$, 413

$\mathcal{F}_0$, 413

C_1 condition, 417

C_2 condition, 417

$r(T)$, 421

$r^*(T)$, 421

G_1', 428

α, 428

α', 428

α'', 428

α''', 428

β, 428

β', 428

$C(M_n)$, 445

$D^a(T)$, 452

$D^{qa}(T)$, 453

$D^{sa}(T)$, 453

$D_S(T)$, 453

$D_S^p(T)$, 453

C^k, 466

S_c class, 471

$\|T\|_c$, 475

S_{Ω}, 476

ρ^{-1}, 484

$J(r, z_0, f)$, 490

$\Delta f(z)$, 493

$S(k)$, 500

c, 504

D_c, 505

M_r, 509

$\theta(f(z))$, 511

$\|T\|_m$, 524

$\theta_W(T)$, 531

α-set, 538

E_p, 543

$P^n(t, A)$, 545

$\|\mu\|$, 545

K, 546

K*, 546

D, 547

D*, 547

SUBJECT INDEX

Accumulation point, 93
Adjoint, 63, 64, 447
σ-algebra, 4
α-set contractions, 538
Analytic element, 467
Analytic functions, 444
Analytic vector, 452
Antidual, 372
Axiom of choice, 5

B*-algebra, 159, 477, 478
Balloon, 27
Banach-Alaoglu theorem, 357
Banach algebra, 142
Banach space, 31
Basis, 13, 115

Calkin algebra, 99, 101, 255, 321, 378, 413, 471, 475
Cartesian product, 5, 6
Cauchy-Schwarz inequality, 110
Cauchy sequence, 8
Cayley transform, 449
C^{∞}-element, 466
Chain, 5
Chebyshev polynomials, 120
Class (c), 307
Class (M,k), 308
Class (N), 266, 308, 332, 333, 341, 455
Class (N,k), 267, 308, 369, 397
Class $\underline{R}$, 296
Class $\bar{R}_1$, 302, 303
Class spectral G_1^{loc}, 295
Closed function, 50
Closed graph theorem, 50
Closed operator, 447
Cluster set, 441
Compact element, 434
Compact operators, 88, 91
Compact set, 16, 84
Complement, 3
Complementary subspace, 25
Complemented subspace, 99
Completely continuous operator, 87
Completely reduced pair, 421
Completely reducible operator, 403
Commutant, 385
Commutative Banach algebra, 142
Commutative monoid, 435
Commutator, 326
Complex extreme points, 499
Compression, 275
Condition (M-W)(+), 359
Cone, 25
Continuous function, 11
Contraction mapping, 252
Convergent, 8
Convex envelope, 545
Convex function, 486, 514
Convex hull, 24
Convexoid, 212, 299, 316, 327
Convolution, 391
Corner, 210, 212
Cramped spectrum, 336
Criteria of subharmonicity, 496
C_0-semigroup, 466
C^{∞}-vector, 452
Cyclic vector, 392

Decomposable element, 241, 264
Densifying mapping, 102
Derivation, 326, 466
Dilation, 275
Dimension, 119
Dirichlet algebra, 281
Disjoint, 3
Dissipative operator, 338, 463
Distribution, 493
Dominant operators, 309
Dual, 34
Dual of a Hilbert space, 126

Essentially self-adjoint operator, 451
Essential numerical range, 217
Essential spectral radius, 527
Exponential function, 233
Extreme point, 38
  of W(T), 222
Extension, 39, 446

Factor, 438
  type I, type II, 435
Filter, 15
Finite, 433
Finite dimensional operator, 86, 87
Finite intersection property, 16
First axiom of countability, 12
First category, 17
Fixed point theorem, 102
Fredholm elements, 434, 436
Fredholm operators, 94, 101, 413
Fredholm region, 413

Garbe operator, 345
Gelfand's lemma, 48
Gelfand-Naimark theorem, 167
Gelfand transform, 155
Generalized Cauchy-Schwarz inequality, 135
Glim, 53, 231
G_1-property, 286, 297, 315, 320, 327, 428, 429
Gram-Schmidt method, 120

Hadamard formula, 61, 72, 161, 484
Hammel basis, 25, 95
Hausdorff space, 14
Hausdorff-Toeplitz theorem, 200, 201, 202, 208
Hellinger-Toeplitz theorem, 130
Herglotz theorem, 471
Hermitian, 130, 159, 238, 315, 338, 448, 478
Hermitian element, 159, 478
Hermitianity, 310
Hilbert-Kaplansky module, 178
Hilbert quadratic form, 369
Hilbert space, 111
Holomorphic functions, 483
Hyperbolic systems, 371
Hyperinvariant subspaces, 382, 387, 391
Hyponormal, 221, 266, 297, 336, 337, 380, 399
Hyponormalizable, 363, 365, 421

Ideal, 151
  left, 151
  right, 151
  two-sided, 151
  maximal, 151
Idempotent, 125, 339
Identity matrix, 139
Index, 413, 435
Injective, 64, 90
Inner product, 109, 352
Interior of a set, 11
Interpolation theory, 372
Intransitive algebras, 382
Invariant subspaces, 93, 380, 381, 396, 397
Invertible operator, 34
Involution, 158
Irreducible operator, 417
Isolated point, 70, 417
Isometry, 45, 230
Isomorphic, 26
*-isomorphism, 167

Jensen's inequality, 497

Left regularizable operator, 99
Leibniz rule, 326
Levi theorem, 496
Linearly dependent set, 21
Linearly independent set, 21
Linear operator, 32
Linear space, 19

Linear subspace, 19
Linear topological space, 29
Liouville's theorem, 62, 73
Local base, 12
Locally convex spaces, 29
Locally m-convex algebra, 262
Locally power α-set contraction, 102, 104, 107, 108, 538, 546
Locally power densifying mapping, 102
Local resolvent, 71
Local spectral radius, 71
Local spectrum, 70, 71

Matrix, 139
  bounded, 139
Maximal element, 5
Maximal numerical range, 219, 220
Maximal orthonormal set, 115
Maximal subspaces, 26
Maximal vectors, 220, 397
Maximum principle, 499, 520
Maximum theorems for the norm, 498
Measurable functions, 179
Measures of noncompactness, 84, 524, 537
Metric spaces, 10
M-hyponormal, 309
Mikhlin conjecture, 100
Minkowski functional, 29, 59
Minimal dilation, 277
Minimal element, 5
Minimal norm, 38, 126
Minimal polynomial of an operator, 408
Mobius function, 248
Modulus:
  of α-contraction, 103
  of the locally power α-contraction, 103
M-power class, 309
Multiplicative functional, 154
Mazur-Gelfand proposition, 148

Negligible function, 179
Neighborhood, 8
Net, 14
Norm, 27, 109
Norm-preserving extension, 39
Normal, 245, 321, 427
Normal eigenvalue, 294
Normality, 321
Normal lifting, 439
Normalizable, 363
Normalized state, 254
Normaloid, 223, 267, 327, 337
Numerical radius, 204, 528
Numerical range, 200

Paranormal, 266
Pell operator, 345
Permutation, 441
Pole, 70, 538
Polynomially compact, 408, 432
Positive functional, 169
Power inequality, 261
Principle of uniform boundedness, 48
Projection, 34, 131, 338

Quasi-analytic element, 467
Quasi-analytic functions, 444
Quasi-analytic vector, 453
Quasi-compact, 103, 537, 546
Quasi-hermitian, 374, 377, 378, 379
Quasi-nilpotent, 326, 381
Quasi-normal, 266
Quasi-normalizable, 363, 364, 367, 374
Quasi-reflexive, 45
Quasi-similar, 387
Quotient algebra, 151
Quotient space, 26

Radical, 157
Radon-Nicodym theorem, 128
Radon-Riesz property, 81
Real extreme point, 500
Reducing subspace, 398
Reductive subspace, 421
Regularizable operators, 99
Reflexive space, 45, 128
Relatively compact set, 84, 85
Representation, 177
Resolvent set, 65
Right regularizable operator, 99
Riesz lemma, 91
Riesz operators, 92, 378

Riesz point, 93
Riesz representation:
  for functionals, 126
  for subharmonic functions, 495
Ring of sets, 4

Schauder fixed point theorem, 302, 303
Schwarz norm, 468
Second axiom of countability, 13
Segment, 23
Self-adjoint, 450
Semi-bare point, 210, 213, 428
Semibounded inferior, 450
Semi-inner product, 55, 226, 478
Seminorm, 27
Seminormal, 266
Semisimple, 157
Semispectral family, 174
Sequential G_1-property, 288, 306
Sequentially weak complete, 58
Separable Hilbert space, 119
Sets:
  absorbing, 23
  affine, 24
  balanced, 30
  bounded, 30
  closed, 8
  convex, 23
  directed, 14
  nowhere dense, 17
  open, 8
  pairwise disjoint, 3
  partially ordered, 4
  symmetric, 23
Shift, 381
Single-valued extension property, 74, 75
Span, 22
Spectrum, 65, 264
  of an element, 147
Spectral radius, 149
Spectral decomposition:
  of identity, 185
  with many dimensions, 186
  with one dimension, 186
Spectral family, 176
Spectral G_1-property, 291
Spectral mapping theorem, 69, 194, 199, 438, 443
  for normal operators, 196, 199
Spectral measure, 187, 281
Spectraloid operator, 303
Spectral sets, 275, 276, 297
Square root, 135
State, pure, 168
Stieltjes vector, 452
Strictly convex space, 38, 508
Strong dilation, 277
Strong maximum principle, 499, 503
Subbase, 12
Subharmonic functions, 485
Subsymmetrizable element, 379
Support, 281
Symmetrizable element, 344, 347, 348, 349, 363, 368
  fully, 346
  left, 344
  strongly, 346

Tietze's extension theorem, 195
Titchmarsh's theorem, 392, 483
Toeplitz operators, 412
Topology, 10, 13
  discrete, 10
  indiscrete, 10
  induced, 11
  linear, 29
  product, 15
  weak, 15
Trigonometric orthonormal system, 121

Unicellular, 389
Uniformly convergent sequence, 56
Uniformly convex space, 361
Unimodular contraction, 232
Unit, 142
Unital Banach algebra, 142
Unitarity, 317
Unitary dilation, 275, 276
Unitary operators, 134, 190
Upper bound, 5
Upper semicontinuous function, 485

V*-algebra, 244
Volterra operator, 288, 391, 396
von Neumann algebra, 433

von Neumann theorem, 439
Weakly convergent sequence, 57, 58
Weakly *-convergent sequence, 58
Weakly holomorphic sequence, 61
Weighted mean, 13
Weyl's spectrum, 412, 413, 433, 527
Weyl's theorem, 420, 429, 431, 432, 436

X-normal dilation, 276

Zero matrix, 139